Remote Sensing Techniques for Environmental Studies

Remote Sensing Techniques for Environmental Studies

Edited by Matt Weilberg

SYRAWOOD
PUBLISHING HOUSE

New York

Published by Syrawood Publishing House,
750 Third Avenue, 9th Floor,
New York, NY 10017, USA
www.syrawoodpublishinghouse.com

Remote Sensing Techniques for Environmental Studies
Edited by Matt Weilberg

International Standard Book Number: 978-1-68286-692-4 (Hardback)

Cataloging-in-Publication Data

Remote sensing techniques for environmental studies / edited by Matt Weilberg.
    p. cm.
Includes bibliographical references and index.
ISBN 978-1-68286-692-4
1. Earth sciences--Remote sensing. 2. Environmental monitoring--Remote sensing.
I. Weilberg, Matt.
QH541.15.M64 R46 2019
550--dc23

TABLE OF CONTENTS

Preface... VII

Chapter 1   **Earth observation archives for plant conservation: 50 years monitoring of Itigi-Sumbu thicket**..1
Susana Baena, Doreen S. Boyd, Paul Smith, Justin Moat and Giles M. Foody

Chapter 2   **Weather radar data correlate to hail-induced mortality in grassland birds**................13
Amber R. Carver, Jeremy D. Ross, David J. Augustine, Susan K. Skagen,
Angela M. Dwyer, Diana F. Tomback and Michael B. Wunder

Chapter 3   **Integrating LiDAR-derived tree height and Landsat satellite reflectance to estimate forest regrowth in a tropical agricultural landscape**................................25
T. Trevor Caughlin, Sami W. Rifai, Sarah J. Graves, Gregory P. Asner and
Stephanie A. Bohlman

Chapter 4   **Bridging disciplines with training in remote sensing for animal movement: an attendee perspective**..38
Bethany L. Clark, Mirjana Bevanda, Eneko Aspillaga and Nicolai H. Jørgensen

Chapter 5   **Investigating animal activity patterns and temporal niche partitioning using camera-trap data: challenges and opportunities**..46
Sandra Frey, Jason T. Fisher, A. Cole Burton and John P. Volpe

Chapter 6   **Will remote sensing shape the next generation of species distribution models?**..........56
Kate S. He, Bethany A. Bradley, Anna F. Cord, Duccio Rocchini,
Mao-Ning Tuanmu, Sebastian Schmidtlein, Woody Turner,
Martin Wegmann and Nathalie Pettorelli

Chapter 7   **Remote sensing training in African conservation**..71
Helen Margaret de Klerk and Graeme Buchanan

Chapter 8   **Building capacity in remote sensing for conservation: present and future challenges**..85
Ilaria Palumbo, Robert A. Rose, Rachel M. K. Headley, Janet Nackoney,
Anthony Vodacek and Martin Wegmann

Chapter 9   **Remote sensing of species dominance and the value for quantifying ecosystem services**..94
Stephanie Pau and Laura E. Dee

Chapter 10   **Observing ecosystems with lightweight, rapid-scanning terrestrial lidar scanners**..105
Ian Paynter, Edward Saenz, Daniel Genest, Francesco Peri, Angela Erb, Zhan Li,
Kara Wiggin, Jasmine Muir, Pasi Raumonen, Erica Skye Schaaf,
Alan Strahler and Crystal Schaaf

Chapter 11   **Satellite remote sensing to monitor species diversity: potential and pitfalls**................................121
Duccio Rocchini, Doreen S. Boyd, Jean-Baptiste Féret, Giles M. Foody,
Kate S. He, Angela Lausch, Harini Nagendra, Martin Wegmann and
Nathalie Pettorelli

Chapter 12   **A simple remote sensing based information system for monitoring sites of
conservation importance**..133
Zoltan Szantoi, Andreas Brink, Graeme Buchanan, Lucy Bastin, Andrea Lupi,
Dario Simonetti, Philippe Mayaux, Stephen Peedell and James Davy

Chapter 13   **Sea turtle nesting patterns in Florida vis-à-vis satellite-derived measures of
artificial lighting**..142
Zachary A. Weishampel, Wan-Hwa Cheng and John F. Weishampel

Chapter 14   **More than counting pixels – perspectives on the importance of remote
sensing training in ecology and conservation**...156
Asja Bernd, Daniela Braun, Antonia Ortmann, Yrneh Z. Ulloa-Torrealba,
Christian Wohlfart and Alexandra Bell

Chapter 15   **A review of camera trapping for conservation behaviour research**................................166
Anthony Caravaggi, Peter B. Banks, A Cole Burton, Caroline M. V. Finlay,
Peter M. Haswell , Matt W. Hayward, Marcus J. Rowcliffe and Mike D. Wood

Chapter 16   **Upland vegetation mapping using Random Forests with optical and radar
satellite data**..180
Brian Barrett, Christoph Raab, Fiona Cawkwell and Stuart Green

Permissions

List of Contributors

Index

PREFACE

Remote sensing techniques are used in acquisition of real-time data from all accessible and inaccessible locations mapped over time and space. The use of remote sensing technology for the topographical mapping of wildlife populations, natural resources, vulnerable ecosystems and pollution provides valuable data for environmental management and monitoring. Remote sensing has also been incorporated in the study of fields like forestry, climatology, glaciology, hydrology and geology. This book explores the applications of remote sensing in environmental studies and conservation efforts. It also presents the effective techniques of incorporating sensor data to make the studies more relevant and informative. It will benefit researchers and students alike.

Various studies have approached the subject by analyzing it with a single perspective, but the present book provides diverse methodologies and techniques to address this field. This book contains theories and applications needed for understanding the subject from different perspectives. The aim is to keep the readers informed about the progresses in the field; therefore, the contributions were carefully examined to compile novel researches by specialists from across the globe.

Indeed, the job of the editor is the most crucial and challenging in compiling all chapters into a single book. In the end, I would extend my sincere thanks to the chapter authors for their profound work. I am also thankful for the support provided by my family and colleagues during the compilation of this book.

Editor

Earth observation archives for plant conservation: 50 years monitoring of Itigi-Sumbu thicket

Susana Baena[1,2], Doreen S. Boyd[2], Paul Smith[3], Justin Moat[1,2] & Giles M. Foody[2]

[1]Royal Botanic Gardens, Kew, Richmond, Surrey TW9 3AB, United Kingdom
[2]School of Geography, University of Nottingham, Nottingham NG7 2RD, United Kingdom
[3]Botanic Gardens Conservation International, Descanso House, 199 Kew Road, Richmond, Surrey TW9 3BW, United Kingdom

Keywords
Accuracy assessment, change detection, deforestation, earth observation data archives, Itigi-Sumbu thicket, plant conservation, red list of ecosystems

Correspondence
Susana Baena, Royal Botanic Gardens, Kew, Richmond, Surrey TW9 3AB, United Kingdom.
E-mail: s.baena@kew.org

Funding Information
No funding information provided.

Editor: Harini Nagendra
Associate Editor: Ned Horning

Abstract

Itigi thicket is a spatially restricted ecosystem only present in Zambia and Tanzania. It is thought to be highly threatened and therefore we need to urgently assess the threats to this ecosystem as well as extent and rates of change to derive its true conservation status. In this study we focus on the Itigi-Sumbu thicket surrounding Lake Mweru Wantipa in Zambia, which occurs both inside and outside a National Park (IUCN category II). Earth observation data archives provide the means to assist the conservation assessment process by allowing the monitoring of the ecosystem over time. In particular, the Landsat archive offers over 40 years of imagery at a resolution suited to the distribution of this ecosystem. In this study we exploit this archive and extend it back 50 years using historical aerial photography. The remote-sensing data were classified according to presence of thicket at five dates across a 50-year period and these outputs were combined to produce a deforestation map. Crucially, this map was assessed for accuracy using a novel approach to expert knowledge, which shows that the resultant map is highly accurate (93% overall accuracy). A confusion matrix was used to provide a confidence interval to the deforestation figures. Results indicate that 64% of the Itigi-Sumbu thicket around Lake Mweru Wantipa has been cleared over the last 50 years and that the largest area of remaining thicket is currently situated within the Mweru Wantipa National Park. This deforestation figure provides the means to assess the conservation status of Itigi-Sumbu thicket as part of the Red List of Ecosystem as Endangered (EN).

Introduction

Itigi thicket is a low, dry forest consisting of a closed stand of bushes and climbers between 3 and 7 m tall (White 1983) and is found only in two small areas in Zambia and Tanzania. This dry, primarily deciduous and almost impenetrable vegetation type is unique in its species composition, featuring nearly one hundred woody plant species, many of them endemic (White 1983; WWF 2014).The entire ecosystem is believed to be threatened, with reported clearance of 50% in the Tanzanian portion (Kideghesho 2001) and as much as 71% in Zambia (Almond 2000). These statistics, however, are not verified and thus the conservation status and level of threat of the ecosystem remain unclear (WWF 2014).

In Zambia, this vegetation type (termed Itigi-Sumbu thicket) is highly sensitive to disturbance and this regression appears to be irreversible (Trapnell 1943; Fanshawe 1971); once the thicket has been cleared it does not regenerate and is replaced by a fire climax wooded grassland, Lake Basin chipya (Trapnell 1943; WWF 2014).The main threat to the Itigi-Sumbu thicket comes from rapid population growth resulting in an increasing demand for land and resources. This results in traditional, shifting 'citemene' slash and burn agriculture in which the thicket is cut down, and its branches piled up and burnt to produce ash which fertilizes the crop. Cassava is the main staple grown in the poor, sandy soils of the area and, after only a few years, the soils are quickly exhausted and the farmer moves on (Trapnell 1943). In addition, the

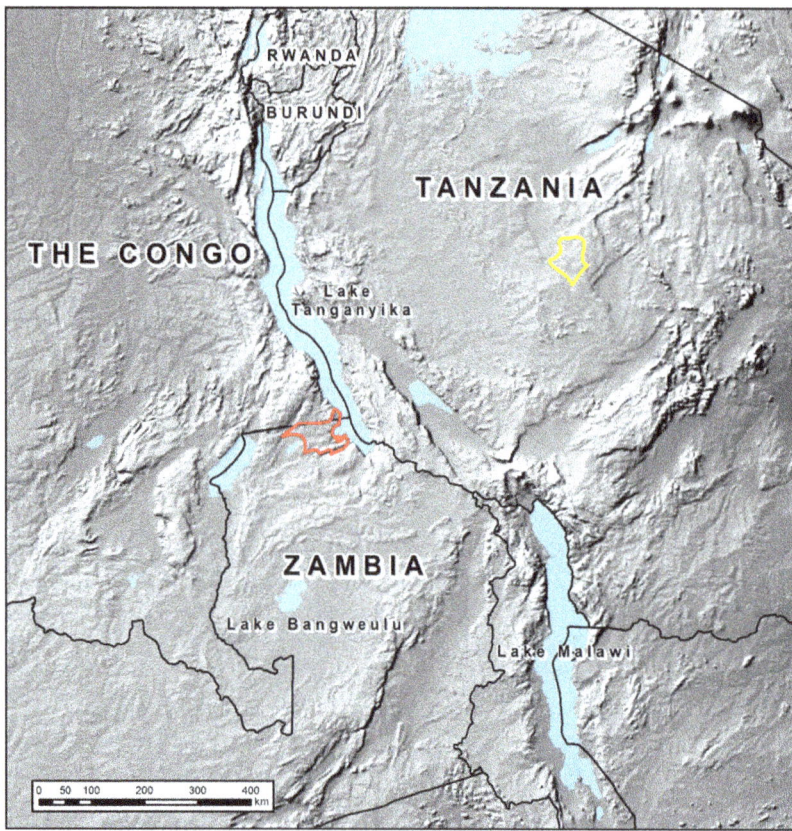

Figure 1. Location map showing areas of known patches of Itigi-Sumbu thicket in Zambia (red) and Tanzania (yellow) based on WWF terrestrial ecoregions (WWF 2014).

close proximity of thicket to Lakes Mweru Wantipa and Tanganyika means that the thicket has come under ever increasing pressure due to immigration associated with the fishing industry on these lakes (NORAD 1989).

The need to conserve the thicket has already been recognized. In 1972 the area along the western shore of Lake Mweru Wantipa, which contained large expanses of thicket, was declared a National Park (IUCN category II). When it was established, the national park had wildlife populations of lion, elephant and the black rhinoceros, all of these species benefiting from the cover provided by the thicket. However, the park's wildlife population has since been much reduced and black rhinoceros have disappeared as a direct consequence of the clearing of the thicket (Almond 2000). It is clear that the need to provide robust data (in terms of quantities and spatial distributions) on the loss of the thicket is pressing. The Red list of Ecosystems provides a scientific tool to document the conservation status of terrestrial ecosystems (http://iucnrle.org/). Five risk assessment criteria have been developed to form the basis of the analysis, one of them being "Decline in Distribution" (Criterion A). This criterion allows for an ecosystem to be assessed (along with other criteria to be fulfilled) based on the reduction in geographic distribution over the past 50 years (Rodriguez et al. 2015).

One approach to deriving data on thicket loss retrospectively is to mine the Earth Observation data (i.e. acquired via remotely sensing) archives. Earth Observation data provide a powerful way to directly observe large-scale ecosystems, consistently across time and space (Boyd 2009), particularly for biodiversity monitoring (Turner et al. 2015). Specifically, these data can contribute to monitoring protected areas and their geographical proximities (Rose et al. 2015). An area can be studied over a period of time using a time series of images with the same coverage acquired at dates of interest (epochs). Here, in this study, the Landsat series of satellites offer the most suitable approach to monitoring Itigi-Sumbu thicket cover over time. They provide repetitive, global coverage, multi-spectral imagery at a spatial scale where natural and human-induced changes can be detected and monitored over time (Markham et al. 2008; Skidmore et al. 2015). The Landsat archive offers over 40 years of imagery (since 1972) freely available to any user (Woodcock et al. 2008) and it will grow in the future given commitments to Landsat-type sensing systems (Committee on Implementation of a Sustained Land Imaging Program 2013). Often a time series of interest can be further extended back in time if complementary remotely sensed data are available. In this study we exploit the fact that

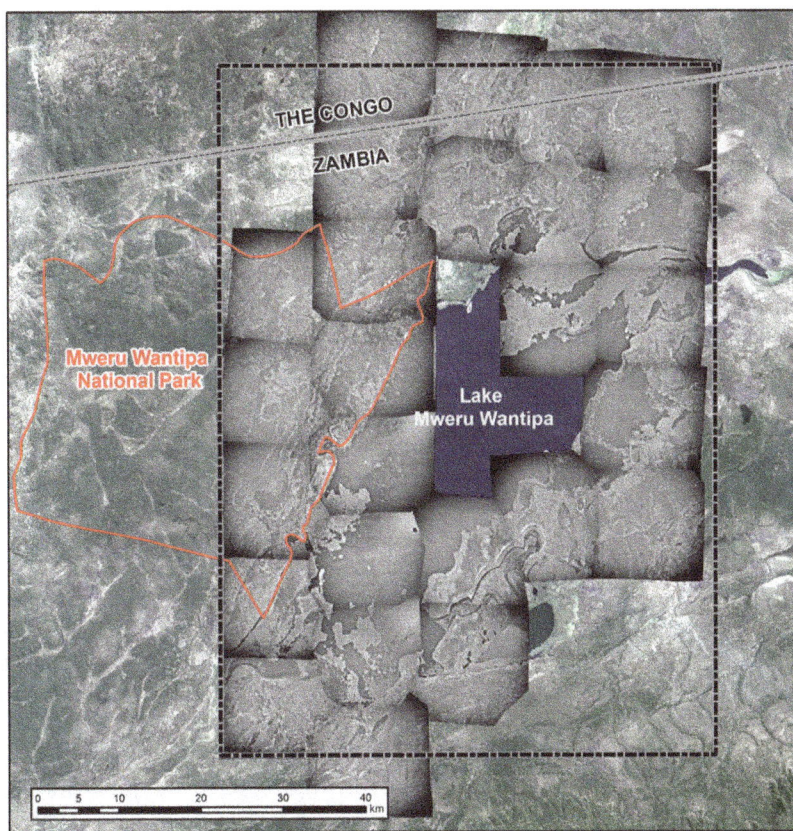

Figure 2. Study area (black box) immediately adjacent the shores of Lake Mweru Wantipa comprising known patches of Itigi-Sumbu thicket in Zambia. The red polygon shows the extent of Mweru Wantipa National Park. Overlapping areas contain the largest block of protected Itigi-Sumbu thicket. Black and White mosaic denotes the extent of aerial photography.

there were aerial photographs captured in this area in 1964 which afforded the extension of the time series of Earth observation data to the ideal of 50 years. Areas where aerial photographs are not available or difficult to obtain could take advantage of the opening up of declassified Corona data (Song et al. 2015) with worldwide spatial coverage and principal focus in Eastern Europe and Asia. This circa 2.5-m spatial resolution dataset would provide a suitable alternative to the aerial photographs used in this study for other areas and ecosystems globally.

It is a key time for the use of remotely sensed data for conservation (Pettorelli et al. 2015), where there is a realization of the intrinsic value of the archive of these data. However, correct practices must be employed in their use and the role of the conservationist is integral to this. The commonly accepted approach to monitoring land cover over time is via a supervised spectral classification which, among other steps, requires an accuracy assessment to be conducted. The remote-sensing science and application communities have developed increasingly reliable, consistent and robust approaches for doing this (Olofsson et al. 2013, 2014) and this ensures integrity in any statistics produced (i.e. for considering the conservation status of an ecosystem) and if used correctly can be extremely informative in ensuring the optimal value of remotely sensed data

(Foody 2015). In this study we determine the extent and rates of change in Itigi-Sumbu thicket around Lake Mweru Wantipa, Zambia, over the last 50 years using historical aerial photography in combination with archived Landsat data and expert knowledge.

Materials and Methods

Study site

Itigi-Sumbu thicket occurs in patches that are immediately adjacent the shores of Lake Mweru Wantipa in Northern Zambia, close to the international borders of the Democratic Republic of Congo and Tanzania (Fig. 1). This comprises an area of 5,000 km^2 in total, with a land cover which is a mosaic of Miombo woodland, Chipya, edaphic grasslands and Itigi-Sumbu thicket. Within the study area is the Mweru Wantipa National Park, comprising a 3134 km^2 area to the west of the lake (Fig. 2). The National Park contains the largest known block of protected Itigi-Sumbu thicket around Lake Mweru Wantipa providing a "best-case" scenario in terms of loss compared with the thicket beyond its protective borders. Expert knowledge alongside remotely sensed data (in particular historical aerial photography) was available in this study area. This was not the case for the other portion of

Figure 3. Itigi-Sumbu thicket in various stages of degradation. (A) Intact thicket; (B) degraded thicket with selective logging of larger trees and shrubs and (C) former thicket area cleared for cassava cultivation.

this area can be divided into three distinct seasons: the cool dry season running from May to August, the hot dry season from August to November and the rainy season from November to April. Rainfall is variable in both amount and distribution over time and location, year-to-year variation being as great as 50%. Itigi-Sumbu thicket is linked to soil type. The soils are typically grey-toned pinkish brown and pinkish buff sandy loams, with a higher clayey content in the subsoil and low humic content (Trapnell 1943). Itigi-Sumbu soils often show impeded drainage and, in some cases, a cement-like duricrust may be present (pers. obs Smith; Trapnell 1943; White 1983). As this unique soil structure is broken down by cultivation (p. Smith, unpubl. data), this is the most likely reason why the thicket does not regenerate after clearance, even if it is protected from fire (Fig. 3).

Physiognomically, the thicket is a two-storeyed forest (Fig. 3) with an open overwood of emergent trees, 6–12 m high over a dense thicket underwood (Fanshawe 1971). The understorey is so dense that very little light penetrates to the ground layer, with the result that ground vegetation is sparse, being confined to a few small herbs and grasses (White 1983). Fanshawe (1971) lists 92 woody species in Zambian Itigi, and notes that it shares only a quarter of its floristic composition with the Tanzanian type. The close proximity of the Zambian thicket to the Mbala local centre of plant endemism, and the lack of botanical exploration carried out in this vegetation type, suggests that Itigi-Sumbu thicket contains many undescribed and undocumented species.

Data

The monitoring of Itigi-Sumbu thicket was carried out using a combination of Landsat images and aerial photography to cover the study area on a frequent basis and over a time period of 50 years. Landsat images are provided as part of the GLS (Global Land Survey) dataset, a collection of images optimally selected for land-cover

protected thicket in Zambia, within Nsumbu National Park and the unprotected thicket in Tanzania (Fig. 1). Therefore, both of these areas of thicket were outside the scope of this study.

The study area lies typically between 950 and 1200 m above sea level in a relatively flat terrain. The climate for

Table 1. Listing of all data sets, including landsat images and aerial photography used in this study.

Collection	Path/row	Acquisition date	Instrument	Sensor	Spatial resolution*	Spectral resolution*
Royal Air Force	V1/543A/512 V2/543A/512	June 1964	Aerial Photography	PAN camera recorded on film	3 m	PAN
GLS1990	172/66	2nd June 1989	Landsat 5	Thematic mapper	30 m	MS: 6 bands
GLS2000	172/66	13th May 2002	Landsat 7	Enhanced Thematic Mapper(ETM+)	15 m / 30 m	PAN / MS: 6 bands
GLS2010	172/66	24th May 2009	Landsat 5	Thematic Mapper	30 m	MS: 6 bands
USGS	172/66	10th June 2015	Landsat 8	Operational Land Imager (OLI)	15 m / 30 m	PAN / MS: 8 bands

*Only those bands used in the analysis.

change studies (Gutman et al. 2008, 2013). The global GLS dataset has been created using the primary Landsat sensor of the time. The earliest GLS of 1975 acquired by the multispectral scanner did not allow for sufficient discrimination (i.e. to distinguish thicket from non-thicket in the supervised classification process) for this study (due to its limited spectral and spatial resolution). Therefore, only images from GLS1990 to 2010 (individual scene dates may range +/− 3 years) were selected. Specifically for this study we used images collected in 1989, 2002 and 2009. In addition, to bring the monitoring up to date one of the latest Landsat 8 images collected in 2015 was added to the dataset (the latter image was downloaded from the USGS).

Due to technological advances over the past 40 years of Landsat missions, the images used were from different sensors; Thematic Mapper (TM), Enhanced Thematic Mapper (ETM) and the Operational Land Imager (OLI). All images had the same spatial resolution (30 m). The accuracy of GLS 2000 data can be expected to be better than 25 m RMSE for most of its constituent scenes (Rengarajan et al. 2015). The GLS 2000 dataset is the geographic reference for the Landsat Data Continuity Mission (LDCM) and the rest of the GLS data products which have been registered to within one pixel geolocation accuracy (Tucker et al. 2004; Gutman et al. 2008). This geometric accuracy ensures any temporal change observed by the time series is a result of land-cover change rather than image rectification issues.

All Landsat images in the time series were acquired during the dry season (May and June), to minimize cloud cover but also changes due to phenological differences. To take the time series of remotely sensed data 50 years back from the present day (2015), a set of aerial photography was acquired. This was captured by the UK's Royal Air Force in June 1964 (series V1/543A/512 and V2/543A/512) by a panchromatic camera and recorded on film. A set of 26 frames divided into five flight line passes were acquired to cover the area. A summary of the full time series of remotely sensed data used is provided in Table 1.

Field data on the species composition and soils of the Itigi-Sumbu thicket were collected in June 1998 and in May 2012 by one of the authors of this study (Smith). GPS data from herbarium collections and soil sample sites were also used as reference data in this study.

Data processing and mapping of Itigi-Sumbu thicket

Change detection commonly uses a time series of remotely sensed data, applying different techniques depending on the nature of the dataset (Singh 1989; Coppin et al. 2004; Lu et al. 2004). For this study, dataset was uncalibrated (i.e. still at digital numbers) and with different spectral and spatial resolution at each time epoch (Thenkabail 2015). In such cases as this, post-classification comparison is the most practical approach as individual images are processed and classified independently based upon each scene's characteristics. Change is then considered at a later stage as the comparative analysis of the output at each time period (Singh 1989). Thus, the remotely sensed data at each epoch were processed and analysed independently to produce a map of Itigi-Sumbu thicket extent. Individual maps were subsequently combined to quantify the loss of Itigi-Sumbu thicket in the study area over the 50 years of study.

Aerial photography

The first step in using the aerial photography captured in 1964 was to apply a geometric correction so that these data could form the base of the time series. This was challenging, due to variation in pixel size across the set of photographs, the presence of strong geometric distortions introduced by the camera and the lack of photogrammetric information on the camera model. The aerial photographs were therefore registered to the 2002 Landsat 7 panchromatic image (spatial resolution of 15 m) using 16 ground control points selected from temporally stable ground features and adjusted using a third-level polynomial transformation. This was the polynomial function that yielded the lowest RMSE of the many tested using a set of eight check points to assess the accuracy of the geometric registration The RMSE obtained was 30 m which was deemed suitable to be used in a time series with the Landsat multispectral data which has a spatial resolution of 30 m.

The panchromatic aerial photographs (Fig. 3) provide limited spectral information to use in the automated mapping of the thicket. This restriction was further exacerbated by the poor quality of the available frames with obvious vignetting on mosaicking of the data to produce a continuous layer. Therefore, the aerial photography (scanned at 3-m resolution) was visually interpreted and digitized after gaining knowledge of the area from ground surveys of the Itigi-Sumbu thicket in 1998 and 2012. The result of this interpretation and digitizing process was a layer which delimited the extent of the thicket in 1964. This layer was then rasterized and resampled to 30-m resolution (to match Landsat imagery) using a majority filter to produce a map of the extent of Itigi-Sumbu thicket in 1964 that could form the basis of the time series.

Landsat multispectral images

The most notable characteristic of Itigi-Sumbu thicket is its dense, mixed evergreen and deciduous composition (Wild and Fernandes 1967; White 1983) which contrasts strongly with the other vegetation types in the area (i.e. Lake Basin Chipya and Miombo woodland) which are deciduous and have a less dense canopy. This feature was key when spectrally discriminating Itigi-Sumbu thicket in the Landsat imagery as the evergreen component of the Itigi-Sumbu thicket remains active in the dry season and thus stands out in its spectral response against the rest of the vegetation in the study area. This enabled successful mapping of the thicket when employing a spectral classifier in a supervised classification (i.e. an automated method of mapping thicket extent from each image at each date).

As Itigi-Sumbu thicket is the priority habitat and the other vegetation types were secondary to the study, there was no need to accurately map all vegetation types. Thus, the supervised classification approach adopted here focussed effort principally on the class of interest, Itigi-Sumbu thicket, at all stages in the classification process (i.e. class training, class allocation and accuracy assessment). Support vector machine (SVM)-based classifications have been demonstrated to be especially suitable for mapping priority habitats as the focus of the classification process is on the vegetation type of interest; resulting in a reduction in the number of training sites required (Sanchez-Hernandez et al. 2007a,b; Li et al. 2011; Stenzel et al. 2014).

Each Landsat image in the time series (i.e. 1989, 2001, 2009 and 2015) was classified to produce a map of thicket extent. The first step in any supervised classification is the training process, in which a dataset that spectrally describes each class to be mapped is produced for machine learning (i.e. the SVM). The challenge here was that coincident information on the thicket for each date of mapping was not available. However, as Itigi-Sumbu thicket has a relict ecology in that once cleared it does not regenerate, the training process exploited this as a rule; if the thicket was present in the study area in 2015 it must have been present throughout the time series, and this was confirmed on the base data of the aerial photography. Thus, a set of 150 training pixel locations for the thicket class was generated from high-resolution imagery (via the Google Earth platform) that was temporally coincident with the 2015 Landsat image.

The Google Earth high-resolution imagery has a horizontal positional accuracy that is sufficient for assessing moderate-resolution remote-sensing data such as those from Landsat (Potere 2008) and thus suitable here. Consequently, the 150 locations were applied to the 1989,

2001, 2009 and 2015 images to form unique spectral training sets for each image to be used by the SVM classifier. For non-thicket areas, comprising all remaining vegetation types and land covers, 150 training sites were randomly selected from the beginning of the time series with the assumption that they remained non-thicket throughout. A SVM supervised classification was performed on each image at each date to produce a thematic map with the extent of Itigi-Sumbu thicket.

The final step in a land-cover classification is to provide an accuracy assessment of a thematic map produced. As contemporary testing data (at each epoch) were not available and could not employ the same rule as used in the training step, it was accuracy of the final thicket deforestation map that was assessed (see below).

Post-classification change detection and thicket deforestation map

The resultant maps showing the extent of Itigi-Sumbu thicket at each date were combined to produce a map of deforestation of Itigi-Sumbu thicket throughout the 50 years (Fig. 4). This produced a final map which has five

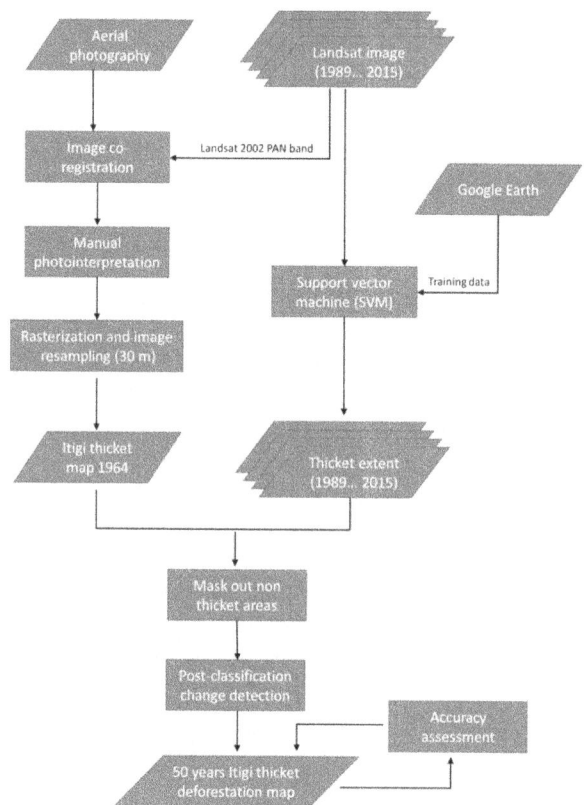

Figure 4. Work flow. Deforestation mapping procedure showing each step of the analysis and the generated final products.

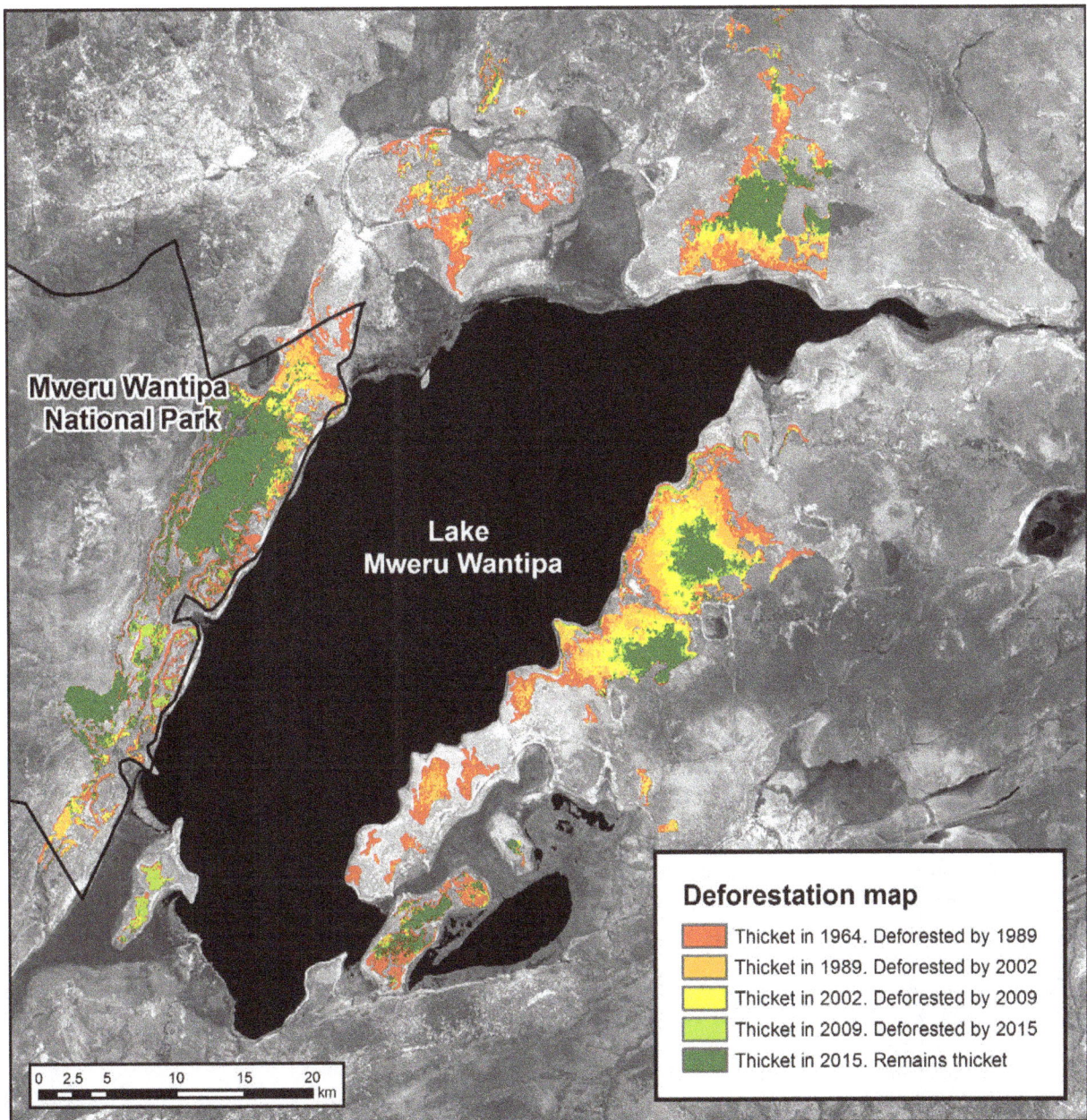

Figure 5. Itigi-Sumbu thicket deforestation map of the study area in and outside of Mweru Wantipa National Park showing the progressive loss of thicket through the 50 years' time span at the five epochs covered in the study (1964, 1989, 2002, 2009 and 2015). Each deforestation class represented in the map provides information on the time span of when the Itigi-Sumbu thicket was lost, e.g. Class 1 shows the extent of the Itigi-Sumbu thicket lost between 1964 and 1989.

classes, each one providing information about where and what period the Itigi-Sumbu thicket was lost. As stated earlier, it was this final map that was assessed for accuracy. A fundamental recommendation of good practice when assessing the accuracy of change maps is that the dataset used for validation is of higher quality change information than that used for training the classification (Olofsson et al. 2014). For change maps going back in time the main issue

is the lack of sources for contemporary validation data; Google Earth provides an extremely useful source of high-resolution data suitable for validation (Knorn et al. 2009; Schneider et al. 2010; Giri et al. 2011), but when the time span of the study goes back beyond the era of high-resolution commercial satellites Google Earth is no longer an option. In those cases historical aerial photography is often used for validation (Lunetta et al. 2006), but for this study,

Table 2. Agreement error matrix – there is full agreement in the labelling procedure by all three experts.

		Reference data					
		Class 1	Class 2	Class 3	Class 4	Class 5	Row total
Classified data	Class 1	86	1	1	0	10	98
	Class 2	1	58	1	0	0	60
	Class 3	0	2	31	0	2	35
	Class 4	0	0	0	19	4	23
	Class 5	0	0	0	2	133	135

	Reference totals	Classified totals	Number correct	Producers accuracy (%)	Users accuracy (%)
Class 1	87	98	86	98.85	87.76
Class 2	61	60	58	95.08	96.67
Class 3	33	35	31	93.94	88.57
Class 4	21	23	19	90.48	82.61
Class 5	149	135	133	89.26	98.52
Totals	351	351	327		
Overall classification accuracy					93.16

the only existing set of aerial photography was used as part of the time series.

High-quality validation data were obtained from three experts directly assessing the final change map rather than the intermediate products (i.e. the maps of Itigi-Sumbu thicket extent at each date). Images were visually enhanced by pan-sharpening methods (where a panchromatic band was available) and/or contrast-stretching techniques to improve visual interpretation. Following best practice (Olofsson et al. 2013, 2014), a probability sample design, here simple random sampling, was used to select 385 validation locations. The sample size selection was informed using sampling theory (Foody 2009; Fleiss et al. 2013). Specifically, for planning purposes it was assumed that the mapping would have an accuracy of 85%, based on a widely used target in remote sensing, and that the half width of the desired 95% confidence interval should be <3% which indicates at least 306 sites be acquired. Each expert was then provided with a project with all enhanced individual images and the 385 validation locations (displayed with black bulls eye symbols with transparent centre for better visibility) and asked to record for each point the date when thicket cover was lost, starting with the 2015 image and working backwards querying out points that had already been assigned. The working scale was set to 1:30,000 to accommodate for differences in image resolution.

Results showed a small amount of uncertainty in the decisions of the three experts (label disagreements in 34 of the 385 locations). Therefore, a confusion matrix was produced using only those points where all three experts agreed on the label, reporting estimates of overall accuracy, user's accuracy (commission error) and producer's accuracy (omission error).

The extent of thicket in each of the epochs was used to calculate the rate of deforestation. This was done temporally, across the 50-year time span, as well as spatially, looking at deforestation rates inside and outside the Mweru Wantipa National Park. The National Park boundaries were downloaded from the World Database on Protected Areas (WDPA) (UNEP-WCMC 2016). Percentage deforestation rates were established by dividing the thicket loss in each epoch (A1−A2) by the original extend (A1), and converted into average annual deforestation rates dividing it by the time difference (t): Average annual deforestation rate (%) = ((A1−A2/A1) 1/t) × 100.

Results

The final map of Itigi-Sumbu thicket deforestation in the study area throughout the 50 years is shown in Figure 5. The accuracy assessment of this map illustrates that the mapping accuracy is extremely high. Taking the error matrix where there is full agreement between each of the three experts (Table 2), it is evident that the overall accuracy of the deforestation map at 93% far exceeds that of the generally accepted accuracy of 85% for land-cover classifications (Foody 2002). Of the 351 validation points finally employed (using only those where all three experts agreed on the label) only 24 were allocated to the wrong deforestation class and in almost every case it was confused with the neighbouring date. All classes in the final deforestation map had producer's accuracies over 89%. Again this exceeded the mapping target.

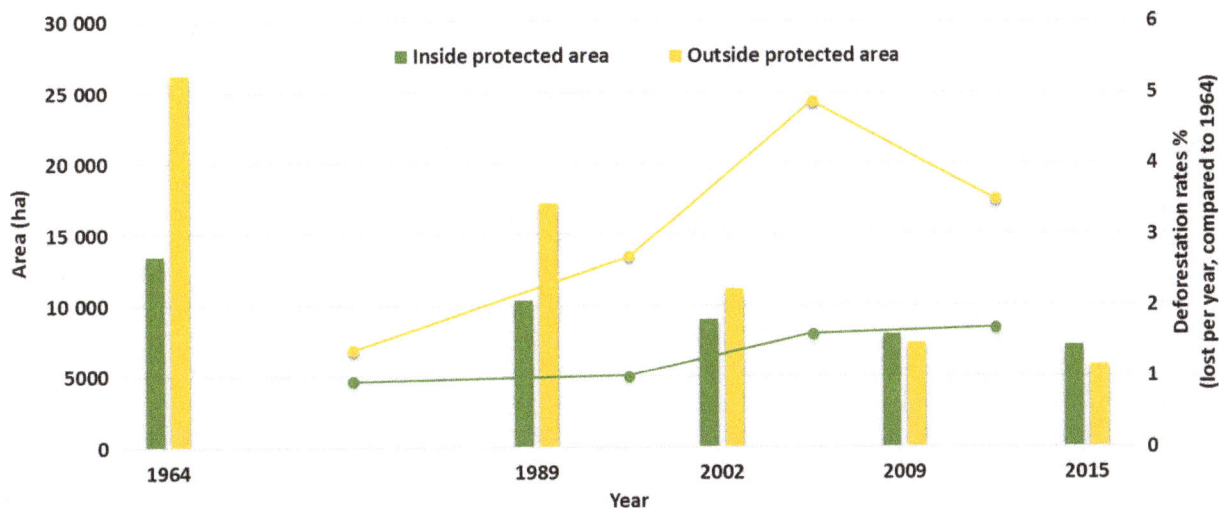

Figure 6. Annual deforestation rates during the 50-year time span of the study.

Employing a rigorous accuracy assessment, as stipulated by Olofsson et al. (2014), allows the adjustment of the amount of thicket mapped at each date. To do so is particularly important here, given the conservation importance of this vegetation. The confusion matrix reports that at the 2015 epoch, of the 351 testing locations 135 had been mapped as remaining thicket by the SVM classifier (i.e. the producer's accuracy). However, 149 of the 351 testing locations were in fact still thicket and therefore in error by ~4%. This error can be applied to the rest of the mapping for that class (thicket remaining in 2015) to adjust the areal estimate of that class and subsequently the deforestation rates. This calculation of required adjustment was applied using the confusion matrix also to the deforestation estimates at each epoch and used to adjust the thicket deforestation rates.

Figure 6 illustrates the extent of remaining thicket at each date along with deforestation rates. The deforestation map (Fig. 5) indicates that 64.8% of the total thicket area has been cleared over the last 50 years. Most of this area (30.6%) was lost before 1989, but the rate of deforestation rapidly increased between 2002 and 2009 reaching annual deforestation rates of 3.4% during this period. When comparing the deforestation patterns inside and outside the reserve, it is evident that Mweru Wantipa National Park experienced the greatest thicket loss during the first period (1964–1989), when 23% of the original thicket extent was lost. By 2015 it is clear that the largest area of the Itigi-Sumbu thicket remaining in the vicinity of Lake Mweru Wantipa lies within the Mweru Wantipa National Park.

These deforestation figures can be used as part of the Red List of Ecosystems framework to provide a conservation assessment for the Itigi-Sumbu thicket. Our results show a 64.8% deforestation rate for the last 50 years in Zambia; this along with the 50% deforestation rate reported on the Tanzanian side (Kideghesho 2001) are within the bounds of 50–80%; this gives Itigi-Sumbu thicket a preliminary ecosystem assessment of Endangered (EN) under criterion A1.

Discussion

The results obtained in this study confirm that the Itigi-Sumbu thicket by Lake Mweru Wantipa in Zambia has been declining over the past 50 years and shows that the thicket is highly threatened in this locality. Given that this area is one of just three areas of this thicket worldwide this is a cause for concern. The thicket deforestation figures agree broadly with the findings of Almond (2000); however, we extended the period of study to afford consideration of this vegetation type by conservation initiatives such as the Red List of Ecosystems. Although our findings indicate lower deforestation rates than those previously reported by Almond (2000), it is evident that removal of thicket has been ongoing for a long time on the shores of Lake Mweru Wantipa and current ecological understanding of this thicket suggests that this decline is irreversible. Lake Mweru Wantipa (as well as Lake Tanganyika) attracts immigrants due to their fishing industries (NORAD 1989) and the thicket in the vicinity of the lake is particularly vulnerable because it occurs on accessible, flat land adjacent the lake shore. Clearing takes two forms: (1) slash and burn 'citemene' agriculture, primarily for the cultivation of cassava (see Fig. 3C) and (2) selective logging of trees and larger shrubs for firewood and charcoal for cooking and smoking fish (Fig. 3B). The peak in deforestation rates recorded between 2002 and

2009 corresponds with a large influx of refugees in the area from political upheavals in the Democratic Republic of Congo. After the second Congo war finished in 2003, many Congolese became naturalized, settled and started farming in the area (Zytkow C. pers. comm.).

With respect to spatial patterns, it is evident that the clearance of thicket has been occurring within, as well as outside the boundaries of the Mweru Wantipa National Park. Inside the park, substantial deforestation (23%) occurred between 1964 and 1989, much of it probably before the park was established in 1972. This cannot be confirmed though, as early Landsat data (i.e. in the period 1972–1982) were considered unsuitable for use due to their poorer spectral resolution (the MSS sensor). If these data could be used then such information would assist in understanding further the patterns of loss immediately prior to and following the establishment of the National Park. More recently, deforestation rates inside the park are substantially lower than those outside, indicating conservation effort is having an effect. The Zambian National Parks and Wildlife Department conducts patrols regularly with the aim of controlling encroachment and illegal farming within the park (Zytkow C. pers. comm.). A visual inspection of the Landsat images (supported by knowledge of the area) does, however, indicate that inside the park there is substantial degradation of thicket canopy cover, indicating that the selective removal of larger trees is an ongoing problem even if encroachment is under control. This occurrence of degraded thicket could also explain the overestimate in deforestation by the spectral classification approach here (i.e. at ~4%). Notwithstanding this problem, Mweru Wantipa National Park probably represents the best opportunity to conserve the Itigi-Sumbu thicket in the environs of Lake Mweru Wantipa. Remote sensing should be used to continue to monitor the thicket; however, a different approach would be required to ensure that mapping of degraded thicket, in addition to intact thicket, is achieved. This would provide further information with which to target conservation effort.

From this study, the power of the Landsat data archive for conservation efforts is clear. In this case the archive was extended back to an ideal of 50 years (as specified by the Red List of Ecosystems) and the spectral classification approach adopted suited the mapping of Itigi-Sumbu thicket in this landscape. The thicket is spectrally distinctive, which when coupled with its relict ecology (i.e. does not regenerate), allowed the exploitation of the Landsat archive without full reference data (i.e. for training and testing of the land-cover classifications at each time stamp). The achievement of an overall classification accuracy of 93% in the final deforestation map demonstrates just how useful this approach was.

The real benefit here is the combination of expert knowledge with a full understanding of how remotely sensed data can be used; something to be further encouraged (Wang et al. 2010). Of particular note when making proper use of remotely sensed data was the use of the best practice approach to estimating area and assessing accuracy of land change, which often is not conducted (Olofsson et al. 2013, 2014). By employing the implicit information in the confusion matrix an adjustment for error could be applied to the deforestation (thicket cleared) estimates which is particularly important within conservation efforts. The requirement that estimates of land cover (and derived data such as clearance, etc.) should reduce any known uncertainties as much as practicably possible has indeed been noted (e.g. UN-IPCC Good Practice Guidance for Land Use, Land Use Change and Forestry) (Penman et al. 2003); but often this has not been the case (Olofsson et al. 2013).

A number of limitations of this study have been noted. The lack of information on deforestation around the time of establishment of the National Park requires future attention to fully understand the patterns in thicket clearance. Focus here would be on overcoming the lack of spectral detail by improving the spatial resolution of the early Landsat MSS data. Approaches such as super-resolution mapping (SRM) (Muad and Foody 2012) could, for instance, be adopted here. Going forward, however, such limitations occurring as a result of the trade-off between spectral, spatial and temporal resolutions and having in this case to emphasize the temporal detail will start to become less important. This will be of great benefit to conservation efforts. The Landsat archive has already seen improved spectral and radiometric resolution with the launch of Landsat 8 (Irons et al. 2012) and indeed the power of its satellites demonstrated for change detection at a global scale (Hansen et al. 2013) and its application to conservation (Tracewski et al. 2016).

Going forward and looking into future conservation assessments we are on the cusp of the Sentinel era with the recently launched Sentinel 2A awaiting its partner satellite (2B due for launch) and Sentinel 3′s launch imminent. These freely available data via the ESA Copernicus programme are assured for beyond 2030 and available to all via the Scientific Data Hub (European Space Agency 2016). Indeed conservation has been identified as a beneficiary here (Turner et al. 2015). There will still be a need, however, for suitable ground data to support the processing of remotely sensed data, but a growing awareness of the role of remote sensing in conservation should afford targeted ground data efforts. The Web 2.0 era should afford the exchange of such data. We have already seen how the rise of Volunteered Geographic Information (VGI – both formal and informal) (Foody and Boyd

2013) data exchange platforms such as Google Earth (Google Earth 2016) could benefit conservation.

Overall, this research contributes to the understanding of the conservation status of and level of threat to the Itigi-Sumbu thicket in the vicinity of Lake Mweru Wantipa, Zambia. Our deforestation results support the assessment of this ecosystem as Endangered, but it is recommended that the remaining Itigi thicket (both in Zambia and Tanzania) be fully surveyed. Doing so will also confirm whether the Zambian thicket is a separate vegetation type from the Tanzanian type as it has been reported that they share only a quarter of their floristic composition (Fanshawe 1971). If a full survey confirms that this is a shared ecosystem between both countries, a deforestation study should be conducted on the Tanzanian side. This assessment is based on reported figures of 50% deforestation (Kideghesho 2001) for Itigi thicket in Tanzania, as of 2001, and if in the Tanzanian side deforestation rates resemble that of Zambia, then the rate of deforestation is likely to have rapidly increased after 2002. Deforestation rates >80% would place this ecosystem under the critically endangered category. Remote sensing affords such a survey, if suitable expert knowledge or supporting information is available.

Acknowledgements

We would like to thank our colleagues Jonathan Timberlake, Tim Wilkinson and Sara Barrios at Royal Botanic Gardens, Kew and Craig Zytkow, CEO, Conservation Lake Tanganyika, for their invaluable knowledge and input in this study.

References

Almond, S., 2000. Itigi thicket monitoring using Landsat TM imagery. MSc. *Remote sensing dissertation*. University College, London.

Boyd, D. S. 2009. Remote sensing in physical geography: a twenty-first century perspective. *Prog. Phys. Geogr.* **33**, 451–456.

Committee on Implementation of a Sustained Land Imaging Program. 2013. *Landsat and beyond: sustaining and enhancing the nation's land imaging program*. National Research Council, Washington DC.

Coppin, P., I. Jonckheere, K. Nackaerts, B. Muys, and E. Lambin. 2004. Review Articledigital change detection methods in ecosystem monitoring: a review. *Int. J. Remote Sens.* **25**, 1565–1596.

European Space Agency. 2016. ESA Sentinel Online. Available at: https://sentinel.esa.int/web/sentinel/home (accessed: 10 June 2016).

Fanshawe, D. B. 1971. *The vegetation of Zambia*. Government Printer, South Africa.

Fleiss, J. L., B. Levin, and M. C. Paik. 2013. *Statistical methods for rates and proportions*. John Wiley & Sons, Hoboken, NJ.

Foody, G. M. 2002. Status of land cover classification accuracy assessment. *Remote Sens. Environ.* **80**, 185–201.

Foody, G. M. 2009. Sample size determination for image classification accuracy assessment and comparison. *Int. J. Remote Sens.* **30**, 5273–5291.

Foody, G. M. 2015. Valuing map validation: the need for rigorous land cover map accuracy assessment in economic valuations of ecosystem services. *Ecol. Econ.* **111**, 23–28.

Foody, G. M., and D. S. Boyd. 2013. Using volunteered data in land cover map validation: mapping west African forests. *IEEE J. Select. Topics Appl. Earth Obs. Remote Sens.* **6**, 1305–1312.

Giri, C., E. Ochieng, L. L. Tieszen, Z. Zhu, A. Singh, T. Loveland, et al. 2011. Status and distribution of mangrove forests of the world using earth observation satellite data. *Glob. Ecol. Biogeogr.* **20**, 154–159.

Google Earth. 2016. Google Earth Outreach. Available at: http://www.google.com/earth/outreach/index.html(accessed: 10 June 2016).

Gutman, B. G., R. Byrnes, J. Masek, S. Covington, C. Justice, S. Franks, et al. 2008. Towards Monitoring Changes at a Global The Global Land Su Land-Cover and Land-Use Scale : rvey (2005). 6–10.

Gutman, G., C. Huang, G. Chander, P. Noojipady, and J. G. Masek. 2013. Assessment of the NASA–USGS Global Land Survey (GLS) datasets. *Remote Sens. Environ.* **134**, 249–265.

Hansen, M. C., P. V. Potapov, R. Moore, M. Hancher, S. a. Turubanova, A. Tyukavina, et al. 2013. High-resolution global maps of 21st-century forest cover change. *Science* (New York, N.Y.)**342**, 850–853.

Irons, J. R., J. L. Dwyer, and J. A. Barsi. 2012. The next Landsat satellite: the landsat data continuity mission. *Remote Sens. Environ.* **122**, 11–21.

Kideghesho, J. R. 2001. The status of wildlife habitats in Tanzania and its implications to biodiversity. *Tanzania Wildl.* **21**, 9–17.

Knorn, J., A. Rabe, V. C. Radeloff, T. Kuemmerle, J. Kozak, and P. Hostert. 2009. Land cover mapping of large areas using chain classification of neighboring landsat satellite images. *Remote Sens. Environ.* **113**, 957–964.

Li, W., Q. Guo, and C. Elkan. 2011. A positive and unlabeled learning algorithm for one-class classification of remote-sensing data. *Geosci. Remote Sens.*, IEEE Transactions on **49**, 717–725.

Lu, D., P. Mausel, E. Brond'izios, and E. Moran. 2004. Change detection techniques. *Int. J. Remote Sens.* **25**, 2365–2407.

Lunetta, R. S., J. F. Knight, J. Ediriwickrema, J. G. Lyon, and L. D. Worthy. 2006. Land-cover change detection using multi-temporal MODIS NDVI data. *Remote Sens. Environ.* **105**, 142–154.

Markham, B. L., P. W. Dabney, N. Gsfc, J. C. Storey, E. J. Knight, and K. Lee. 2008. Landsat data continuity mission calibration and validation. *Pecora* **17**, 18–20.

Muad, A. M., and G. M. Foody. 2012. Super-resolution mapping of lakes from imagery with a coarse spatial and fine temporal resolution. *Int. J. Appl. Earth Obs. Geoinf.* **15**, 79–91.

NORAD. 1989. Environmental effects of agricultural change and development in the Northern Province, Zambia. A report prepared for the Provincial Planning Unit, Northern Province, Govt. of Republic of zambia and the Norwegian Agency for International Development (NORAD).

Olofsson, P., G. M. Foody, S. V. Stehman, and C. E. Woodcock. 2013. Making better use of accuracy data in land change studies: estimating accuracy and area and quantifying uncertainty using stratified estimation. *Remote Sens. Environ.* **129**, 122–131.

Olofsson, P., G. M. Foody, M. Herold, S. V. Stehman, C. E. Woodcock, and M. Wulder. 2014. Good practices for estimating area and assessing accuracy of land change. *Remote Sens. Environ.* **148**, 42–57.

Penman, J., M. Gytarsky, T. Hiraishi, T. Krug, D. Kruger, R. Pipatti, et al. 2003. *Good practice guidance for land use, land-use change and forestry.* Institute for Global Environmental Strategies, Kanagawa, Japan.

Pettorelli, N., H. Nagendra, R. Williams, D. Rocchini, and E. Fleishman. 2015. A new platform to support research at the interface of remote sensing, ecology and conservation. *Remote Sen. Ecol. Conserv.* **1**, 1–3.

Potere, D. 2008. Horizontal Positional accuracy of google earth's high-resolution imagery archive. *Sensors* **8**, 7973–7981.

Rengarajan, R., A. Sampath, J. Storey, and M. Choate. 2015. Validation of geometric accuracy of global land survey (GLS) 2000 data. *Photogramm. Eng. Remote Sensing* **81**, 131–141.

Rodriguez, J. P., D. A. Keith, K. M. Rodriguez-Clark, J. Murray, E. Nicholson, T. J. Regan, et al. 2015. A practical guide to the application of the IUCN Red List of Ecosystems criteria. *Philos. Trans. R. Soc. B: Biol. Sci.* **370**, 20140003.

Rose, R. A, D. Byler, J. R. Eastman, E. Fleishman, G. Geller, S. Goetz, et al. 2015. Ten ways remote sensing can contribute to conservation. *Conserv. Biol.* **29**, 350–359.

Sanchez-Hernandez, C., D. S. Boyd, and G. M. Foody. 2007a. Mapping specific habitats from remotely sensed imagery: support vector machine and support vector data description based classification of coastal saltmarsh habitats. *Ecol. Inf.* **2**, 83–88.

Sanchez-Hernandez, C., D. S. Boyd, and G. M. Foody. 2007b. One-class classification for mapping a specific land-cover class : SVDD classification of Fenland. *IEEE Trans. Geosci. Remote Sens.* **45**, 1061–1073.

Schneider, A., M. A. Friedl, and D. Potere. 2010. Mapping global urban areas using MODIS 500-m data: new methods and datasets based on "urban ecoregions". *Remote Sens. Environ.* **114**, 1733–1746.

Singh, A. 1989. Digital change detection techniques using remotely-sensed data. *Int. J. Remote Sens.* **10**, 989–1003.

Skidmore, A. K., N. Pettorelli, N. C. Coops, G. N. Geller, M. Hansen, R. Lucas, et al. 2015. Environmental science: agree on biodiversity metrics to track from space. *Nature* **523**, 403–405.

Song, D. X., C. Huang, J. O. Sexton, S. Channan, M. Feng, and J. R. Townshend. 2015. Use of Landsat and Corona data for mapping forest cover change from the mid-1960s to 2000s: Case studies from the Eastern United States and Central Brazil. *ISPRS J. Photogramm. Remote Sens.* **103**, 81–92.

Stenzel, S., H. Feilhauer, B. Mack, A. Metz, and S. Schmidtlein. 2014. Remote sensing of scattered Natura 2000 habitats using a one-class classifier. *Int. J. Appl. Earth Obs. Geoinf.* **33**, 211–217.

Thenkabail, P. S.. 2015. *Remotely Sensed Data Characterization, Classification, and Accuracies.* CRC Press, Boca Raton, FL.

Tracewski, L., S. H. M. Butchart, P. F. Donald, M. Evans, L. D. C. Fishpool, and M. Graeme. 2016. Patterns of twenty-first century forest loss across a global network of important sites for biodiversity. *Remote Sens. Ecol. Conserv.* **2**, 37–44.

Trapnell, C. G. 1943. The soils, vegetation and agriculture of North-Eastern Rhodesia. Report of the Ecological Survey. Govt. Printer, Lusaka.

Tucker, C. J., D. M. Grant, and J. D. Dykstra. 2004. NASA 's Global Orthorectified Landsat Data Set. *Photogramm. Eng. Remote Sens.* **70**, 313–322.

Turner, W., C. Rondinini, N. Pettorelli, B. Mora, K. Leidner, Z. Szantoi, et al. 2015. Free and open-access satellite data are key to biodiversity conservation. *Biol. Conserv.* **182**, 173–176.

UNEP-WCMC, 2016. The World Database on Protected Areas (WDPA). Cambridge, UK: UNEP-WCMC. Available at: www.protectedplanet.net.(accessed: 10 June 2016).

Wang, K., S. E. Franklin, X. Guo, and M. Cattet. 2010. Remote sensing of ecology, biodiversity and conservation: a review from the perspective of remote sensing specialists. *Sensors* (Basel, Switzerland) **10**, 9647–9667.

White, F. 1983. *The vegetation of Africa, natural resources research 20.* United Nations Scientific and Cultural Organization, Paris.

Wild, H., and A. Fernandes. 1967. *Flora Zambesiaca supplement.* M.O. Collins, Salisbury, Rhodesia.

Woodcock, C. E., et al. 2008. Free Access to Landsat Imagery. *Science* (New York, N.Y.)**320**, 1011.

WWF. 2014. Itigi-sumbu thicket. Terrestrial Ecoregions. World Wildlife Fund. Available at: http://www.worldwildlife.org. (accessed 10 June 2016)

Weather radar data correlate to hail-induced mortality in grassland birds

Amber R. Carver[1], Jeremy D. Ross[2,3], David J. Augustine[4], Susan K. Skagen[5], Angela M. Dwyer[6], Diana F. Tomback[1] & Michael B. Wunder[1]

[1]Department of Integrative Biology, University of Colorado Denver, Campus Box 171, P.O. Box 173364, Denver, Colorado 80217-3364
[2]Sutton Avian Research Center, P.O. Box 2007, Bartlesville, Oklahoma 74005
[3]Oklahoma Biological Survey, University of Oklahoma, 111 E. Chesapeake St, Norman, Oklahoma 73019
[4]Rangeland Resources Research Unit, USDA-Agricultural Research Service, 1701 Centre Ave., Fort Collins, Colorado 80526
[5]U.S. Geological Survey, Fort Collins Science Center, 2150 Centre Ave, Bldg. C, Fort Collins, Colorado 80526
[6]Bird Conservancy of the Rockies, 230 Cherry St. Fort Collins, Colorado 80521

Keywords

Extreme weather events, grassland birds, hail, nest mortality, nest survival, nest vegetation, NEXRAD, severe weather

Correspondence

Amber R. Carver, Department of Integrative Biology, University of Colorado Denver, Campus Box 171, P.O. Box 173364, Denver, Colorado 80217-3364. E-mail: amber.carver@ucdenver.edu

Funding Information

This project was funded by a United States Fish and Wildlife Service State Wildlife Grant, administered through the Nebraska Game and Parks Commission. The project was made possible in part by research grants from the Denver Field Ornithologists and the Colorado Field Ornithologists

Editor: Nathalie Pettorelli
Associate Editor: Alienor Chauvenet

Abstract

Small-bodied terrestrial animals such as songbirds (Order Passeriformes) are especially vulnerable to hail-induced mortality; yet, hail events are challenging to predict, and they often occur in locations where populations are not being studied. Focusing on nesting grassland songbirds, we demonstrate a novel approach to estimate hail-induced mortality. We quantify the relationship between the probability of nests destroyed by hail and measured Level-III Next Generation Radar (NEXRAD) data, including atmospheric base reflectivity, maximum estimated size of hail and maximum estimated azimuthal wind shear. On 22 June 2014, a hailstorm in northern Colorado destroyed 102 out of 203 known nests within our research site. Lark bunting (*Calamospiza melanocorys*) nests comprised most of the sample ($n = 186$). Destroyed nests were more likely to be found in areas of higher storm intensity, and distributions of NEXRAD variables differed between failed and surviving nests. For 133 ground nests where nest-site vegetation was measured, we examined the ameliorative influence of woody vegetation, nest cover and vegetation density by comparing results for 13 different logistic regression models incorporating the independent and additive effects of weather and vegetation variables. The most parsimonious model used only the interactive effect of hail size and wind shear to predict the probability of nest survival, and the data provided no support for any of the models without this predictor. We conclude that vegetation structure may not mitigate mortality from severe hailstorms and that weather radar products can be used remotely to estimate potential for hail mortality of nesting grassland birds. These insights will improve the efficacy of grassland bird population models under predicted climate change scenarios.

Introduction

The damaging effect of large hail (≥2.5 cm diameter) is well documented in literature related to agriculture (Wille and Kleinkopf 1992; Vorst 1991) and commerce (e.g. Hohl et al. 2002; Das et al. 2009; Saa Requejo et al.

2011), but its contribution to wildlife mortality is understudied. This is due partly to economics and partly to the relative difficulty in measuring hail-induced animal mortality. Carcasses of hail victims may be difficult to detect or may be quickly scavenged after a severe hail event. This complicates measurement of hailstorm effects on populations and requires researchers to act quickly after mass mortality events. The best-documented cases of mass mortality during hailstorms pertain to birds (Hall and Harvey 2007; Saunders et al. 2011; Diehl et al. 2014), suggesting that birds are highly vulnerable to hail. Small-bodied, non-avian taxa may be similarly vulnerable, but hail-induced mortality of avian species that form large flocks may be more obvious due to the unusual concentration of carcasses. Information derived from opportunistic studies on avian hail mortality are of limited utility to demographic modeling without spatially comprehensive hail incidence data across a population's range. Our study demonstrates the value of remotely sensed weather data for overcoming this obstacle.

Breeding birds are among the taxa most vulnerable to hail-related mortality because of their physical attributes and life history. Although skeletal hollowness in birds is offset by increased bone tissue density and tensile strength (Dumont 2010), most birds are small-bodied (Bokma 2004), making it unlikely that they could survive the impact of large hail. Even relatively large-bodied birds experience broken bones and other internal damage when hit by large hail (Heflebower and Klett 1980). A compounding factor is the rarity of shelter use by birds. Most species do not use burrows or cavities for nesting but rather build open-cup or scrape nests with little overhead structure (Collias 1997). Nest-monitoring studies have documented nest destruction by hail in both shorebirds (Graul 1975; Mabee and Estelle 2000; Saalfeld et al. 2011) and songbirds (La Rivers 1944; Conway and Johnson 2011). Nests may serve as a proxy for population-wide mortality. Relative to mobile adults, nests are easier to locate and monitor. Thus, it can be easier to measure the impact of hail on nests than on adults. High adult mortality concurrent with nest destruction (*ARC, personal observation*) suggests that bird mobility does not reduce hail risk in breeding areas. The correlation between hail-induced nest failure and adult mortality is unknown, but the density of destroyed nests could serve as an index of adult mortality.

The North American Great Plains is an ecoregion where relatively low annual precipitation and high evaporative potential (Rosenberg 1987), combined with cyclical grazing and fire (Fuhlendorf and Engle 2001; Derner et al. 2009), favor grasses over taller, woody vegetation. These environmental drivers generate a suite of unique ecological niches, and several bird species have evolved to breed only on the Great Plains (Knopf 1996). Populations of these species have declined due to habitat loss (Brennan and Kuvlesky 2005). As of 2004, 70% of the Great Plains had been converted to agriculture or other land cover types (Samson et al. 2004). Other environmental changes likely contribute to population declines, but may be more difficult to quantify than habitat loss. Developing metrics

of other sources of mortality is important for estimating demographic parameters for grassland bird populations, which in turn is key to predicting population change and assessing population viability.

Hail is a common weather phenomenon in North America (Changnon 2008), and it can cause high bird mortality in areas where individuals are concentrated (e.g. Smith and Webster 1955; Johnson 1979; Diehl et al. 2014). Large-diameter hail regularly occurs on the Great Plains (Allen et al. 2015; Fig. 1). Further, the Great Plains breeding bird community is dominated by songbirds (Veech 2006), most of which construct open-cup nests (Collias 1997) that have little protection from hail. The frequency of hail mortality during grassland birds' evolutionary history may have been insufficient to drive an adaptive response. Increased hail frequency, as predicted under several climate change models (Trapp et al. 2007; Kapsch et al. 2012; Allen et al. 2015; Fischer and Knutti 2015), could lead to greater mortality in species that are not adapted to avoid hail impact. In combination with other threats such as habitat loss and fragmentation, higher frequency of severe hail could impair the ability of isolated populations to recover from these combined events (Saunders et al. 2011; McKechnie et al. 2012). Developing methods to quantify hail-induced bird mortality will be important for population monitoring if hail frequency increases.

Localized storm paths and intensity are randomly distributed (Morgan and Towery 1975) and not readily predicted. Thus, it would be challenging to design studies evaluating the impacts of naturally occurring extreme weather events; often the zone of highest intensity within a weather system may occur where populations are not under scientific observation. Quantifying the impact of hail on animal populations requires remote sensing technology that can be applied at large spatial scales and capture local variability in storm location and intensity. The metrics produced must distinguish between rain and hail, as well as between small and large hail, as these different hydrometeor classes are likely to cause very different outcomes for breeding birds.

Remote sensing technology that can provide metrics to quantify hailstorms is now available in the USA. The network of Next Generation Radars (NEXRAD) employed by the U.S. National Weather Service provides high-coverage data on atmospheric conditions. These stationary Doppler radars transmit pulses of electromagnetic radiation, and then coupled receiving sensors measure the intensity of returning radiation as decibels of reflectivity (dBZ). Output data are provided at fixed radial widths (measured in degrees) over five to fourteen elevations above the horizon, depending on operational mode (Lakshmanan et al. 2006). The base reflectivity (BR) represents the lowest elevational swath, with a beam centered 0.5° above the horizon and

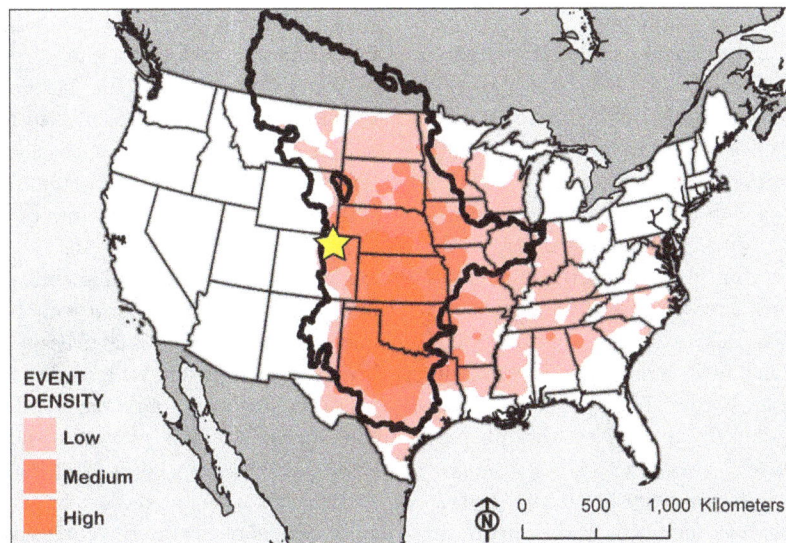

Figure 1. Distribution of United States large hail events in relation to The Great Plains ecoregion is indicated with the black polygon, derived from North America Level I ecoregion classification by the U.S. Environmental Protection Agency (https://www.epa.gov/eco-research/ecoregions). We obtained georeferenced records of U.S. hail events from the National Oceanic and Atmospheric Administration Storm Prediction Center Severe Weather GIS website (http://www.spc.noaa.gov/gis/svrgis/). This included 326 455 events documented in 1955–2015. We selected records with hail ≥2.5 cm, corresponding with the threshold for what is defined as damaging hail (Kapsch et al. 2012). This yielded 13 960 results. The region of highest large hail frequency (indicated in red) includes much of the Great Plains. The location of the study site is indicated with the yellow star.

measuring approximately $1° \times 1°$. Reflectivity is positively correlated with the density of precipitation and can be used to reliably distinguish between pure rain and rain mixed with hail (Kunz and Kugel 2015). Reflectivity and hail diameter, however, are not strongly correlated, although reflectivity readings in the upper end of the spectrum usually indicate large hail (Depue et al. 2007). NEXRAD also measures relative velocity of airborne objects, calculated from the Doppler shift within the reflected radar signal. Secondary products, such as the maximum estimated size of hail (MESH) (Witt et al. 1998; Stumpf et al. 2004) or maximum estimated azimuthal wind shear (AWS); (Smith and Elmore 2004), can then be extracted by applying algorithms to the reflectivity and relative velocity as part of the Warning Decision Support System – Integrated Information (WDSS-II; Lakshmanan et al. 2007). In this study, we demonstrate how these publicly accessible weather data (i.e. BR, MESH, and AWS) provide the means to estimate local storm intensity, which we relate to nest loss in a population of grassland-breeding birds.

This study leveraged a rare opportunity to study ecological impacts after a severe weather event on 22 June 2014 passed over an ongoing grassland songbird nest-monitoring project at the Central Plains Experimental Range (CPER) – a shortgrass steppe grazing research facility in northeastern Colorado. Similar to previous studies of hail-induced mortality in birds, we quantified hail-related mortality opportunistically, incidental to other research. However, unlike other studies, we went beyond anecdotal reporting and investigated the extent to which remotely sensed weather data could be used to predict the spatial distribution of nest destruction. Our specific objective was to assess the accuracy of NEXRAD data products in explaining the distribution of hail-induced nest mortality attributed to this single event. As part of this, we tested the hypothesis that the microhabitat at the nest site (i.e. surrounding vegetation) has an ameliorative factor in hail risk. We hypothesized that woody vegetation (shrubs and subshrubs) would shelter nests from hail impact and that vegetation cover directly over the nest bowl would be important to the probability of nest survival. We also hypothesized that vertical structure would be important to nest fate, since the buffering effect of vegetation against hail impact should increase with the height of the vegetation.

Materials and Methods

Study Area

We conducted this study at CPER, Weld County, Colorado, on the western edge of the Pawnee National Grassland. The climate is semi-arid, with most precipitation occurring in the summer months (Sala et al. 1992). Severe weather such as extreme rain and hail commonly occurs during summer, but its spatiotemporal distribution is heterogeneous and unpredictable (Rosenberg 1987). Vegetation at CPER is dominated by C4 shortgrasses (primarily *Bouteloua gracilis* and *B. dactyloides*), with smaller proportions of cacti, midgrasses and forbs (Lauenroth and Sala 1992). The woody plant community includes both tall shrubs and low subshrubs (Lee and Lauenroth 1994). Plant community composition and structure at the study site are heterogeneous due to grazing treatments and topoedaphic factors. Average vegetation height and percent cover by shrubs increase along a northeasterly axis. The CPER breeding bird community is primarily comprised of songbirds (Order Passeriformes). These small organisms are unlikely to survive the physical impact of large hail (i.e. ≥2.5 cm diameter), based on previous observations of mass mortality (Hall and Harvey 2007; Diehl et al. 2014). The dominant nest structure at CPER among these species is a partially recessed cup in the ground with little or no woody vegetation over the nest (*ARC, personal observation*). For ground-nesting species, nest-site vegetation may afford protection from wind, but the extent of overhead structure varies within and among species (With and Webb 1993).

Nest fates and measurements

From May to July 2014, we located nests by rope-dragging and completed high-frequency (2–4 day interval) nest checks as part of a larger study examining the relationship between nest survival and livestock grazing regime. As of 20 June 2014, we were monitoring 203 active nests, divided among species as follows: 186 lark bunting (*Calamospiza melanocorys*), seven horned lark (*Eremophila alpestris*), three Brewer's sparrow (*Spizella breweri*), two mourning dove (*Zenaida macroura*), two McCown's longspur (*Rhynchophanes mccownii*), two grasshopper sparrow (*Ammodramus savannarum*) and one western meadowlark (*Sturnella neglecta*). Lark buntings construct an open-cup nest on the ground, similar to most other grassland-breeding species; so, we treated it as representative of grassland open-cup ground-nesters.

On 23 June, the day after the hail event, we checked all previously active nests and recorded the fate of each; nests were classified as 'failed' if all eggs were destroyed or all chicks were killed, and as 'survived' if at least one undamaged egg or living chick remained. After the storm, any failed nest was attributed to the storm, although the interval since the previous nest check was either 3 or 4 days for every nest. In most cases (n = 74), this assumption was supported by the presence of crushed eggs or dead chicks. In many cases, a dead parent bird was also found on or near the nest. We classified nests with live chicks, whole eggs and/or a live adult on the nest as still active.

As part of our larger project, we measured and identified vegetation at as many nest sites as possible. This provided us with vegetation data for 133 ground nests and three shrub nests within the hail-affected nest sample. We omitted shrub nests from subsequent analyses due to the small sample size. To quantify vegetation cover, we used a modified version of the line-point intercept (LPI) method (Canfield 1941). We centered a 0.5 m × 0.5 m quadrat over the nest and took measurements within the quadrat at intersections of a 5 × 5 grid with 10-cm spacing. At each intersection, we aimed a laser beam at the ground directly beneath the point, and we recorded the number of contacts with any above-ground vegetation (by vegetation type), and whether the laser contacted bare ground, litter, or the basal portion of a plant at ground level. The laser projector, which was mounted on a tripod for stability, produced a 1.6 mm diameter beam (equivalent to the diameter of pin flags used by U.S. land management agencies for line-point intercept monitoring; Herrick et al. 2005). We quantified shrub cover as the number of intercepts by a shrub or subshrub. We quantified nest bowl cover as the number of intercepts by any vegetation type directly over the nest bowl. To quantify vertical structure, we used the Robel Pole method (Robel et al. 1970). We obtained low and high visual obstruction readings in each cardinal direction and calculated the means of low and high readings.

Hail event and NEXRAD data

The severe thunderstorm for this analysis passed over the CPER area between ~1600 and 1900 MDT on 22 June 2014. It produced large hail as it crossed our study area on a southeasterly trajectory. It was part of a larger system of storms that day that produced hailstones with a maximum diameter in the range of 1.6–2.8 cm (SPC 2016). We obtained base reflectivity data from the Cheyenne, Wyoming NEXRAD station (KCYS) for the period from 2135 GMT on 22 June to 0051 GMT on 23 June from the NOAA Climate Data Online website (https://www.ncdc.noaa.gov/cdo-web/). This station was located 27–37 km from nests in the study area, and the data obtained represented 29 sweeps of the area; 18 intra-sweep intervals were 6 min long, four were 12 min long and three were 46 min long. These reflectivity files are provided in binary format. We used the NOAA Weather and Climate Toolkit (WCT; http://www.ncdc.noaa.gov/wct/) to convert these data to vector format. The WCT can be used to convert binary files directly to raster format, but reflectivity data are collected as polygons, so conversion into vector format preserves more information than conversion into raster. Furthermore, the WCT produces relatively coarse rasters. Export to vector format with the WCT, followed by conversion into raster (res:

0.0069°×0.0069°) with Esri ArcMap v10.2.2 resulted in better raster resolution and higher fidelity between the binary file and the resulting raster. We used the ArcMap Extract Values to Points tool to identify the maximum BR across all radar sweeps for each specific nest location. We downloaded additional NEXRAD weather radar data products associated with our focal severe hail event from the National Severe Storm Laboratory's (NSSL) WDSS-II portal (http://ondemand.nssl.noaa.gov). The WDSS-II system incorporates reflectivity data and relative velocity data from all available elevational sweeps of all available NEXRAD data and runs these through separate algorithms to calculate the MESH and the maximum azimuthal wind shear (AWS). The NSSL pre-calculates MESH data by first combining all available NEXRAD data onto a three-dimensional grid (i.e. latitude, longitude and height; Lakshmanan et al. 2006) and then applying a modified Hail Detection Algorithm to produce MESH patterns over the selected landscape at 60-sec intervals (Witt et al. 1998; Stumpf et al. 2004). The composite grid that we ultimately were able to download is based upon MESH data accumulated over successive radar sweeps to produce areal swaths of hailfall. Azimuthal wind shear, which is a metric of abrupt horizontal wind shifts (e.g. downdrafts or tornadoes), is pre-calculated by the NSSL through application of a linear least squares derivative filter on relative velocity data (Smith and Elmore 2004). The maximum value of shear, in units of per-seconds (s^{-1}), is calculated across multiple radars and the lower 2 km of the atmosphere and then assigned to georeferenced pixels measuring 0.005° for both latitude and longitude. We downloaded the georeferenced MESH and AWS rasters from the WDSS-II portal for a broad time span (15:00–23:00 MDT on June 22, 2014) which fully encompassed the period of severe weather passage over the CPER study site (~1800–MDT).

Statistical analyses

For all nests in our sample, we compared distributions for failed and surviving nests for each of the NEXRAD variables using a two-sample non-parametric Kolmogorov–Smirnov test (K-S Test), which tests for equality of continuous probability distributions for two samples of the same measurement variable. We selected the K-S test because it is non-parametric and does not assume the data conform to a specific distribution; it merely tests for equality of two empirically derived distributions. We used K-S tests for the subset of nests where vegetation was measured, to assess whether that subsample might be representative of the full sample. We also used a K-S test to compare the distributions of vegetation variables between failed and surviving nests.

For the subset of ground nests where vegetation was measured ($n = 133$), we fit a set of 21 logistic regression models to predict the probability of nest survival as functions of NEXRAD and vegetation variables. As part of the process of constructing these models, we evaluated the potential for collinearity among variables by estimating the absolute value of the pairwise correlation coefficient $|r|$. Because $|r|$ was greater than 0.59 for all pairs of NEXRAD variables, we did not consider any model that included these variables independently. Base reflectivity measures only atmospheric density, whereas maximum estimated size of hail and maximum estimated azimuthal wind shear are more advanced metrics of estimated hail size and angular velocity respectively. We therefore omitted base reflectivity as a covariate and considered only models with the effects of hail size, wind shear and the interaction between the two. Mean low and high visual obstruction readings were moderately correlated ($r = 0.58$); so, we considered only mean low visual obstruction reading in subsequent analyses. Correlation between other vegetation variables was low (range of $|r|$: 0.20–0.41). Because the NEXRAD data provided such strong separation between failed and surviving nests (quasi-complete separation), classical maximum likelihood estimators for logistic regression (e.g. function glm in R) did not converge. We therefore fit all models using the function bayesglm (package:arm; Gelman and Su 2015) in R ver. 3.2.3 (R Development Core Team 2015), specifying the binomial family to model the error distribution. This function uses the Estimation Maximization (EM) algorithm (rather than maximum likelihood) to fit the data, and we specified the default cauchy distribution as a vague prior (Gelman et al. 2008). To compare relative parsimony among models, we evaluated BIC values, associated weights of evidence, $w = e_i^{-\frac{1}{2}\Delta BIC} / \sum_i e_i^{-\frac{1}{2}\Delta BIC}$ and evidence e. r. = ratios $e^{\frac{1}{2}\Delta BIC}$ to compare the relative support in the data for pairs of models.

Results

Hail impact

Of the 203 monitored nests active at last check prior to the 22 June 2014 hailstorm, 102 (50.2%) appeared to have failed because of the storm, as evidenced by broken eggs and crushed chicks (Table 1). The average lark bunting nesting cycle (from the start of incubation to the fledging of young) is 19 days (Jehle et al. 2004), which is comparable to the nesting cycle durations for the horned lark (Verbeek 1967), McCown's longspur (Mickey 1943), Brewer's sparrow (Reynolds 1981) and grasshopper sparrow (Wray et al. 1982). Jehle et al. (2004) estimated

Table 1. Sample composition for nests located prior to the 22 June 2014 hailstorm, active at most recent nest check before storm, and checked the day after the storm.

Species	All nests		Nests with vegetation data	
	Total sample	Failed	Total sample	Failed
LARB	186	94	123	49
HOLA	7	5	5	4
BRSP	3	1	2	1
MCLO	2	2	1	1
GRSP	2	0	2	0
MODO	2	0	1	0
WEME	1	0	1	0
Sum	203	102 (50.2%)	135	55 (40.7%)

LARB, lark bunting; HOLA, horned lark; BRSP, brewer's sparrow; MCLO, McCown's longspur; GRSP, grasshopper sparrow; MODO, mourning dove; WEME, western meadowlark; see Methods for scientific names.

survival probability for lark bunting nests at three locations near our study site over the course of 3 years without severe hailstorms. Those estimates assumed a constant daily survival probability across nesting stages (i.e. nest survival probability did not vary between the egg and chick stages of nesting) and were fit using estimators in program MARK (Dinsmore et al. 2002). Average survival estimates for lark bunting nests for the entire nesting period ranged from 0.13 ± 0.05 to 0.45 ± 0.07, yielding average daily survival ranging from 0.90 to 0.96. Assuming daily nest survival events are constant over time and independent, total nest survival probability for any interval can be estimated as the product of daily survival probabilities over that interval (Dinsmore et al. 2002). Using Jehle et al.'s (2004) minimum daily survival estimate of 0.90 over the maximum of 4 days between last nest check and the storm, the most conservative average background nest survival rate is 0.66; this translates to an expected 134 of 203 nests surviving the 4-day interval. We recorded 101 surviving nests over that interval. If the only process at work were background nest survival (i.e. no loss due to storm), we can use the binomial distribution to compute the probability of our result as follows: $Pr(X \leq k) = \sum_{i}^{[k]} \binom{n}{i} p^i (1-p)^{n-i}$, where $n = 203$, $k = 101$ and $P = 0.66$. The probability of 100 or fewer of 203 nests surviving 4 days if the true survival rate were 0.66 (the most conservative estimate for intervals without hailstorms) is $1.4 * 10^{-6}$. In other words, a conservative estimate of background nest loss represents an extremely improbable explanation for our data.

Mourning dove and western meadowlark nest cycles are comparatively longer (~29 days), such that under the same background daily survival probability, these species

dBZ (dB)

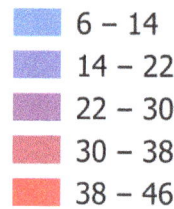
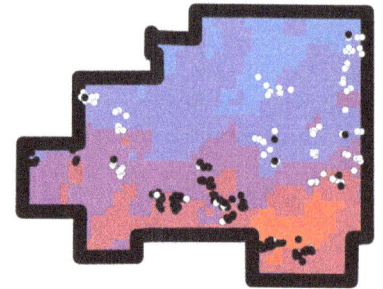

MESH (mm)

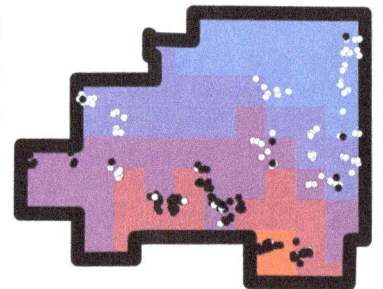

AWS (m s⁻¹)

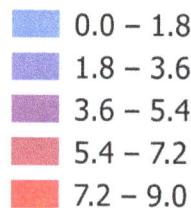
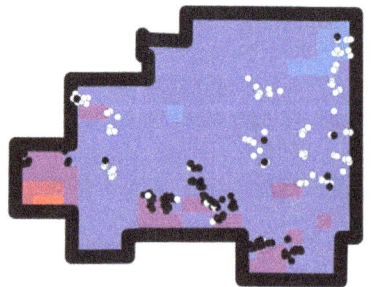

Figure 2. Distribution of nest fates at CPER relative to maximum NEXRAD BR, estimated size of hail (MESH) and AWS for the period of 21:35–00:51 GMT on June 22/23, 2014. Both BR and MESH rose from moderate levels in northwestern CPER (blue) to high levels in the area's southeast (red), which corresponded to an observed rise in nest mortality rates. Black dots represent nests that failed, and white dots represent nests that survived. AWS, azimuthal wind shear; BR, base reflectivity; CPER, Central Plains Experimental Range; MESH, maximum estimated size of hail.

would experience lower total nest survival probability. However, none of their nests failed, so elevated nest mortality cannot be attributed to including nests from these species in our dataset .

Mortality and hail-related NEXRAD data values were each heterogeneously distributed across the study area (Fig. 2). For the full nest sample ($n = 203$), values for estimated hail size and wind shear were not normally distributed and included multiple modes (Fig. 3). The distributions of values for MESH and AWS differed between failed and surviving nests (K-S test MESH: $D = 0.74$, $P < 0.001$; AWS: $D = 0.61$, $P < 0.001$). The distribution of estimated hail size at failed nest locations was to the right of that for locations with nests that survived, indicating that larger hail was estimated for locations where nests failed. The same comparative pattern was observed for wind shear; locations where nests failed experienced greater wind shear force than locations where nests survived. Our sample of nests where vegetation was measured ($n = 133$) included 55 (41.3%) that had failed, and the comparative differences between distributions of weather variable values for failed and surviving nests were the same as for the full dataset.

Vegetation data

For the subset of nests where vegetation was measured ($n = 133$), distributions of values for shrub cover did not differ between failed and surviving nests (K-S test

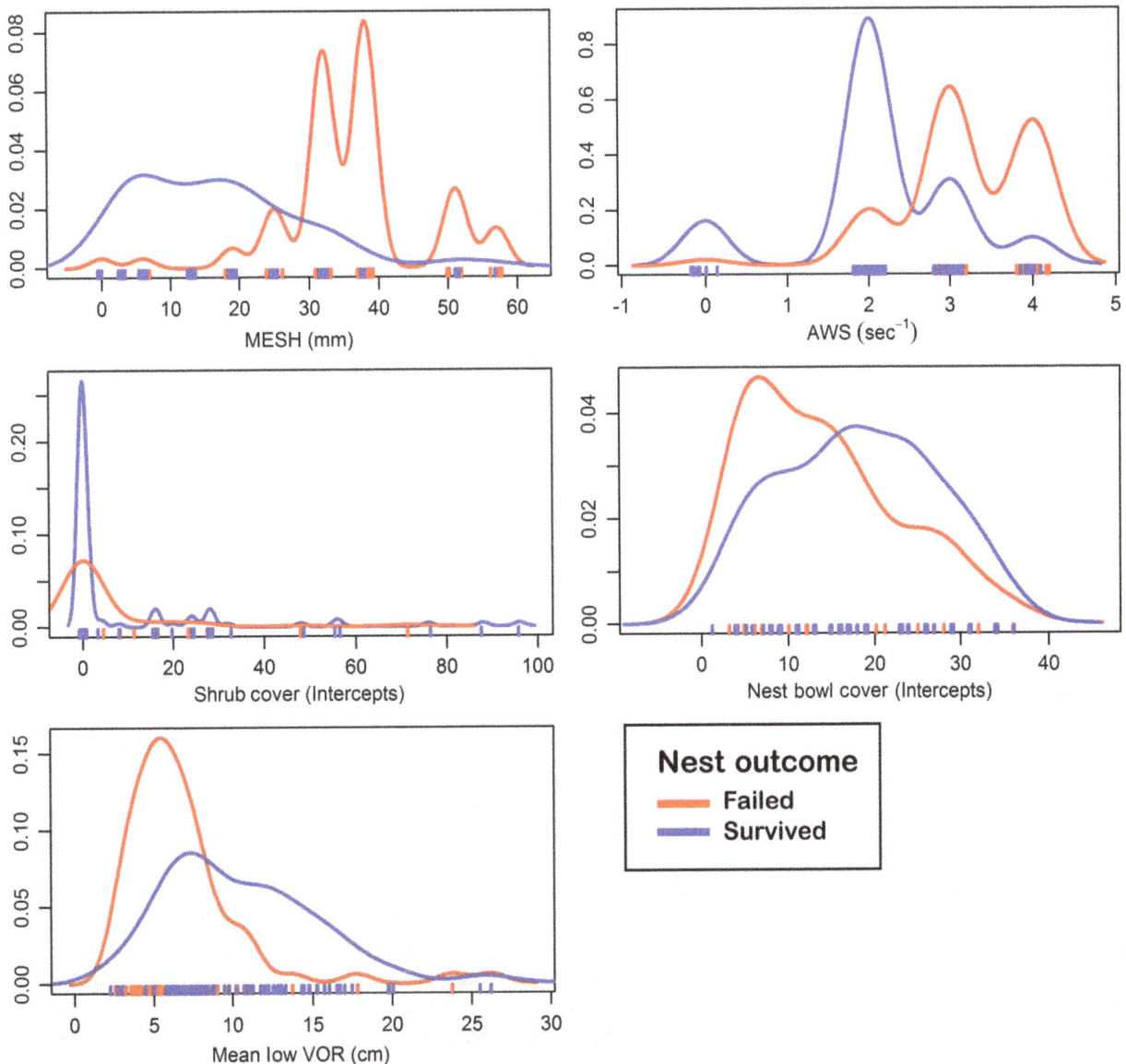

Figure 3. Density distribution of environmental covariates. Distributions of values for MESH and maximum estimated azimuthal wind shear differed between failed and surviving nests. Distributions of values for shrub abundance did not differ between failed and surviving nests, but distributions of nest cover and mean low visual obstruction reading (VOR) differed between failed and surviving nests. See text for test statistics. MESH, maximum estimated size of hail.

$D = 0.14$, $P = 0.57$; Fig. 3). Shrubs were uncommon nest-site vegetation components: only 35 nests had any cover by shrubs. The distribution of nest bowl cover ($D = 0.36$, $P < 0.001$) and mean low visual obstruction reading ($D = 0.43$, $P < 0.001$) did differ between samples of failed and surviving nests. For both of these variables, the distribution at failed nest locations was to the left of that for locations with nests that survived, indicating more nest bowl cover and higher visual obstruction readings for surviving nests than failed nests.

Model fit

A model including the effects of MESH and AWS, and the interaction between the two covariates was best supported by the data (Table 2). The parameterized linear component for that model is ~$2.59-0.08*$MESH$-1.1*$AWS$-0.10*$MESH$*$AWS. The coefficient for the interaction term in the fitted model was negative (-0.10) and non-zero (z-test for H_0: X = 0: $z = -4.1$, $P < 0.00001$). This indicates that as the product of hail size and wind shear increases (increase in either hail size, wind force or both), predicted failure probability increases (Fig. 4).

The second most parsimonious model added the effect of nest cover to the model above; the evidence against this second model as the one responsible for generating the data we observed was over 10 times greater than that for the top model. All models without the effects of MESH and AWS received no support from the data; the cumulative weight in favor of models including MESH and AWS was 100%.

Discussion

Severe storms can cause animal mortality through drowning or hail impact. Despite numerous other storms during our 2014 field season, each involving high-volume rainfall and small hail, no other storm caused perceptible mortality in nests under observation. No cases of birds drowning during severe storms have been recorded at our study site. Birds nesting near waterbodies may drown or abandon their nests (e.g. Sexson and Farley 2012), but there are no rivers or lakes in our study area. The apparent absence of drowning allowed us to focus on hail impact. Our focal storm was the only large-diameter (>2.5 cm) hail event in 2014. The apparent absence of mortality associated with small-diameter hail suggests estimation of hail incidence alone would be insufficient to gauge impact on breeding birds. Doppler weather data quantify various storm attributes, and our objective was to assess the accuracy of these data in predicting hail impact on breeding birds. We thought that combining multiple NEXRAD datasets might better explain nest mortality risk from hail. We further

Table 2. Models for the probability of nest survival as functions of NEXRAD and vegetation covariates were fit using Bayesian generalized linear models.

Model: Fate~	BIC	ΔBIC	Model weight	Evidence ratio
MESH*AWS	87.19	—	0.81	—
MESH*AWS + nest cover	91.85	4.66	0.08	10.26
MESH*AWS + VOR	91.86	4.67	0.08	10.30
MESH*AWS + shrubs	93.42	6.22	0.04	22.44
Global Model	99.28	12.08	0.00	420.47
MESH + VOR	102.81	15.61	0.00	$2.46*10^3$
MESH	103.49	16.30	0.00	$3.46*10^3$
MESH + cov	106.89	19.69	0.00	$1.89*10^4$
MESH + shrubs	108.38	21.19	0.00	$3.98*10^4$
VOR	164.13	76.93	0.00	$5.07*10^{16}$
Nest cover	172.49	85.30	0.00	$3.32*10^{18}$
Intercept only	182.42	95.23	0.00	$4.77*10^{20}$
Shrubs	183.76	96.57	0.00	$9.31*10^{20}$

Bayesian Information Criterion (BIC), a measure of model parsimony; ΔBIC is the difference between the BIC for a given model and that of the most parsimonious model and is used to quantify model selection uncertainty among models in the set; model weight (w) is the proportional support for each model in the total set (weights normalized so that they sum to 1 for the set); evidence ratio uses the ratio of ΔBIC to quantify relative degree of evidence against each model; for example, an evidence ratio of 10.26 indicates that the model is only one tenth as likely as the top model to have generated the data.
All models including the MESH*AWS interaction term also included the main effect. MESH, maximum estimated size of hail; AWS, maximum estimated azimuthal wind shear; VOR, mean low visual obstruction reading; AWS, azimuthal wind shear

hypothesized that nest-site vegetation would have an ameliorative effect on mortality risk.

Efficacy of NEXRAD datasets

We observed high efficacy of NEXRAD BR, maximum estimated size of hail (MESH) and maximum estimated azimuthal wind shear (AWS) for predicting hail-related mortality. That is, distributions of values for each measure were generally distinct between failed and surviving nests. These three NEXRAD measurements covaried along a spatial gradient in the study area (Fig. 2). We attribute the resulting collinearity to parallel variance associated with storm cloud dynamics. We subsequently focused on the interactive effect of MESH and AWS on hail-related mortality risk. Strong support (as measured by w and the evidence ratios) for a nest failure model including the additive effect of MESH, AWS and the MESH*AWS interaction term indicates an important relationship between hail size and wind shear. It suggests that hailstones' potential for damage depends in large part on the force with which they strike the ground. This makes

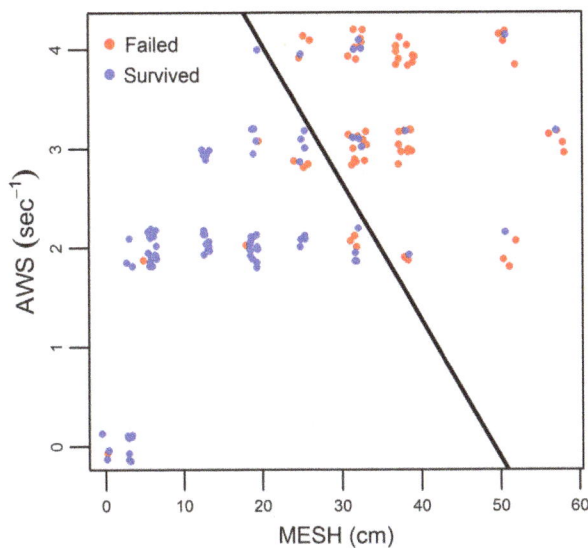

Figure 4. Effect of MESH and AWS on nest survival, based on the best supported logistic regression model. As maximum estimated AWS and MESH increased, hail-induced nest mortality became likely. The strong support for the model including only the additive and interactive effects of MESH and AWS ($w = 0.81$) indicates that wind shear augmented the impact of hail, inducing greater impact than would have been predicted based on hail size alone. AWS, azimuthal wind shear; MESH, maximum estimated size of hail.

intuitive sense, but to the best of our knowledge, we are the first to demonstrate the accuracy of Level-III data through a study on known wildlife mortality. This is a powerful indication that, given the knowledge of animal distributions and life histories, NEXRAD data can inform population modeling at large spatiotemporal scales.

Ameliorative effect of nest environment

We quantified the impact of shrub cover, subshrub cover, vegetation cover directly over the nest bowl and vertical vegetation structure on hail-related mortality risk. None of these variables strongly influenced mortality risk. Shrub cover was rare in our nest sample, and the distribution of values did not differ between failed and surviving nests. Models including shrub cover were not well-supported by the data. Partial nest bowl cover by vegetation was ubiquitous, and there was a difference between distributions of this variable for failed and surviving nests; nests that failed tended to have less cover than those that survived. However, models including this variable received little support from the data, suggesting that nest bowl cover did not mitigate the impacts of hail size and wind shear on the probability of nest survival. There was a similar difference in vertical vegetation structure between failed and surviving nests but effectively no support for models including this variable. These observed differences in nest

bowl cover and vertical structure between failed and surviving nests probably reflect the study site's gradient in vegetation height and density, which both increased along the same axis as increasing storm intensity (Fig. 2).

Nest-site vegetation associations vary with species life history (e.g. Cody 1981; Martin 1993; Powell and Steidl 2000), and the vegetation used by some species would likely play a larger role in offsetting risk of hail impact. On the other hand, in areas of less frequent severe hailstorm occurrence (Fig. 1), the biological community may lack protective behavior [e.g. shielding clutches (Thompson 1934; Johnson 1966), nest placement (Hadley 1969), sheltering in dense vegetation (Rockwell 1909), or physical attributes, e.g. egg shell thickness (Rahn and Paganelli 1989)]. We expect that hail-induced mortality likely affects open-cup nesters differently than birds that use other nest types, and warrants further examination across guilds.

Ecological implications

Large hail occurs more frequently on the Great Plains than in most other U.S. ecoregions, and our study site is located in one of the areas of highest frequency (Fig. 1). Hail may have particularly strong effects on grassland bird population viability because a single event not only destroys nests but also kills breeding adults. A nest failure rate of 50% is not extreme for grassland birds. Lark buntings, which dominated our sample, typically lose >40% of nests to predation (Jehle et al. 2004; Skagen et al. 2005). However, the heterogeneous distribution of mortality (Fig. 2) was highly atypical, as was the widespread adult mortality observed in this event. All species in this study have the capacity to re-nest, so nests destroyed by hail may be replaced by surviving adults within the same season if sufficient breeding opportunity remains following a severe hail event. However, even in the absence of additional storms, the probability of reproductive failure remains moderate to high due to increased nest predation pressure in grassland ecosystems (Vickery et al. 1992; Pietz and Granfors 2000; Murray 2015). Multiple severe hailstorms within the same season could prevent adults from re-nesting successfully, and the loss of breeding adults could depress population productivity for multiple seasons. For demographically marginal populations, this could result in extirpation.

Conservation implications

Our study supports the utility of Doppler weather radar data for estimating the impact of hail events on nesting birds across a larger area than that typically encompassed by direct monitoring efforts. The frequency of days conducive to severe storm formation is projected to increase (Kapsch et al. 2012). Remotely sensed weather radar data make it possible to estimate the extent and severity of severe hail impacts for nesting birds and, as such, can serve as an important tool in monitoring bird populations. Combined with known ranges of breeding birds, it should be possible to estimate hail-induced mortality at a regional or local scale. Further, based on climate models and the expected increase in hail frequency, adding hail threats to population viability projections should allow us to generate more accurate estimates across possible management scenarios. Although the incidence of large hail cannot be controlled, estimating its prevalence and contribution to mortality may help prioritize actions, especially pertaining to land use, to offset the effect of hail on nesting bird populations. These estimates can help identify potential changes to management practices that will most benefit the conservation of bird populations.

Acknowledgments

This project was funded by a United States Fish and Wildlife Service State Wildlife Grant, administered through the Nebraska Game and Parks Commission. The Bird Conservancy of the Rockies provided both material and intellectual support. The project was made possible in part by research grants from the Denver Field Ornithologists and the Colorado Field Ornithologists. Scientific collecting permits were obtained for each year of the project from the U.S. Fish and Wildlife Service Region 6 and Colorado Parks and Wildlife. Egg handling methods were approved by the Institutional Animal Care and Use Committee of the USGS Fort Collins Science Center and the University of Colorado Denver. Any use of trade, product or firm names is for descriptive purposes only and does not imply endorsements by the United States Government. We thank Tom Stanley and two anonymous reviewers for comments on earlier drafts of this manuscript.

References

Allen, J. T., M. K. Tippett, and A. H. Sobel. 2015. An empirical model relating U.S. monthly hail occurrence to large-scale meteorological environment. *J. Adv. Model. Earth Sys.* **7**, 226–243.

Bokma, F. 2004. Differential rates of morphological divergence in birds. *J. Evol. Biol.* **17**, 933–940.

Brennan, L. A., and W. P. Kuvlesky. 2005. North American grassland birds: an unfolding conservation crisis? *J. Wildl. Manage.* **69**, 1–13.

Canfield, R. H. 1941. Application of the line interception method in sampling range vegetation. *J. Forest.* **39**, 388–394.

Changnon, S. A. 2008. Temporal and spatial distributions of wind storm damages in the United States. *Clim. Change.* **94** (3–4), 473–482.

Cody, M. L. 1981. Habitat selection in birds: the roles of vegetation structure, competitors, and productivity. *Bioscience* **31**, 107–113.

Collias, N. E. 1997. On the origin and evolution of nest building by passerine birds. *Condor* **99**, 253–270.

Conway, W. C., and E. Johnson. 2011. Cassin's Sparrows nesting on the southern high plains of Texas. *Bull. Tex. Ornithological Soc.* **44**(1–2), 90–94.

Das, M., E. Oterkus, E. Madenci, and H. Razi. 2009. Residual strength of sandwich panels with hail damage. *Compos. Struct.* **88**, 403–412.

Depue, T. K., P. C. Kennedy, and S. A. Rutledge. 2007. Performance of the hail differential reflectivity (H DR) polarimetric radar hail indicator. *J. Appl. Meteorol. Climatol.* **46**, 1290–1301.

Derner, J. D., W. K. Lauenroth, P. Stapp, and D. J. Augustine. 2009. Livestock as ecosystem engineers for grassland bird habitat in the western Great Plains of North America. *Rangeland Ecol. Manage.* **62**, 111–118.

Diehl, R. H., J. M. Bates, D. E. Willard, and T. P. Gnoske. 2014. Bird mortality during nocturnal migration over Lake Michigan: a case study. *Wilson J. Ornithol.* **126**, 19–29.

Dinsmore, S. J., G. C. White, and F. L. Knopf. 2002. Advanced techniques for modeling avian nest survival. *Ecology* **83**, 3476–3488.

Dumont, E. R. 2010. *Bone density and the lightweight skeletons of birds.* Proceedings of the Royal Society of London B, Biological Sciences. doi:10.1098/rspb.2010.0117.

Fischer, E. M., and R. Knutti. 2015. Anthropogenic contribution to global occurrence of heavy-precipitation and high-temperature extremes. *Nat. Clim. Chang.* **5**, 560–564.

Fuhlendorf, S. D., and D. M. Engle. 2001. Restoring heterogeneity on rangelands: ecosystem management based on evolutionary grazing patterns. *Bioscience* **51**, 625–632.

Gelman, A., A. Jakulin, M. G. Pittau, and Y.-S. Su. 2008. A weakly informative default prior distribution for logistic and other regression models. *Ann. App. Sci.* **2**, 1360–1383.

Gelman, A. and Y.-S. Su. 2015. Arm: Data Analysis using Regression and Multilevel/Hierarchical Models. R package version 1.8–03 Available at http://CRAN.R-project.org/package=arm. (accessed 4 August 2016).

Graul, W. D. 1975. Breeding biology of the Mountain Plover. *Wilson Bulletin* **87**, 6–31.

Hadley, N. F. 1969. Microenvironmental factors influencing the nesting sites of some subalpine Fringillid birds in Colorado. *Arct. Alp. Res.* **1**, 121–126.

Hall, D. W., and T. M. Harvey. 2007. Mortality at a night roost of great-tailed grackles and European starlings during a spring hail storm. *Wilson J. Ornithol.* **119**, 309–312.

Heflebower, C. C., and E. V. Klett. 1980. A killer hailstorm at the washita refuge. *Bulletin Okla. Ornithological Soc.* **13**, 26–28.

Herrick, J. E., J. W. Van Zee, K. M. Havenstad, L. M. Burkett, and W. G. Whitford. 2005. *Monitoring manual for grassland, shrubland and savanna ecosystems*. USDA - ARS Jornada Experimental Range, Las Cruces, New Mexico.

Hohl, R., H. H. Schiesser, and D. Aller. 2002. Hailfall: The relationship between radar-derived hail kinetic energy and hail damage to buildings. *Atmos. Res.* **63**, 177–207.

Jehle, G., A. A. Yackel Adams, J. A. Savidge, and S. K. Skagen. 2004. Nest survival estimation: a review of alternatives to the Mayfield estimator. *The Condor* **106**, 472–484.

Johnson, R. E. 1966. Alpine birds of the Little Belt Mountains, Montana. *Wilson Bulletin* **78**, 225–227.

Johnson, D. H. 1979. Effect of a summer storm on bird populations. *Prairie Naturalist* **11**, 78–82.

Kapsch, M. L., M. Kunz, R. Vitolo, and T. Economou. 2012. Long-term trends of hail-related weather types in an ensemble of regional climate models using a Bayesian approach. *J. Geophys. Res.: Atmos.* **117**(D15107), 1–16.

Knopf, F. L. 1996. Prairie Legacies–Birds. Pp. 135–148 In: F. B. Samson, F. L. Knopf, eds. *Prairie Conservation*. Island Press, Washington, D.C.

Kunz, M., and P. I. S. Kugel. 2015. Detection of hail signatures from single-polarization C-band radar reflectivity. *Atmos. Res.* **153**, 565–577.

La Rivers, I. 1944. Observations on the nesting mortality of the Brewer Blackbird, *Euphagus cyanocephalus*. *Am. Midl. Nat.* **32**, 417–437.

Lakshmanan, V., T. Smith, K. Hondl, G. J. Stumpf, and A. Witt. 2006. A real-time, three-dimensional, rapidly updating, heterogeneous radar merger technique for reflectivity, velocity, and derived products. *Weather Forecast.* **21**, 802–823.

Lakshmanan, V., Fritz, A., Smith, T., Hondl, K. & Stumpf, G. 2007. An automated technique to quality control radar reflectivity data. *J. Appl. Meteorol. Clim.*, **46**, 288–305.

Lauenroth, W. K., and O. E. Sala. 1992. Long-term forage production of North American shortgrass steppe. *Ecol. Appl.* **2**, 397–403.

Lee, C. A., and W. K. Lauenroth. 1994. Spatial distributions of grass and shrub root systems in the Shortgrass Steppe. *Am. Midl. Nat.* **132**, 117–123.

Mabee, T. J., and V. B. Estelle. 2000. Assessing the effectiveness of predator exclosures for plovers. *Wilson Bulletin* **112**, 14–20.

Martin, T. E. 1993. Nest predation and nest sites new perspectives on old patterns. *Bioscience* **43**, 523–532.

McKechnie, A. E., P. A. R. Hockey, and B. O. Wolf. 2012. Feeling the heat: Australian landbirds and climate change. *Emu* **112**:i–vii.

Mickey, F. W. 1943. Breeding habits of McCown's Longspur. *Auk* **60**, 181–209.

Morgan, G. M. J., and N. G. Towery. 1975. Small-scale variability of hail and its significance for hail prevention experiments. *J. Appl. Meteorol.* **14**, 763–770.

Murray, L. 2015. Success and predation of bird nests in grasslands at Valley Forge National Historical Park. *Northeastern Naturalist* **22**, 10–19.

Pietz, P. J., and D. A. Granfors. 2000. Identifying predators and fates of grassland passerine nests using miniature video cameras. *J. Wildl. Manage.* **64**, 71–87.

Powell, B. E., and R. J. Steidl. 2000. Nesting habitat and reproductive success of southwestern riparian birds. *Condor* **102**, 823–831.

R Development Core Team. 2015. R: A Language and Environment for Statistical Computing. Vienna, Austria: the R Foundation for Statistical Computing. ISBN: 3-900051-07-0. Available at http://www.R-project.org/. (accessed 29 October 2016)

Rahn, H., and C. V. Paganelli. 1989. Shell mass, thickness and density of avian eggs derived from the tables of schönwetter. *J. für Ornithologie* **130**, 59–68.

Reynolds, T. D. 1981. Nesting of the sage thrasher, sage sparrow, and brewer's sparrow in Southeastern Idaho. *Condor* **83**, 61–64.

Robel, R. J., J. N. Briggs, A. D. Dayton, and L. C. Hulbert. 1970. Relationships between visual obstruction measurements and weight of grassland vegetation. *J. Range Manag.* **23**, 295–297.

Rockwell, R. B. 1909. The use of magpies' nests by other birds. *The Condor* **11**, 90–92.

Rosenberg, N. J. 1987. Climate of the great plains region of the United States. *Great Plains Q.*. **7**(Winter):344:22–32.

Saa Requejo, A., R. García Moreno, M. C. Díaz Alvarez, F. Burgaz, and M. Tarquis. 2011. Analysis of hail damages and temperature series for peninsular Spain. *Nat. Hazard. Earth Syst. Sci.* **11**, 3415–3422.

Saalfeld, S. T., W. C. Conway, D. A. Haukos, and W. P. Johnson. 2011. Nest success of Snowy Plovers (*Charadrius nivosus*) in the southern high plains of Texas. *Waterbirds* **34**, 389–399.

Sala, O. E., W. K. Lauenroth, and W. J. Parton. 1992. Long-term soil water dynamics in the shortgrass steppe. *Ecology* **73**, 1175–1181.

Samson, F. B., F. L. Knopf, and W. R. Ostlie. 2004. Great Plains ecosystems: past, present, and future. *Wildl. Soc. Bull.* **32**, 6–15.

Saunders, D. A., P. Mawson, and R. Dawson. 2011. The impact of two extreme weather events and other causes of death on Carnaby's Black Cockatoo: a promise of things to come for a threatened species? *Pac. Conserv. Biol.* **17**, 141–148.

Sexson, M. G., and G. H. Farley. 2012. Snowy plover nest survival in Kansas and effective management to counter negative effects of precipitation. *J. Wildl. Manage.* **76**, 1587–1596.

Skagen, S. K., A. A. Yackel Adams, and R. D. Adams. 2005. Nest survival relative to patch size in a highly fragmented shortgrass prairie landscape. *Wilson Bulletin* **117**, 23–34.

Smith, T. M., and K. L. Elmore. 2004. The use of radial velocity derivatives to diagnose rotation and divergence. Preprints, 11th Conf. on Aviation, Range, and Aerospace, Hyannis, MA, American Meteorological Society P5.6. Available: http://ams.confex.com/ams/pdfpapers/81827.pdf. (accessed 24 February 2016)

Smith, A. G., and H. R. Webster. 1955. Effects of hail storms on waterfowl populations in Alberta, Canada: 1953. *J. Wildl. Manage.* **19**, 368–374.

SPC. 2016. SPC Storm Reports for 06/22/14. Storm Prediction Center, National Oceanic and Atmospheric Administration National Weather Service. Available: http://www.spc.noaa.gov/climo/reports/140622_rpts.html .

Stumpf, G. J., T. M. Smith, and J. Hocker. 2004. New hail diagnostic parameters derived by integrating multiple radars and multiple sensors. In: Preprints, 22d Conference on Severe Local Storms. Hyannis, MA: American Meteorological Society.

Thompson, B. H. 1934. A wilderness-use technique. *The Condor* **36**, 153–157.

Trapp, R. J., N. S. Diffenbaugh, H. E. Brooks, M. E. Baldwin, E. D. Robinson, and J. S. Pal. 2007. Changes in severe thunderstorm environment frequency during the 21st century caused by anthropogenically enhanced global radiative forcing. *Proc. Natl Acad. Sci.* **104**, 19719–19723.

Veech, J. A., ed. 2006. A probability based analysis of temporal and spatial co-occurrence in grassland birds. *J. Biogeogr.* **33**:2145–2153.

Verbeek, N. A. 1967. Breeding biology and ecology of the horned lark in alpine tundra. *Wilson Bulletin* **79**, 208–218.

Vickery, P. D., M. L. Hunter, and J. V. Wells. 1992. Evidence of incidental nest predation and its effects on nests of threatened grassland birds. *Oikos* **63**, 281–288.

Vorst, J. J. 1991. *Assessing hail damage to corn*. Cooperative Extension Institute of Agricultural and Natural Resources, University of Nebraska, Lincoln.

Wille, J. J., and G. E. Kleinkopf. 1992. Effect of simulated hail damage on yield and quality of Russet Burbank potatoes. *Am. Potato J.* **69**, 705–714.

With, K. A., and D. R. Webb. 1993. Microclimate of ground nests: the relative importance of radiative cover and wind breaks for three grassland species. *Condor* **95**, 401–413.

Witt, A., M. D. Eilts, G. J. Stumpf, J. T. Johnson, E. D. W. Mitchell, and K. W. Thomas. 1998. An enhanced hail detection algorithm for the WSR-88D. *Weather Forecast.* **13**, 286–303.

Wray, T. II, K. A. Strait, and R. C. Whitmore. 1982. Reproductive success of grassland sparrows on a reclaimed surface mine in West Virginia. *Auk* **99**, 157–164.

3

Integrating LiDAR-derived tree height and Landsat satellite reflectance to estimate forest regrowth in a tropical agricultural landscape

T. Trevor Caughlin[1], Sami W. Rifai[1], Sarah J. Graves[1], Gregory P. Asner[2] & Stephanie A. Bohlman[1]

[1]University of Florida, School of Forest Resources and Conservation, Gainesville, Florida 32601
[2]Department of Global Ecology, Carnegie Institution for Science, 260 Panama Street, Stanford, California, 94305

Keywords

Continuous change detection, forest landscape restoration, Landsat–LiDAR integration, reforestation, spectral mixture analysis, tropical secondary forest

Correspondence

T. Trevor Caughlin, PO Box 110410, Gainesville, FL 321611. E-mail: trevor.caughlin@gmail.com

Funding Information

We thank D. Knapp, C. Anderson, T. Kennedy-Bowdoin and the entire Carnegie Airborne Observatory staff for data collection and processing support. T. T. Caughlin was supported by the NSF grant #1415297 Science, Engineering and Education for Sustainability Fellow program. Airborne data collection, processing and analysis was funded by the Grantham Foundation for the Protection of the Environment and William R. Hearst III. The Carnegie Airborne Observatory has been made possible by grants and donations to G.P. Asner from the Avatar Alliance Foundation, Margaret A. Cargill Foundation, David and Lucile Packard Foundation, Gordon and Betty Moore Foundation, Grantham Foundation for the Protection of the Environment, W. M. Keck Foundation, John D. and Catherine T. MacArthur Foundation, Andrew Mellon Foundation, Mary Anne Nyburg Baker and G. Leonard Baker Jr, and William R. Hearst III.

Editor: Harini Nagendra
Associate Editor: Lola Fatoyinbo

Abstract

Remotely sensed data have revealed ongoing reforestation across many tropical landscapes. However, most studies have quantified changes between discrete land cover categories that are difficult to relate to the continuous changes in forest structure that underlie reforestation. Here, we demonstrate how generalized linear models (GLMs) can predict tree height and tree canopy cover from Landsat satellite reflectance in a 109 882 ha tropical agricultural landscape of western Panama. We derived tree canopy cover and tree height from airborne Light Detection and Ranging (LiDAR) data, and related these variables to the fraction of photosynthetic vegetation (PV) in Landsat pixels. We found large gains in predictive accuracy from modeling tree canopy height with a gamma GLM and tree canopy cover with a binomial GLM, relative to modeling these variables using linear regression. Adding social and environmental covariates to our GLMs, including topography and parcel membership (representing different land owners), increased predictive accuracy, resulting in best-fit models with an R^2 of 55.68% and RMSE of 23.69% for tree canopy cover, and an R^2 of 51.24% and RMSE of 3.42 m for tree height. Finally, we applied the GLMs to predict tree height and tree canopy cover in Landsat images from c. 2000 to 2012, and used results to quantify changes in forest structure during this 12-year period. We found that >60% of pixels in our study area had increased in tree height and tree canopy cover, suggesting widespread forest regrowth. These increases were spatially widespread across the study area, yet subtle, with most pixels increasing <2 m in tree height. Our results suggest ecological and agricultural changes that could be overlooked if measuring land cover change with discrete forest and non-forest categories. Overall, we show the advantages of linking LiDAR and Landsat data to quantify forest regrowth in an agricultural landscape.

Introduction

Reforestation is occurring throughout the tropics with consequences for human livelihood, carbon storage and biodiversity conservation (Chazdon 2014). Remotely sensed data has played an important role in quantifying the spatial extent of this land cover change at landscape to continental scales (Nagendra and Southworth 2009; Aide et al. 2013). Many of the remotely sensed data classifications used to assess reforestation involve discrete land cover categories, often just forest and non-forest categories (Hansen et al. 2013) and occasionally more detailed categories, such as 'primary forest' and 'secondary forest' (Sloan et al. 2015). In reality, reforestation is not a sudden transition from a non-forest state to a forest state, but a gradual change in forest structure and composition (Van Breugel et al. 2006; Chazdon et al. 2007; Norden et al. 2015). Binning these continual changes into multiple discrete categories, for example, early, mid and late succession, is possible but may require reflectance data with fine spatial (Arroyo-Mora et al. 2005) or spectral (Kalacska et al. 2007) resolution. In contrast to discrete land cover categories, continuous metrics from remote sensing related to leaf chlorophyll, such as photosynthetic vegetation fraction (Negrón-Juárez et al. 2011) and NDVI (Campo-Bescós et al. 2013), can more accurately represent gradual landscape changes, including reforestation (Southworth et al. 2004; Vogelmann et al. 2016). Recent work has revealed the potential for time series of remotely sensed data to measure a wide variety of change metrics with ecological relevance, including number of years to recover from a disturbance (Pickell et al. 2016), non-linear trends (Pflugmacher et al. 2014) and persistence of spectral changes (Hermosilla et al. 2015). Relating these change metrics to absolute measures of forest structure, such as tree height, will play an important role in scaling up field data on tree demography to landscapes and regions (Antonarakis et al. 2011). Ultimately, remote sensing detection of tree cover and tree height will improve our ability to link widespread gains in forest cover with the ecological processes that drive reforestation.

Airborne light detection and ranging (LiDAR) technology has emerged as a data source that can measure forest canopy structure at the scale of individual trees. Models fit with LiDAR data can quantify aboveground biomass (Asner et al. 2013), tree height and canopy cover (Kellner et al. 2009) and even tree leaf area (Stark et al. 2012) with comparable accuracy to on-the-ground field data. Landscape estimates of aboveground biomass with LiDAR models and forest inventory data can be more accurate and less biased than estimates generated from forest inventory data alone, because the spatial configuration of forest type in field plots is seldom representative of the variation in the larger landscape (Marvin et al. 2014). In tropical secondary forests, LiDAR data have been used to classify forest successional stage from forest vertical structure (Castillo et al. 2012) and to relate forest structure to successional mechanisms, such as seed dispersal (Castillo-Núñez et al. 2011). Time series of LiDAR data from repeated flyovers are rare, but can measure ecological processes including gap dynamics (Réjou-Méchain et al. 2015) and competition for canopy space between neighboring trees (Kellner and Asner 2014). Repeat LiDAR surveys over larger areas show potential for measuring regional forest dynamics, including biomass change and carbon flux (Hudak et al. 2012; Englhart et al. 2013; Cao et al. 2016). While LiDAR data show great promise for quantifying ecological process over large spatial extents, these data are currently limited to a tiny fraction of the earth's surface and to a small number of acquisition dates (Mascaro et al. 2014).

One solution to the limited spatial and temporal coverage of LiDAR data is to predict LiDAR-derived metrics of forest structure from medium resolution satellite imagery, including the Landsat archive, which provides open access to 30×30 m resolution data over a global extent from 1972 to present (Wulder et al. 2016). Although these satellite data represent spectral reflectance, not direct measurements of forest structure, predicting LiDAR-derived tree canopy cover and tree height from satellite reflectance data is possible (Hudak et al. 2002; Wulder and Seemann 2003; Chen and Hay 2011; Ahmed et al. 2015; Zald et al. 2016). Successful prediction of forest structure from Landsat data has largely taken place in relatively homogenous forest where human impacts are either rare (Ota et al. 2014) or are limited to industrial logging (Hudak et al. 2002; Ahmed et al. 2015; Zald et al. 2016). Recent studies have identified 'time since disturbance' as an important predictor variable for tree height and canopy cover at the time of LiDAR acquisition (Ota et al. 2014; Ahmed et al. 2015; Zald et al. 2016). This approach assumes homogenous successional trajectories, with time since disturbance having a similar effect on forest structure across the entire landscape. Tropical agricultural landscapes present an even wider range of successional trajectories than young forest stands embedded in continuous forest (Arroyo-Rodríguez et al. 2015), and quantifying time since disturbance in these landscapes is not straightforward.

Mixed-use agricultural landscapes, where potential for reforestation is highest (Latawiec et al. 2015), present challenges for remote measurement of forest structure relative to more homogeneously forested landscapes. First, agricultural landscapes are dominated by privately owned land. Heterogeneous landowner decision making creates a patchwork of land use history that can lead to divergent successional trajectories between secondary forest patches (Manson and Evans 2007; Arroyo-Rodríguez et al. 2015). Consequently, even neighboring patches with a similar age since agricultural abandonment can have different times to forest recovery, species composition and forest structure (Mesquita et al. 2015). Second, tree cover in tropical agricultural landscapes varies on fine spatial scales, including large isolated trees in pastures (Slocum 2001), narrow strips of short trees planted as live fences (Harvey et al. 2005) and patches of native secondary forest (van Breugel et al. 2013). These differences are subtle at the scale of a 30-m × 30-m pixel, but have major consequences for the ecological dynamics of forest succession (Zahawi and Augspurger 2006). For example, a shift from monoculture pasture to silvopasture systems with a combination of grass, trees and shrubs can lead to significant increases in soil carbon, biodiversity potential and milk production (Murgueitio et al. 2011).

We integrate LiDAR and Landsat data to predict tree cover in a tropical agricultural landscape of western Panama, the Azuero Peninsula. Similar to many other landscapes in Latin America, land at our study site is predominantly privately owned cattle pasture. Previous remote sensing studies have documented positive net forest cover change from 2000 to 2010 in the Azuero, potentially indicating that the region is undergoing a forest transition (Metzel and Pacala 2010; Sloan 2015). Our ultimate goal is to quantify the ecological dynamics of this forest transition by predicting tree canopy cover and tree height from Landsat data. We focus on these forest structural variables because both can be related to individual tree demography in models of forest dynamics (Purves et al. 2008; Caughlin et al. 2016). We predict LiDAR-derived tree height and tree canopy cover using fraction of photosynthetic vegetation in Landsat pixels as a predictor variable in generalized linear models (GLMs). We address the question of how well these models can scale up to predict forest structure in heterogeneous landscapes by determining the minimum number of pixels required to accurately predict forest structure for a test dataset. Finally, we apply our models to quantify differences in tree height and tree canopy cover between Landsat images from c. 2000 to 2012. Overall, we demonstrate an approach that can provide continuous estimates of forest regrowth from spectral data in a tropical agricultural landscape.

Materials and Methods

Site description

Our study takes place in Los Santos Province in Southwestern Panama (Fig. 1), a region historically covered by

tropical dry forest that was cleared for cattle pasture in the early to mid-20th century (Heckadon-Moreno 2009). Currently, the mosaic of habitats in the landscape includes actively managed cattle pasture, shrubland in fallow pasture, strips of riparian forest and a few secondary forest patches (Griscom et al. 2011; Fig. 2). The study site has a pronounced dry season from December through March, with an average of 1700 mm of rainfall mostly falling from late April to late November. The region is hilly with an elevational range from 0 to 468 m. Our research is motivated by previous studies that have suggested that Los Santos province is undergoing a forest transition, likely driven by changes in agricultural labor that lead to tree encroachment on cattle pastures (Metzel and Pacala 2010; Sloan 2015). Quantifying the spatial extent of this land cover change could help restore Mesoamerican tropical dry forest, a critically endangered ecosystem, to Los Santos.

LiDAR data

LiDAR data were collected in January 2012 by the Carnegie Airborne Observatory-2, using a dual-laser waveform light detection and ranging (LiDAR) scanner with a density of 2 points per square meter. Classified point-cloud data were used to develop a canopy height model (CHM) with a pixel size of 1.13 m covering a 13 333 ha area (Asner et al. 2012; additional details in Appendix S1).

We resampled the CHM pixels to the same spatial extent as Landsat pixels using bilinear interpolation, resulting in a total of exactly 676 CHM pixels of 1.25 m^2 within each 30 m^2 Landsat pixel. We then calculated tree canopy cover as the number of CHM pixels in a single Landsat pixel with height >1.5 m out of a total of 676 pixels. We obtained similar results from height thresholds for tree canopy cover ranging from >0.3 to 3 m (Appendix S2). We calculated tree height as the mean value of CHM pixels within each Landsat pixel. These operations resulted in two response variables: tree canopy cover (number of pixels with height >1.5 m out of a total of 676) as a proportional variable and tree height (m) as a continuous and positive response variable.

Landsat data

We selected Landsat 7 data with low cloud cover during the dry season between December 2011 and March 2012. Surface reflectance data were retrieved from the USGS Climate Data Record where it was processed using the Landsat Ecosystem Disturbance Adaptive Processing System (LEDAPS) algorithm (Masek et al. 2006). The cloud mask band from the LEDAPS product was used to remove cloud contaminated pixels. This resulted in a total

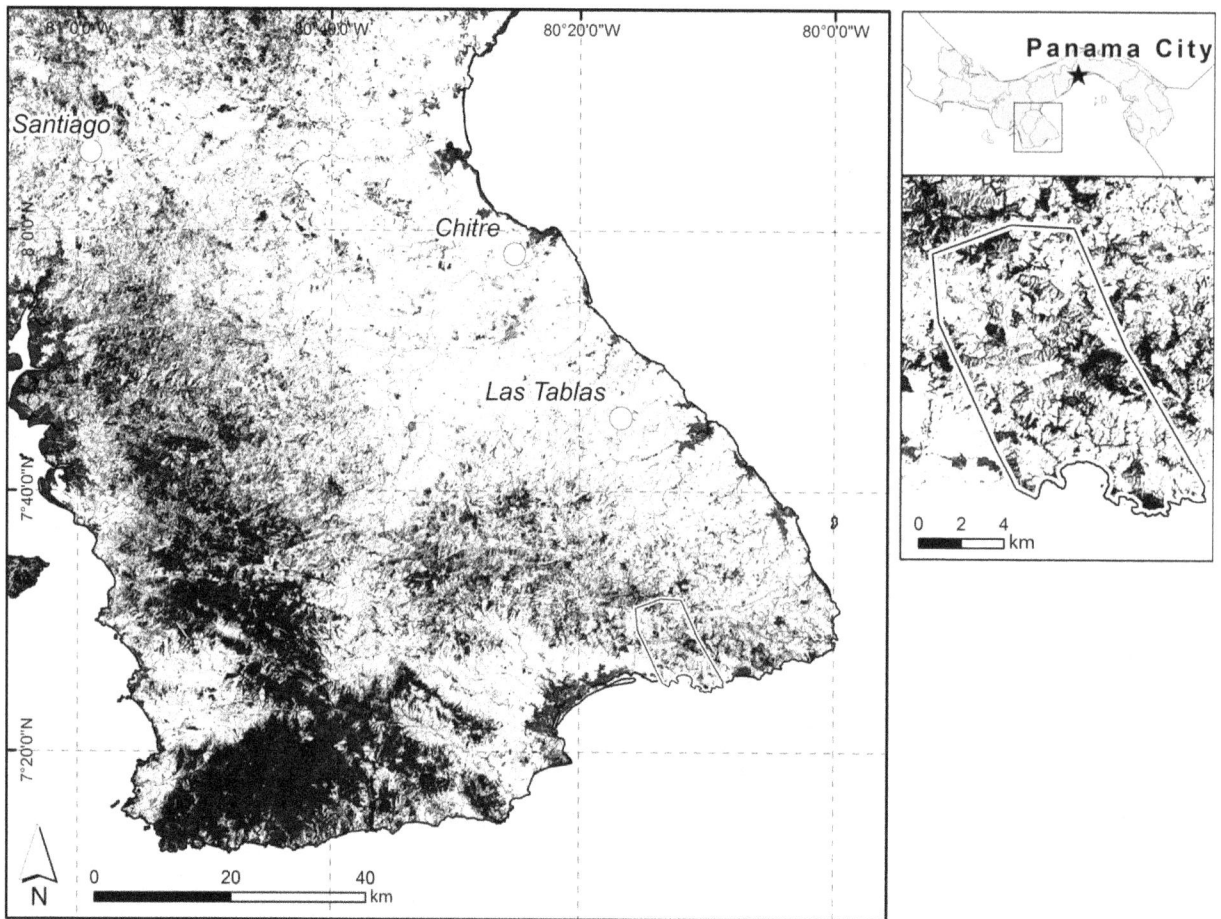

Figure 1. Map of study region. The black outline, shown enlarged in the right panel, represents the boundaries of the study region. The black shading in the figure represents forest cover from Hansen et al. (2013).

of four Landsat 7 scenes (Path/Row 012/055), taken on 31 December 2011, 16 January 2012, 1 February 2012 and 20 March 2012. The performance of our models was likely enhanced by the stability of the modeled surface reflectance data, which can ameliorate the influence of atmospheric and aerosol contamination on illumination (Masek et al. 2006). We used spectral mixture analysis to estimate fractional cover of photosynthetic vegetation from the Landsat bands. Spectral mixture analysis is a physical modeling approach that 'unmixes' the fractional composition of pixels into components with known reflectance, or endmembers (Adams et al. 1995; Asner et al. 2009). Fractional components are biologically meaningful because they have a physical interpretation that can be directly linked to field measurements (Asner et al. 2004). Spectral mixture analysis has previously been used to classify forest and non-forest habitat in tropical dry forest (Mayes et al. 2015), estimate gap mortality at the scale of individual trees from Landsat data (Rifai et al.

2016) and quantify subpixel disturbances in tropical forest habitats (Negrón-Juárez et al. 2011). We used the software package CLASlite to unmix pixels into three components: photosynthetic vegetation (PV), non-photosynthetic vegetation (NPV) and soil. CLASlite applies an automated Monte Carlo procedure that draws off of an extensive library of endmembers to unmix components (Asner et al. 2009). We unmixed each separate Landsat image, resulting in fractional composition of pixels on 31 December 2011, 16 January 2012, 1 February 2012 and 20 March 2012. In preliminary analyses, we compared the predictive power of PV, NPV and soil fractions for each date separately and for the mean of all the Landsat images. We found highest predictive accuracy was achieved using PV as the sole predictor variable; the soil fraction contributed little to predictive power, and NPV was highly collinear with PV but performed slightly worse as a sole predictor variable. Mean PV (PV_{mean}) across all dates from December 2011 to March 2012 resulted in

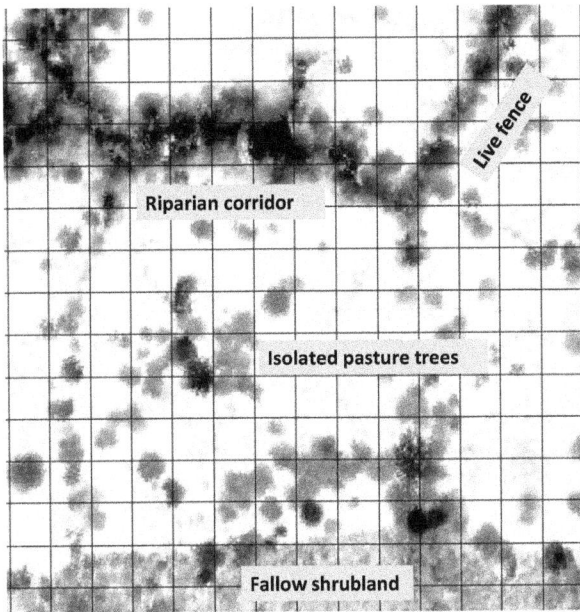

Figure 2. Complex patterns of tree cover in a tropical agricultural landscape. This figure illustrates a subsection of the LiDAR-derived canopy height model (CHM) data from the study region. Darker colors indicate higher tree height. Grid cells indicate the spatial scale and position of 30 × 30 m Landsat pixels. Each Landsat pixel contains 676 CHM pixels.

significantly higher predictive accuracy than PV for any single date, so we used PV_{mean} as a predictor variable in subsequent analyses (Table S1).

Environmental and social covariates

Because topography can capture biophysical properties that influence differences in forest structure, including topographic metrics in remote sensing models can improve prediction of forest structural attributes (Anderson et al. 2008), we included topography from the Shuttle Radar Topography Mission (SRTM) at 30-m × 30-m resolution as a predictor variable. We calculated six different topographic variables (elevation, slope, aspect, terrain ruggedness index, topographic position index and roughness) in R using the package 'raster' (Hijmans et al. 2015). Aspect resulted in the greatest increase in predictive accuracy when added to the univariate GLMs with photosynthetic fraction as a predictor variable (Table S2).

Because the majority of the land in the study site is privately owned, and landowner decision making results in heterogeneous patterns of land cover, we also included landownership as a predictor variable using cadastral data (mapped boundaries of individual properties). Cadastral data comes from a census conducted in 2000 by the National Authority for the Administration of Lands

(ANATI – Autoridad Nacional de Administración de Tierras). Although the cadastral data are older than the LiDAR data, the parcel boundaries corresponded closely to boundaries of 30 farms in the region whose boundaries were ground-truthed in 2013 (Valencia, unpubl. data). We removed parcels that contained <3 Landsat pixels from the dataset prior to analyses. This resulted in 685 parcels with a mean (±SD) parcel area of 19 (±27) ha.

Modeling

Because tree canopy cover is proportional data (number of pixels >1.5 m out of a total of 676), we modeled tree canopy cover using a binomial GLM. Because tree canopy height is continuous and positive, we modeled tree canopy height using a gamma GLM. Generalized linear models use a link function to ensure that predictor variables vary linearly with the expected value of the response (Bolker et al. 2009). For the gamma regression, we used a log-link function to constrain the expected value of the response to be positive, and for the binomial regression we used a complementary log-log link ('cloglog') to constrain the expected value of the response to be between 0 and 1. For our data, the cloglog function resulted in higher predictive accuracy than common alternatives, including probit and logit link functions, indicating zero-skewed data (Hardin and Hilbe 2007). All models were fit in base R, using the GLM and predict functions (R Core Team 2015). To assess the gain in predictive accuracy from our GLM approach, we compared the fit of GLMs to linear regression with log-transformed response variables. We separated the data into a training set, consisting of 9607 randomly selected pixels, and a validation set consisting of the remaining 50 992 pixels. We assessed predictive accuracy as out-of-sample R^2 and RMSE (root mean squared error) of predictions from a model fitted to the training data and applied to the test data. For tree canopy cover, rather than log transforming the percentage of tree canopy cover, we used a log transformation of the total number of LiDAR scale pixels with height >1.5 m.

Next, we determined the minimum number of training pixels required to obtain accurate predictions of tree canopy cover and tree height from Landsat data. This analysis addresses the question of how well univariate GLMs with a single predictor variable (photosynthetic vegetation fraction) can scale up to predict forest structure in a heterogeneous landscape. If a sampling scheme that captures a wide range of environmental conditions is required to accurately predict forest structure in our agricultural landscape, we would expect models developed with small training sets to perform poorly. On the other hand, if the relationship between photosynthetic fraction and forest structure is consistent across many different environmental conditions, we would expect that only a minimal sample size would be required to predict forest structure across the entire landscape. For this analysis, we randomly sampled between 5 and 3000 pixels of training data from the dataset, representing from 0.0055% to 3.3% of the entire dataset. We note that our sampling strategy was completely random (i.e. we did not stratify sampling to sample a range of habitats). We then fit the gamma and binomial GLMs to each subsample of training data and quantified predictive accuracy using all pixels not used in the training data. At each level of pixel sample size, we replicated sampling five times and calculated out-of-sample RMSE and R^2 as metrics of predictive accuracy.

We tested how including social and environmental covariates improves predictive accuracy, relative to the univariate GLMs. First, we evaluated predictive accuracy of GLMs with aspect as an additive predictor variable in the generalized linear models. Second, we evaluated the predictive accuracy of generalized linear mixed models (GLMMs) including parcel identity as an additive random effect. Including this term as a random effect enables parcels with few replicate pixels to borrow strength from parcels with many replicate pixels (Gelman and Hill 2006).

Finally, we applied our GLMs to compare tree canopy cover and tree height between c. 2000 and 2012 for a 109 882 ha region of the study area centered on the LiDAR image. To avoid effects of inter-annual phenology unrelated to forest regrowth, we averaged multiple dry season images, including 5 scenes from 2000 to 2001 and 12 scenes from 2011 to 2013. Results were nearly identical when using only images from 2000 and 2012 (Appendix S3). We then applied GLM models with PV and aspect as covariates to predict tree canopy and tree height from the two composite images. We produced change metrics for each pixel by subtracting the 2000–2001 epoch predictions from the 2011–2013 epoch predictions. The scan line corrector (SLC) failure in 2003 on board Landsat 7 results in data gaps on the sides of the scene (Wulder et al. 2008). Although the streaking is visually evident, even after compositing over the time series of Landsat 7 images (Fig. 5), we found little impact of SLC failure on forest regrowth detection (Appendix S3). We then compared our measures of land cover change with the Global Forest Change dataset v1.2 (Hansen et al. 2013), a widely used remote sensing product that categorizes pixels into 'forest' and 'non-forest' categories. The Global Forest Change dataset is also derived from Landsat 7 data and has approximately the same resolution as our final product (30 m^2), properties that ease comparison with our forest gain metrics.

Table 1. Predictive accuracy of different models for tree canopy cover.

Tree canopy cover		
Model	R^2 (%)	RMSE (%)
Linear regression	16.3	49.38
Binomial GLM	44.31	26.55
Binomial GLM + aspect	44.77	26.44
Binomial GLMM	55.39	23.77
Binomial GLMM + aspect	55.68	23.69

GLM, generalized linear models.

Table 2. Predictive accuracy of different models for tree canopy height.

Tree height		
Model	R^2 (%)	RMSE (m)
Linear regression	33.99	3.98
Gamma GLM	44.61	3.65
Gamma GLM + aspect	45.60	3.62
Gamma GLMM	50.70	3.45
Gamma GLMM + aspect	51.24	3.42

GLM, generalized linear models.

Results

For both tree canopy cover and tree height, GLMs provided a better fit than linear regression (Tables 1 and 2). This difference was more pronounced for tree canopy cover than for tree height, with an increase in R^2 of 28.01% from the log-transformed linear regression to the binomial GLM for tree canopy cover, compared to an increase in R^2 of 10.62% from the log-transformed linear regression to the gamma GLM for tree height (Tables 1 and 2). The advantage of the GLM models is apparent in plots of the relationship between PV and tree canopy height and tree canopy cover (Fig. 3). The gamma GLM, with a log-link function, accurately captures the nonlinear relationship between tree height and PV. Similarly, the binomial GLM with a cloglog link function accurately captures the sigmoidal relationship between tree canopy cover and PV, including saturation at high values of PV.

Fewer than 200 randomly selected pixels were required to reach a maximum level of predictive accuracy for both tree canopy cover and tree height using univariate GLMs (Fig. S1). We found sharp increases in predictive accuracy when increasing from 5 pixels to 100 pixels, weaker increases from 100 pixels to 200 pixels and near-constant predictive accuracy after sample sizes of 200 pixels. This result shows that GLMs using only PV can be trained to predict forest structure with fair accuracy in a

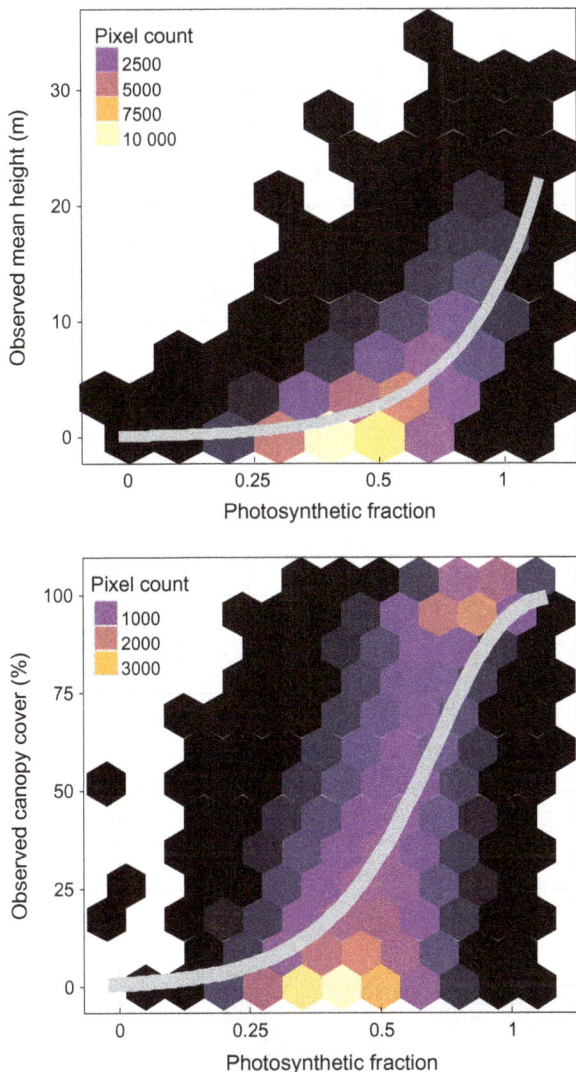

Figure 3. Photosynthetic fraction from spectrally unmixed Landsat data is correlated with mean tree height and tree canopy cover. Hexagon colors represent pixel density (n = 80 000 pixels). The gray line represents predicted relationship between photosynthetic fraction and forest structure from a univariate binomial GLM (tree canopy cover) or a gamma GLM (tree height).

heterogeneous landscape using <0.1% of the study region's total area.

The increases in predictive accuracy when aspect was added to the GLM were relatively minor, with increases in R^2 of 1.46 and 3.3% for tree canopy cover and tree canopy height respectively. Including parcel ID as a random effect in a GLMM had a larger impact, with an increase in R^2 of 10.18 and 4.42% for tree canopy cover and tree canopy height respectively. The best model for both tree height and tree canopy cover was the GLMM with parcel ID as a random effect and PV and aspect as fixed effects.

For the best-fitting models, plots of predicted versus observed values suggest that maximum predictive accuracy is achieved at 20% tree canopy cover and 5 m tree height (Fig. S2). The binomial GLMM overpredicted tree canopy cover at low tree cover levels (<20% tree canopy cover) and both the binomial and gamma GLMMs underpredicted tree canopy cover and tree height at high tree cover levels (>60% tree canopy cover and >20 m of tree canopy height). Consequently, although the models reproduce the overall structure of the heterogeneous landscape fairly well, including riparian corridors and live fences, maps of predicted tree canopy cover and tree height are somewhat less distinctly resolved compared to maps of observed tree canopy cover and tree height (Fig. 4).

Spectral mixture analysis of Landsat images from $c.$ 2000 to 2012 suggests widespread increases in photosynthetic vegetation across the landscape during this time period (Fig. 5). Our GLMs enabled us to translate changes in endmember fractions into changes in tree height and tree canopy cover. We found that 60.67% of pixels increased in tree height and 61% of pixels increased in tree canopy cover (Fig. 6). However, most of these increases were small, with a median increase of 0.2 m in height (95% CI: −5.34, 3.07) and 3% in tree cover (95% CI: −0.43, 0.33). 7.2% of pixels increased in tree height more than 2 m, and only 4.4% of pixels increased more than 5 m in tree height. In comparison, the discrete classification of forest gain in the study area from $c.$ 2000 to 2012 only detected forest gain in 1.4% of pixels, most of which were clustered together in only a few parcels (Fig 7).

Discussion

Spatially extensive yet ecologically detailed measurements of forest regrowth will be necessary to understand the successional mechanisms that drive landscape scale forest transitions. We demonstrate that LiDAR–Landsat data integration can measure continuous changes in tree canopy cover and tree height in an agricultural landscape. For both variables, GLMs outperformed linear regression models, indicating the value of modeling these data with the appropriate probability distribution and link function. An advantage of the relatively simple GLMs is the ability to rapidly fit models with minimum data requirements. In our case, training datasets representing <0.1% of the total landscape area were sufficient to maximize predictive accuracy for tree canopy height and tree canopy cover. When applied to predict tree canopy cover and tree height change between $c.$ 2000 and 2012, our method reveals a widespread increase in both, suggesting an ongoing land cover change. Although spatially extensive, these structural changes were subtle and may be overlooked by

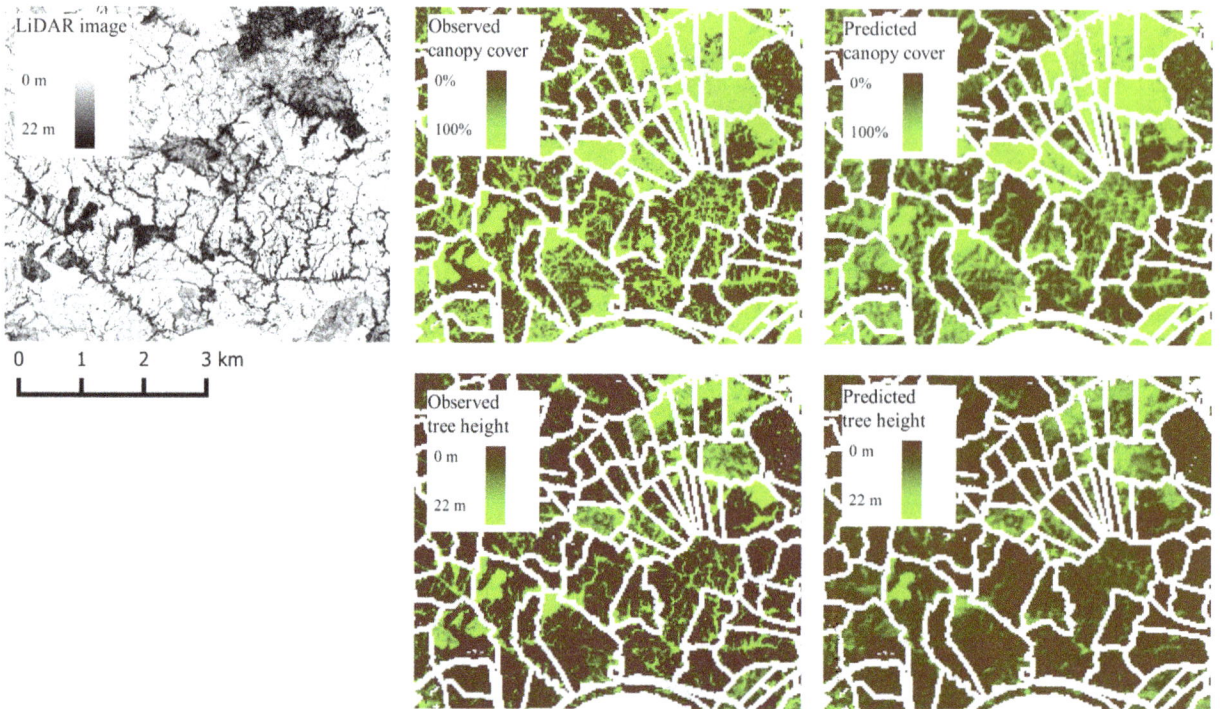

Figure 4. Map of predicted versus observed tree canopy cover and tree height for a subset of the landscape. The black and white LiDAR image show the original LiDAR-derived canopy height model (CHM) at 1.25 m resolution. The observed tree height and tree canopy cover panels show these forest structure variables derived from the CHM at 30 m resolution. Predictions are from the best-fit GLMMs. White lines indicate boundaries between parcels.

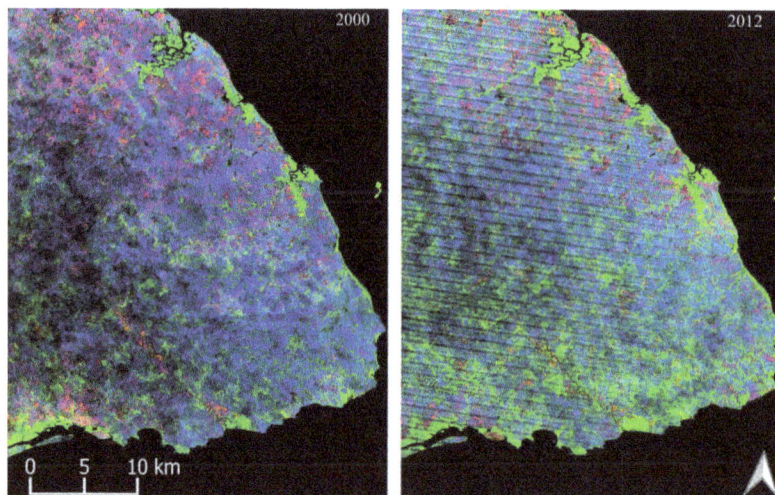

Figure 5. Endmember fractions (photosynthetic vegetation, non-photosynthetic vegetation and soil) derived through spectral unmixing of Landsat 7 data for the Azuero Peninsula in epochs 2000–2001 and 2011–2013. Green color indicates photosynthetic vegetation, red indicates soil and blue indicates non-photosynthetic vegetation.

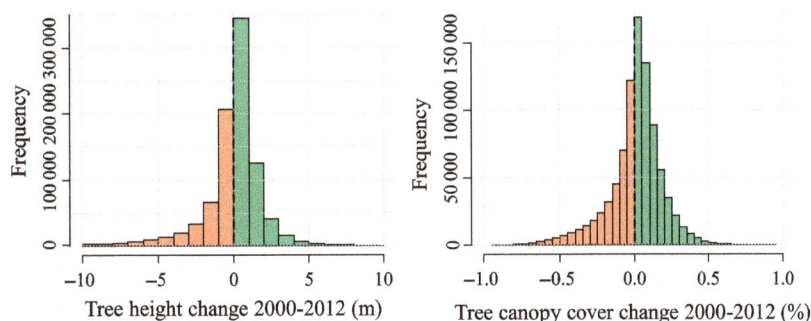

Figure 6. Distribution of pixel-level tree height and canopy cover changes. The green bars indicate gain while the red bars indicate loss. This figure shows an overall increase in tree height and tree canopy cover during the 12-year period, largely driven by increases in height of <3 m, and in canopy cover of <20%.

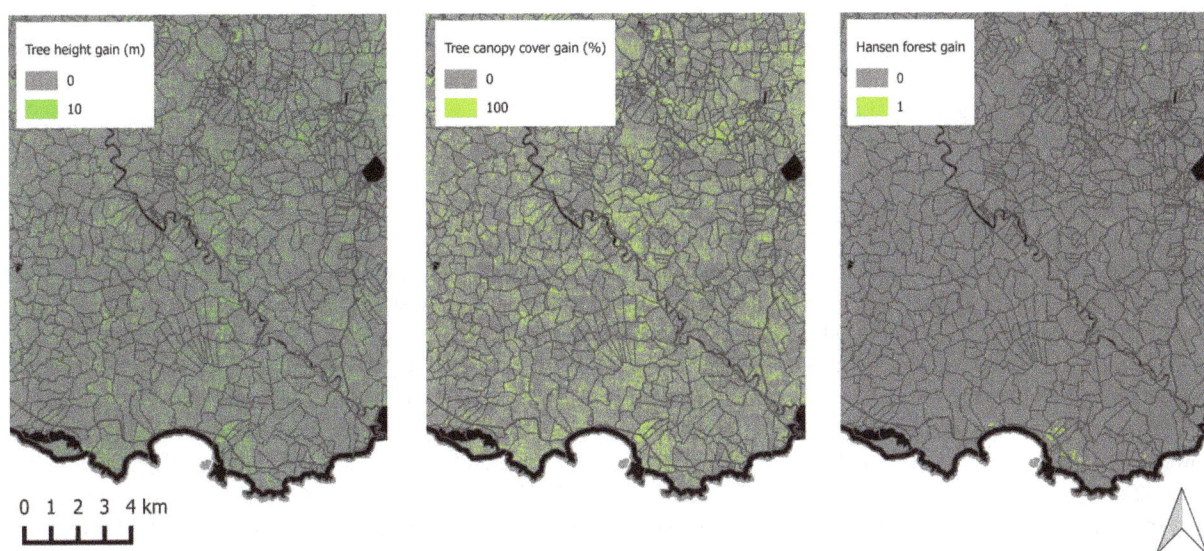

Figure 7. Comparison of forest gain from continuous changes in tree height and tree canopy cover to discrete non-forest to forest pixels. These maps illustrate how the discrete classification fails to detect spatially extensive gains in tree height across the landscape. Gray lines indicate parcel boundaries and black areas indicate urban or beach habitat.

discrete classification of pixels in 'forest' or 'non-forest' categories. Overall, we demonstrate an approach that could unlock the potential of the open access, multi-decadal and global Landsat database to predict secondary forest dynamics – even in landscapes with complex and spatially variable land use history.

Our approach has the potential to improve predictions of tropical reforestation. Studies that use field data to compare secondary forest dynamics across multiple landscapes reveal high variability between sites (Norden et al. 2015). For example, across 45 Neotropical sites, biomass of 20-year-old secondary forest varied by 11-fold (Poorter et al. 2016). Forest inventory plots may even underestimate variability in successional rates since they are generally placed in sites where secondary succession has already

occurred and where there are multiple trees >5 cm DBH (diameter at breast height) that can be measured. Thus, sites with low tree cover where secondary succession has failed or is proceeding very slowly (Duncan and Chapman 1999) are underrepresented. The Landsat archive, and other widely available remotely sensed data products, could improve our understanding of tropical reforestation by increasing the temporal and spatial scale of field studies. Our analysis will likely be reproducible on other calibrated optical remote sensing platforms that can be processed to surface reflectance and have a similar spatial resolution (e.g. Landsat 8, ASTER, Sentinel 2, and SPOT 1-4). Our models are most accurate for quantifying the early stages of secondary succession, when tree cover is 20–40% and tree height is ≤10 m, and could fill the data gap on a critical early successional stage (Holl et al. 2000) that is often classified as 'non-forest' in categorical remote sensing studies (Aide et al. 2013; Hansen et al. 2013; Sloan et al. 2015). Our statistical models were not very sensitive to the choice of a height threshold for tree canopy cover (Appendix S2), suggesting that our approach could be adapted to a wide range of ecologically relevant height thresholds for tree cover, such as the height of a succession-inhibiting grass layer (Caughlin et al. 2016), or the minimum size of trees measured in field plots.

Our main predictor variable was photosynthetic vegetation fraction (PV) from spectrally unmixed Landsat pixels. This variable was able to capture a wide range of variation in tree height and tree canopy cover across our study area, including a model trained with a small proportion (<0.01%) of pixels. While our study area represents a relatively small area of Southwestern Panama, our findings add to a growing body of evidence that subpixel fractions of photosynthetic and non-photosynthetic material capture a general relationship between forest structure and satellite reflectance (Asner et al. 2009; Caviglia-Harris et al. 2015; Mayes et al. 2015). These relationships may be due to greater leaf area index, greater moisture at the beginning of the dry season in forest patches (Kalacska et al. 2007) or other mechanisms. For some secondary forests, fraction of non-photosynthetic vegetation (NPV) may provide more information than PV on forest type (Mayes et al. 2015). However, in this landscape, the addition of NPV as a covariate did not add any predictive power. Ultimately, the generality of our models will need to be evaluated over larger spatial extents, including different forest types and disturbance regimes.

Environmental and social covariates could improve the ability of regression models to predict forest structure from Landsat data. We found that parcel membership improved predictive accuracy for both tree canopy cover and tree height. In tropical agricultural landscapes, differential land use can create spatial differences in secondary forest composition and structure (Manson and Evans 2007; Caviglia-Harris et al. 2015; Mesquita et al. 2015). Our results demonstrate that remote sensing techniques that account for the patchwork of land history in agricultural landscapes have the potential to increase predictive accuracy of secondary forest characterization. Incorporating landownership in remote sensing classifications can also lead to opportunities to engage stakeholders (Vergara-Asenjo et al. 2015). Cadastral data, mapped boundaries of parcels, are ideal for this purpose as they are spatially explicit and can be easily merged with remotely sensed datasets. As more cadastral data comes online (e.g. Brazil's Cadastro Ambiental Rural program; Gibbs et al.

2015), we expect the utility of parcel membership as a predictor variable in remote sensing to increase.

Previous remote sensing studies in our study region have revealed small increases (<5%) in net forest cover from 2000 to 2010 (Metzel and Pacala 2010; Aide et al. 2013; Sloan 2015). In contrast, our results suggest a more dynamic landscape that is undergoing widespread forest regrowth, with increased tree height and tree canopy cover in a majority of pixels. A likely reason for this discrepancy is that previous land cover change analyses have used discrete land cover classifications with conservative thresholds for forest change. For example, the Global Forest Change dataset (Hansen et al. 2013), produced with the same Landsat 7 data used in our study, shows only a 1% increase in pixels classified as 'forest' from c. 2000 to 2012. In the Azuero, many pixels have tree cover below the 25% tree cover threshold used to classify 'forest' versus 'non-forest' pixels in Hansen et al., and thus would not be counted as 'forest.' While the Global Forest Change dataset is a highly automated global product, in contrast to our more local classification, its discrete forest categories are similar to many other remote sensing studies focused on detecting reforestation (Aide et al. 2013; Sloan 2015). Definitions of forest that overlook fine-scale patterns of tree cover are problematic because small differences in tree cover in agricultural landscapes can be ecologically significant (Slocum 2001; Harvey et al. 2005; Murgueitio et al. 2011). Integrating fine-scale LiDAR data with Landsat data provides a solution to represent fine-scale tree cover differences in medium resolution satellite data with pixels larger than the area of most individual tree crowns. However, we note that different definitions of reforestation are useful for different purposes, and for some applications, defining forest categories based on a cover threshold may be desirable (Chazdon et al. 2016). While we measured changes in forest structure, reforestation also involves changes in species composition (Van Breugel et al. 2006; Chazdon et al. 2007). An important next step will be to integrate remote detection of tree community composition (e.g. Fagan et al. 2015; Graves et al. 2016) into forest structural models.

Possible explanations for the widespread but small increases in tree cover and tree height in our study area include an agricultural regime shift, from monoculture pasture to silvopastoral systems (Murgueitio et al. 2011), lengthened fallow cycles due to labor scarcity (Rudel 2012) or the early stages of native secondary forest succession (Chazdon 2014). Distinguishing between these possibilities will require additional ecological and social data. Change metrics and trend analyses that utilize time series of remotely sensed data will also be necessary to explain the ecological processes behind land cover change (Kennedy et al. 2014). Spatially and temporally extensive measurements of forest structure derived from satellite data will play a key role in scaling up field measurements to understand regional land cover change.

Acknowledgments

We thank D. Knapp, C. Anderson, T. Kennedy-Bowdoin and the entire Carnegie Airborne Observatory staff for data collection and processing support. T. T. Caughlin was supported by NSF grant #1415297 Science, Engineering and Education for Sustainability Fellow program. Airborne data collection, processing and analysis was funded by the Grantham Foundation for the Protection of the Environment and William R. Hearst III. The Carnegie Airborne Observatory has been made possible by grants and donations to G.P. Asner from the Avatar Alliance Foundation, Margaret A. Cargill Foundation, David and Lucile Packard Foundation, Gordon and Betty Moore Foundation, Grantham Foundation for the Protection of the Environment, W. M. Keck Foundation, John D. and Catherine T. MacArthur Foundation, Andrew Mellon Foundation, Mary Anne Nyburg Baker and G. Leonard Baker Jr, and William R. Hearst III.

References

Adams, J. B., D. E. Sabol, V. Kapos, R. Almeida Filho, D. A. Roberts, M. O. Smith, et al. 1995. Classification of multispectral images based on fractions of endmembers: application to land-cover change in the Brazilian Amazon. *Remote Sens. Environ.* **52**, 137–154.

Ahmed, O. S., S. E. Franklin, M. A. Wulder, and J. C. White. 2015. Characterizing stand-level forest canopy cover and height using landsat time series, samples of airborne LiDAR, and the random forest algorithm. *ISPRS J. Photogramm. Remote Sens.* **101**, 89–101.

Aide, T. M., M. L. Clark, H. R. Grau, D. López-Carr, M. A. Levy, D. Redo, et al. 2013. Deforestation and reforestation of Latin America and the Caribbean (2001–2010). *Biotropica* **45**, 262–271.

Anderson, J. E., L. C. Plourde, M. E. Martin, B. H. Braswell, M.-L. Smith, R. O. Dubayah, et al. 2008. Integrating waveform lidar with hyperspectral imagery for inventory of a northern temperate forest. *Remote Sens. Environ.* **112**, 1856–1870.

Antonarakis, A. S., S. S. Saatchi, R. L. Chazdon, and P. R. Moorcroft. 2011. Using Lidar and Radar measurements to constrain predictions of forest ecosystem structure and function. *Ecol. Appl.* **21**, 1120–1137.

Arroyo-Mora, J. P., G. A. Sánchez-Azofeifa, M. E. Kalacska, B. Rivard, J. C. Calvo-Alvarado, and D. H. Janzen. 2005. Secondary forest detection in a Neotropical dry forest landscape using Landsat 7 ETM+ and IKONOS Imagery1. *Biotropica* **37**, 497–507.

Arroyo-Rodríguez, V., F. P. L. Melo, M. Martínez-Ramos, F. Bongers, R. L. Chazdon, J. A. Meave, et al. 2015. Multiple successional pathways in human-modified tropical landscapes: new insights from forest succession, forest fragmentation and landscape ecology research. *Biol. Rev.* doi: 10.1111/brv.12231.

Asner, G. P., M. Keller, J. Rodrigo Pereira, J. C. Zweede, and J. N. M. Silva. 2004. Canopy damage and recovery after selective logging in Amazonia: field and satellite studies. *Ecol. Appl.* **14**, 280–298.

Asner, G. P., D. E. Knapp, A. Balaji, and G. Páez-Acosta. 2009. Automated mapping of tropical deforestation and forest degradation: CLASlite. *J. Appl. Remote Sens.* **3**, 033543.

Asner, G. P., D. E. Knapp, J. Boardman, R. O. Green, T. Kennedy-Bowdoin, M. Eastwood, et al. 2012. Carnegie Airborne Observatory-2: increasing science data dimensionality via high-fidelity multi-sensor fusion. *Remote Sens. Environ.* **124**, 454–465.

Asner, G. P., J. Mascaro, C. Anderson, D. E. Knapp, R. E. Martin, T. Kennedy-Bowdoin, et al. 2013. High-fidelity national carbon mapping for resource management and REDD+. *Carbon balance and management* **8**, 7.

Bolker, B. M., M. E. Brooks, C. J. Clark, S. W. Geange, J. R. Poulsen, M. H. H. Stevens, et al. 2009. Generalized linear mixed models: a practical guide for ecology and evolution. *Trends Ecol. Evol.* **24**, 127–135.

van Breugel, M., J. S. Hall, D. Craven, M. Bailon, A. Hernandez, M. Abbene, et al. 2013. Succession of ephemeral secondary forests and their limited role for the conservation of floristic diversity in a human-modified tropical landscape.

Campo-Bescós, M. A., R. Muñoz-Carpena, J. Southworth, L. Zhu, P. R. Waylen, and E. Bunting. 2013. Combined spatial and temporal effects of environmental controls on long-term monthly NDVI in the Southern Africa Savanna. *Remote Sens.* **5**, 6513–6538.

Cao, L., N. C. Coops, J. L. Innes, S. R. J. Sheppard, L. Fu, H. Ruan, et al. 2016. Estimation of forest biomass dynamics in subtropical forests using multi-temporal airborne LiDAR data. *Remote Sens. Environ.* **178**, 158–171.

Castillo, M., B. Rivard, A. Sánchez-Azofeifa, J. Calvo-Alvarado, and R. Dubayah. 2012. LIDAR remote sensing for secondary Tropical Dry Forest identification. *Remote Sens. Environ.* **121**, 132–143.

Castillo-Núñez, M., G. A. Sánchez-Azofeifa, A. Croitoru, B. Rivard, J. Calvo-Alvarado, and R. O. Dubayah. 2011. Delineation of secondary succession mechanisms for tropical dry forests using LiDAR. *Remote Sens. Environ.* **115**, 2217–2231.

Caughlin, T. T., S. Elliott, and J. W. Lichstein. 2016. When does seed limitation matter for scaling up reforestation from patches to landscapes? *Ecol. Appl.* doi: 10.1002/eap.1410.

Caviglia-Harris, J. L., M. Toomey, D. W. Harris, K. Mullan, A. R. Bell, E. O. Sills, et al. 2015. Detecting and interpreting secondary forest on an old Amazonian frontier. *J. Land Use Sci.* **10**, 442–465.

Chazdon, R. L. 2014. *Second growth: the promise of tropical forest regeneration in an age of deforestation.* University of Chicago Press, Chicago, IL.

Chazdon, R. L., S. G. Letcher, M. Van Breugel, M. Martínez-Ramos, F. Bongers, and B. Finegan. 2007. Rates of change in tree communities of secondary Neotropical forests following major disturbances. *Philos. Trans. R. Soc. Lond. B Biol. Sci.* **362**, 273–289.

Chazdon, R. L., P. H. S. Brancalion, L. Laestadius, A. Bennett-Curry, K. Buckingham, C. Kumar, et al. 2016. When is a forest a forest? Forest concepts and definitions in the era of forest and landscape restoration. *Ambio* **45**, 538

Chen, G., and G. J. Hay. 2011. A support vector regression approach to estimate forest biophysical parameters at the object level using airborne lidar transects and quickbird data. *Photogramm. Eng. Remote Sensing* 77, 733–741.

Duncan, R. S., and C. A. Chapman. 1999. Seed dispersal and potential forest succession in abandoned agriculture in tropical Africa. *Ecol. Appl.* **9**, 998–1008.

Englhart, S., J. Jubanski, and F. Siegert. 2013. Quantifying dynamics in tropical peat swamp forest biomass with Multi-temporal LiDAR datasets. *Remote Sens.* **5**, 2368–2388.

Fagan, M. E., R. S. DeFries, S. E. Sesnie, J. P. Arroyo-Mora, C. Soto, A. Singh, et al. 2015. Mapping species composition of forests and tree plantations in Northeastern Costa Rica with an integration of hyperspectral and multitemporal landsat imagery. *Remote Sens.* **7**, 5660–5696.

Gelman, A., and J. Hill. 2006. *Data analysis using regression and multilevel/hierarchical models.* Cambridge University Press, Cambridge.

Gibbs, H. K., J. Munger, J. L'Roe, P. Barreto, R. Pereira, M. Christie, et al. 2015. Did ranchers and slaughterhouses respond to zero-deforestation agreements in the Brazilian Amazon? *Conserv. Lett.* **9**, 32–42.

Graves, S. J., G. P. Asner, R. E. Martin, C. B. Anderson, M. S. Colgan, L. Kalantari, et al. 2016. Tree species abundance predictions in a tropical agricultural landscape with a supervised classification model and imbalanced data. *Remote Sens.* **8**, 161.

Griscom, H. P., A. B. Connelly, M. S. Ashton, M. H. Wishnie, and J. Deago. 2011. The structure and composition of a tropical dry forest landscape after land clearance; Azuero Peninsula, Panama. *J. Sustain. Forest.* **30**, 756–774.

Hansen, M. C., P. V. Potapov, R. Moore, M. Hancher, S. A. Turubanova, A. Tyukavina, et al. 2013. High-resolution global maps of 21st-century forest cover change. *Science* **342**, 850–853.

Hardin, J. W., and J. Hilbe. 2007. *Generalized linear models and extensions,* 2nd ed. Stata Press, College Station, Tex.

Harvey, C. A., C. Villanueva, J. Villacís, M. Chacón, D. Muñoz, M. López, et al. 2005. Contribution of live fences to the ecological integrity of agricultural landscapes. *Agric. Ecosyst. Environ.* **111**, 200–230.

Heckadon-Moreno, S. 2009. Exedra Books, Panamá. *De selvas a potreros: la colonización santeña en Panamá, 1850–1980.* Exedra Books, Panamá.

Hermosilla, T., M. A. Wulder, J. C. White, N. C. Coops, and G. W. Hobart. 2015. An integrated Landsat time series protocol for change detection and generation of annual gap-free surface reflectance composites. *Remote Sens. Environ.* **158**, 220–234.

Hijmans, R. J., van Etten J., M. Mattiuzzi, M. Sumner, J. A. Greenberg, O. P. Lamigueiro, et al. 2015. Package "raster."

Holl, K. D., M. E. Loik, E. H. V. Lin, and I. A. Samuels. 2000. Tropical Montane forest restoration in Costa Rica: overcoming barriers to dispersal and establishment. *Restor. Ecol.* **8**, 339–349.

Hudak, A. T., M. A. Lefsky, W. B. Cohen, and M. Berterretche. 2002. Integration of lidar and Landsat ETM+ data for estimating and mapping forest canopy height. *Remote Sens. Environ.* **82**, 397–416.

Hudak, A. T., E. K. Strand, L. A. Vierling, J. C. Byrne, J. U. H. Eitel, S. Martinuzzi, et al. 2012. Quantifying aboveground forest carbon pools and fluxes from repeat LiDAR surveys. *Remote Sens. Environ.* **123**, 25–40.

Kalacska, M., G. A. Sanchez-Azofeifa, B. Rivard, T. Caelli, H. P. White, and J. C. Calvo-Alvarado. 2007. Ecological fingerprinting of ecosystem succession: estimating secondary tropical dry forest structure and diversity using imaging spectroscopy. *Remote Sens. Environ.* **108**, 82–96.

Kellner, J. R., and G. P. Asner. 2014. Winners and losers in the competition for space in tropical forest canopies. *Ecol. Lett.* **17**, 556–562.

Kellner, J. R., D. B. Clark, and S. P. Hubbell. 2009. Pervasive canopy dynamics produce short-term stability in a tropical rain forest landscape. *Ecol. Lett.* **12**, 155–164.

Kennedy, R. E., S. Andréfouët, W. B. Cohen, C. Gómez, P. Griffiths, M. Hais, et al. 2014. Bringing an ecological view of change to Landsat-based remote sensing. *Front. Ecol. Environ.* **12**, 339–346.

Latawiec, A. E., B. B. Strassburg, P. H. Brancalion, R. R. Rodrigues, and T. Gardner. 2015. Creating space for large-scale restoration in tropical agricultural landscapes. *Front. Ecol. Environ.* **13**, 211–218.

Manson, S. M., and T. Evans. 2007. Agent-based modeling of deforestation in southern Yucatán, Mexico, and reforestation in the Midwest United States. *Proc. Natl Acad. Sci.* **104**, 20678.

Marvin, D. C., G. P. Asner, D. E. Knapp, C. B. Anderson, R. E. Martin, F. Sinca, et al. 2014. Amazonian landscapes and the bias in field studies of forest structure and biomass. *Proc. Natl Acad. Sci.* **111**, E5224–E5232.

Mascaro, J., G. P. Asner, S. Davies, A. Dehgan, and S. Saatchi. 2014. These are the days of lasers in the jungle. *Carbon Balance Manage.* **9**, 7.

Masek, J. G., E. F. Vermote, N. E. Saleous, R. Wolfe, F. G. Hall, K. F. Huemmrich, et al. 2006. A Landsat surface reflectance dataset for North America, 1990-2000. *IEEE Geosci. Remote Sens. Lett.* **3**, 68–72.

Mayes, M. T., J. F. Mustard, and J. M. Melillo. 2015. Forest cover change in Miombo Woodlands: modeling land cover of African dry tropical forests with linear spectral mixture analysis. *Remote Sens. Environ.* **165**, 203–215.

Mesquita, R. C. G., P. E. Dos Santos Massoca, C. C. Jakovac, T. V. Bentos, and G. B. Williamson. 2015. Amazon Rain forest succession: stochasticity or land-use legacy? *Bioscience* **65**, 849–861.

Metzel, R. N. B., and S. Pacala. 2010. From "Finca" to forest: forest cover change and land management in Los Santos, Panama.

Murgueitio, E., Z. Calle, F. Uribe, A. Calle, and B. Solorio. 2011. Native trees and shrubs for the productive rehabilitation of tropical cattle ranching lands. *For. Ecol. Manage.* **261**, 1654–1663.

Nagendra, H., and J. Southworth. 2009. *Reforesting landscapes: linking pattern and process.* Springer, Dordecht.

Negrón-Juárez, R. I., J. Q. Chambers, D. M. Marra, G. H. P. M. Ribeiro, S. W. Rifai, N. Higuchi, et al. 2011. Detection of subpixel treefall gaps with Landsat imagery in Central Amazon forests. *Remote Sens. Environ.* **115**, 3322–3328.

Norden, N., H. A. Angarita, F. Bongers, M. Martínez-Ramos, I. Granzow-de la Cerda, M. van Breugel, et al. 2015. Successional dynamics in Neotropical forests are as uncertain as they are predictable. *Proc. Natl Acad. Sci.* **112**, 8013–8018.

Ota, T., O. S. Ahmed, S. E. Franklin, M. A. Wulder, T. Kajisa, N. Mizoue, et al. 2014. Estimation of Airborne Lidar-Derived Tropical Forest Canopy Height Using Landsat Time Series in Cambodia. *Remote Sens.* **6**, 10750–10772.

Pflugmacher, D., W. B. Cohen, R. E. Kennedy, and Z. Yang. 2014. Using Landsat-derived disturbance and recovery history and lidar to map forest biomass dynamics. *Remote Sens. Environ.* **151**, 124–137.

Pickell, P. D., T. Hermosilla, R. J. Frazier, N. C. Coops, and M. A. Wulder. 2016. Forest recovery trends derived from Landsat time series for North American boreal forests. *Int. J. Remote Sens.* **37**, 138–149.

Poorter, L., F. Bongers, T. M. Aide, A. M. Almeyda Zambrano, P. Balvanera, J. M. Becknell, et al. 2016. Biomass resilience of Neotropical secondary forests. *Nature* **530**, 211–214.

Purves, D. W., J. W. Lichstein, N. Strigul, and S. W. Pacala. 2008. Predicting and understanding forest dynamics using a simple tractable model. *Proc. Natl Acad. Sci.* **105**, 17018.

R Core Team. 2015. *R: a language and environment for statistical computing.* R Foundation for Statistical Computing, Vienna, Austria.

Réjou-Méchain, M., B. Tymen, L. Blanc, S. Fauset, T. R. Feldpausch, A. Monteagudo, et al. 2015. Using repeated small-footprint LiDAR acquisitions to infer spatial and temporal variations of a high-biomass Neotropical forest. *Remote Sens. Environ.* **169**, 93–101.

Rifai, S. W., J. D. Urquiza Muñoz, R. I. Negrón-Juárez, F. R. Ramírez Arévalo, R. Tello-Espinoza, M. C. Vanderwel, et al. 2016. Landscape-scale consequences of differential tree mortality from catastrophic wind disturbance in the Amazon. *Ecol. Appl.* **26**, 2225–2237.

Rudel, T. K. 2012. The Human Ecology of Regrowth in the Tropics. *J. Sustain. Forest.* **31**, 340–354.

Sloan, S. 2015. The development-driven forest transition and its utility for REDD+. *Ecol. Econ.* **116**, 1–11.

Sloan, S., M. Goosem, and S. G. Laurance. 2015. Tropical forest regeneration following land abandonment is driven by primary rainforest distribution in an old pastoral region. *Landscape Ecol.* **31**, 601–618.

Slocum, M. G. 2001. How tree species differ as recruitment foci in a tropical pasture. *Ecology* **82**, 2547–2559.

Southworth, J., D. Munroe, and H. Nagendra. 2004. Land cover change and landscape fragmentation—comparing the utility of continuous and discrete analyses for a western Honduras region. *Agric. Ecosyst. Environ.* **101**, 185–205.

Stark, S. C., V. Leitold, J. L. Wu, M. O. Hunter, C. V. de Castilho, F. R. C. Costa, et al. 2012. Amazon forest carbon dynamics predicted by profiles of canopy leaf area and light environment. *Ecol. Lett.* **15**, 1406–1414.

Van Breugel, M., M. Martínez-Ramos, and F. Bongers. 2006. Community dynamics during early secondary succession in Mexican tropical rain forests. *J. Trop. Ecol.* **22**, 663–674.

Vergara-Asenjo, G., D. Sharma, and C. Potvin. 2015. Engaging stakeholders: assessing accuracy of participatory mapping of land cover in Panama. *Conserv. Lett.* **8**, 432–439.

Vogelmann, J. E., A. L. Gallant, H. Shi, and Z. Zhu. 2016. Perspectives on monitoring gradual change across the continuity of Landsat sensors using time-series data. *Remote Sens. Environ.* **185**, 258–270.

Wulder, M. A., and D. Seemann. 2003. Forest inventory height update through the integration of lidar data with segmented Landsat imagery. *Can. J. Remote Sens.* **29**, 536–543.

Wulder, M. A., S. M. Ortlepp, J. C. White, and S. Maxwell. 2008. Evaluation of Landsat-7 SLC-off image products for forest change detection. *Can. J. Remote Sens.* **34**, 93–99.

Wulder, M. A., J. C. White, T. R. Loveland, C. E. Woodcock, A. S. Belward, W. B. Cohen, et al. 2016. The global Landsat archive: status, consolidation, and direction. *Remote Sens. Environ.* **185**, 271–283.

Zahawi, R. A., and C. K. Augspurger. 2006. Tropical forest restoration: tree islands as recruitment foci in degraded lands of Honduras. *Ecol. Appl.* **16**, 464–478.

Zald, H. S. J., M. A. Wulder, J. C. White, T. Hilker, T. Hermosilla, G. W. Hobart, et al. 2016. Integrating Landsat pixel composites and change metrics with lidar plots to predictively map forest structure and aboveground biomass in Saskatchewan, Canada. *Remote Sens. Environ.* **176**, 188–201.

4

Bridging disciplines with training in remote sensing for animal movement: an attendee perspective

Bethany L. Clark[1], Mirjana Bevanda[2], Eneko Aspillaga[3] & Nicolai H. Jørgensen[4]

[1]Environment and Sustainability Institute, University of Exeter, Penryn Campus, TR10 9FE, Penryn, Cornwall, United Kingdom
[2]Animal Ecology and Tropical Biology - Ecological Field Station, Institute of Biology, University of Würzburg, 97074 Würzburg, Germany
[3]Departament d'Ecologia, Universitat de Barcelona, 08029 Barcelona, Spain
[4]Department of Animal and Aquacultural Science, Norwegian University of Life Science, 1430 Ås, Norway

Keywords
Animal movement, bio-logging, satellite imagery, survey, tracking, training

Correspondence
Bethany L. Clark, Environment and Sustainability Institute, University of Exeter, Penryn Campus, TR10 9FE Penryn, Cornwall, United Kingdom.
E-mail: b.l.clark@exeter.ac.uk
Nicolai H. Jørgensen, Department of Animal and Aquacultural Science, Norwegian University of Life Science, 1430 Ås, Norway.
E-mail: nicolai.jorgensen@nmbu.no

Funding Information
The APC will be funded by Norwegian University of Life Sciences (NMBU).

Editor: Duccio Rocchini
Associate Editor: Martin Wegmann

Abstract

Remote sensing and animal movement datasets are increasingly used to answer key questions in ecology and conservation. Collecting and accessing this data is becoming ever cheaper and easier, but limited analytical expertise limits its wider use. Working at the interface between these two disciplines is challenging as there are no standard techniques for handling the complex spatial data, so specific and in-depth training is required. Higher education programs rarely cover remote sensing for animal movement, so external courses play a major role in training newcomers and creating a more unified global community. We conducted an online survey to investigate the views of previous attendees of four training courses that involve remote sensing and animal location data. These courses provided subject-specific knowledge, practical and coding skills, networking, collaboration opportunities, insightful discussions and transferable research skills. Our survey highlighted the importance of real-world examples, practical sessions, time for participants to work with their own data, preparatory material and open source software. Despite the value of interdisciplinary training in remote sensing and animal movement, it reaches few ecology and conservation practitioners outside of academia. We advocate more funding for underrepresented participants to attend existing course and the development of new courses.

Introduction

Observing the changing status of the planet is key to understanding and predicting animal distributions (O'Connor et al. 2015; Skidmore et al. 2015). With increased habitat loss, invasive species, climate change and overexploitation, it is imperative that future conservationists understand how these changes may affect species (Mace and Purvis 2008). Advances in remote sensing technology are rapidly providing new variables and higher resolution data, which

could be harnessed to address key questions in ecology and conservation (Rose et al. 2014). For example, Landsat data can identify habitat degradation caused by grazing (Karnieli et al. 2013), and sea surface temperature patterns explain seabird foraging behavior (Scales et al. 2014). Similarly, animal movement is an emerging field driven by rapid developments in technology, specifically miniaturized bio-loggers, which are sensors attached to animals to record their behavior (Ropert-Coudert and Wilson 2005). Satellite imagery, in particular, is becoming increasingly important for investigating animal movement patterns (Pettorelli et al. 2014a, 2016; Wikelski et al. 2007).

The importance of earth observation for ecology and conservation was postulated over a decade ago (Roughgarden et al. 1991; Kerr and Ostrovsky 2003; Nagendra 2001; Turner et al. 2003), but various challenges still limit its application (Pettorelli et al. 2014a). Among obstacles such as data access (Turner et al. 2015) and software availability (Rocchini and Neteler 2012), the data generated by remote sensing and animal tracking technology is not straightforward to handle. Remote sensing and animal tracking data are spatially and temporally autocorrelated, and therefore cannot be analyzed using standard statistical methods (Legendre 1993; Dray et al. 2010). A key challenge is to keep up with the rapid development of analytical approaches (Nathan et al. 2008), which has led to a lack of standard methods as many similar tools have been developed without a cohesive strategy (Nathan 2008). Additionally, datasets from both fields are becoming rapidly large and more complex. The spatial and temporal resolution of satellite images is increasing and new technology, including multispectral imaging, hyperspectral imaging, RADAR and LiDAR, allow additional variables to be measured (Pettorelli et al. 2014a; He et al. 2015). Bio-logging devices are recording at a higher frequency and for longer, as well as logging additional measurements such as acceleration, temperature and sound. Furthermore, devices are becoming smaller and cheaper, allowing more species and individuals to be tagged (Ropert-Coudert and Wilson 2005; Wilson et al. 2008). Skills in analysis, coding and data management are needed to overcome the challenges posed by such complex data. Moreover, standard approaches using open access methods allow for better comparison between studies and facilitate collaboration (Hampton et al. 2013), and open access datasets allow different methods to be directly compared using the same data.

Despite the value of remote sensing in ecology and conservation, and the numerous challenges in its application, there is little training that bridges the gap between these disciplines (Pettorelli et al. 2014a). These topics are rarely covered in higher education programs (Pettorelli et al. 2014b), due to the small number of lecturers within biological science departments with such expertise. Furthermore, research institutions, universities and funding bodies rarely facilitate interdisciplinary research or teaching between departments (Cech and Rubin 2004). As such, there is a strong need to train newcomers including postgraduate students, postdoctoral researchers and ecology or conservation practitioners (Wegmann et al. 2016). Furthermore, courses have a key role in establishing connections between communities and encouraging consistent methods (Pettorelli et al. 2014a).

This interdisciplinary perspective has three main aims. Firstly, to outline the multiple benefits of attending specialist training in remote sensing and animal movement. Secondly, to demonstrate the challenges faced by those running courses, and provide information from an attendee viewpoint to aid the planning of future courses. Finally, to encourage the training of a larger number and wider diversity of users in this rapidly developing research area. We surveyed previous attendees of relevant training courses to assess their needs and preferences, and having attended the AniMove Animal Movement Analysis for Conservation in 2014 or 2015, we are equipped to present a student perspective.

Materials and Methods

Few interdisciplinary courses covering remote sensing for ecology and conservation are available (Pettorelli et al. 2014a,b), especially in relation to animal movement. We conducted an online survey of people who had previously attended a 5- or 10-day training course that involved using remote sensing and animal movement data (surveyplanet.com/56aa68df1fb2579657c33e50). We used open questions to ask what the course's main benefits and drawbacks were, and multiple choice questions to capture more detail. Between 9 March and 25 April 2016, we received 49 responses relating to four courses, which varied in the amount of animal movement material covered (Table 1).

Value of attending training courses

Training courses in remote sensing and animal movement are highly valued by participants; 98% would recommend the course they attended. We identified a number of key benefits of training courses. Firstly, new skills and

Table 1. Survey responses by course and year (1 person attended both the AniMove and CAnMove course).

Course title and host organization	2009–13	2014	2015	2016	Total
Spatial analysis of ecological data using R, PR statistics	–	1	10	9	20
Animal movement analysis for ecology & conservation, AniMove	2	6	8	–	16
ECO 304 animal movement ecology, University of Zurich	–	–	8	–	8
Ecology of migration, CAnMove, Lund University	3	–	3	–	6

knowledge acquired through the course curriculum. Secondly, a broader exposure to different approaches and problem solving through discussions with tutors and other attendees. And finally, career ideas and opportunities for networking and starting collaborative projects.

New skills and knowledge

Attendees reported improved coding and practical skills, as well as transferable research and communication skills (Fig. 1). As these courses deal with rapidly developing research areas without standard techniques (Nathan 2008; Nathan et al. 2008), providing attendees with the latest technical and computational tools is key. Learning state-of-the-art methods was the most commonly stated main benefit in response to our open question (14 participants), which is particularly important in such a fast-moving field, where published articles or books may not yet be available. Additionally, participants found it useful to learn about current methodological debates and priority areas for future development. The courses generally covered the main area of the attendees' research, and resulted in new analyses within attendees' own projects (Fig. 1). For example, Figure 2 illustrates the application of remote sensing and animal movement analysis skills learnt on the AniMove course to a participant's own data. The environmental correlates of the home range of sheep

were investigated using satellite imagery and the Normalized Difference Vegetation Index in combination with dynamic Brownian Bridge Movement Models of home range from Global Positioning System (GPS) tracking (Kranstauber et al. 2012). In the era of big data, research groups and employers highly value skills in handling large datasets and analyzing complex patterns (Hampton et al. 2013).

Besides the direct acquisition of theoretical and technical knowledge, more informal discussions between the course teachers and participants proved to be highly enriching and were specifically stated as a main benefit by six respondents. Meeting both the teachers and the other attendees was beneficial to all course participants, with meeting teachers benefitting more people in relation to their studies or research and meeting other attendees benefitting more people with respect to personal aspects such as comparing lifestyles (Fig. 3). In discussions, attendees dealt with common concerns from different points of view, leading to the emergence of new ideas and approaches. Moreover, resolving specific problems is easier in diverse groups, as solutions may arise from different disciplines or people who have previously overcome similar obstacles. Additionally, explaining projects to others with relevant knowledge is a good way to clarify ideas and improve communication skills, which is particularly valuable when such opportunities are not available at home institutions.

Networking and collaboration

Courses create networks of researchers and promote future synergies between both teachers and attendees. When our survey asked an open question on what the main benefits of the course were, 11 attendees mentioned the opportunity to make useful contacts and develop a network, and three mentioned meeting collaborators. Teachers and other attendees provided a similar number of participants with benefits relating to networking, career ideas and potential employment (Fig. 3). Interdisciplinary collaborations between remote sensing researchers and ecologists enhance the potential of both disciplines (Pettorelli et al. 2016, 2014b). Survey respondents reported three collaborations that arose from courses, including the development of a method for predicting birth events in moose, deer and elk from GPS collar data (E. Fuller, pers. comm. 2016), and a study of cougar movement in the Southern Yellowstone Ecosystem (A. Kusler, pers. comm. 2016). A further 13 attendees were planning collaborations with attendees or tutors after meeting through a course (Fig. 3). We expect more collaborations stemming from these courses to develop as 78% responses were registered less than a year and 18% less than a month after the courses.

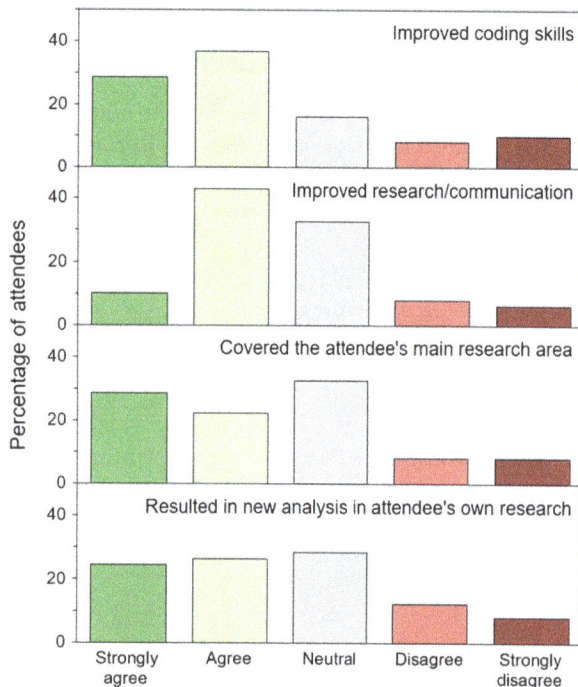

Figure 1. Survey responses from the 49 participants relating to the benefits of attending courses on a Likert scale. Answers to each question total 100%.

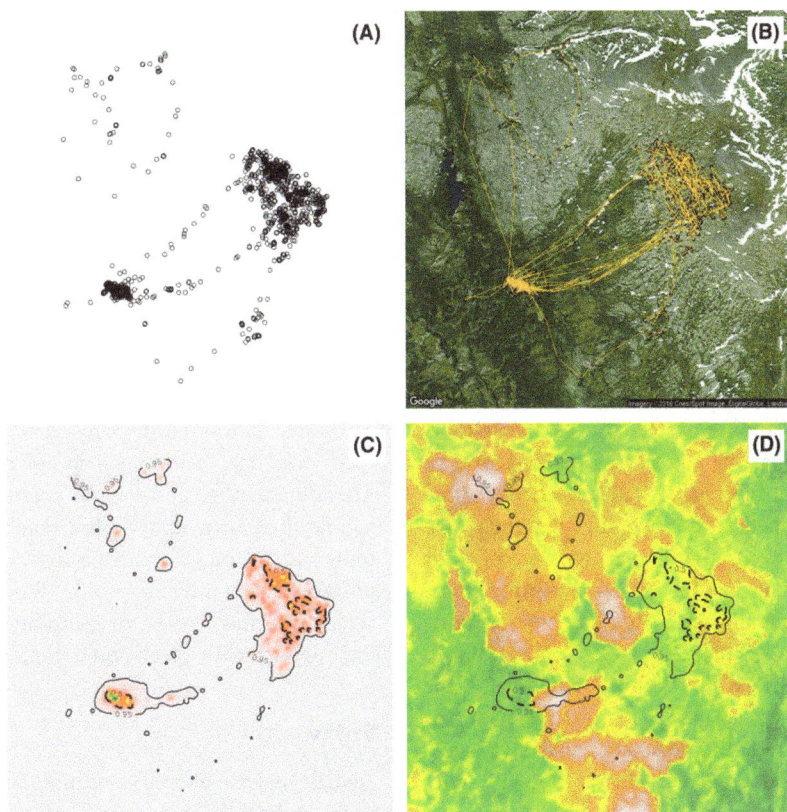

Figure 2. Hourly GPS from one sheep *Ovis aries* in Spekedalen, Norway, 23rd Jun to 3rd Sep 2013. (A) Locations plotted in R. (B) Track plotted onto Google satellite map using the 'ggmap' R package. (C) A dynamic Brownian Bridge Movement Model (dBBMM) to estimate the home range with 50 and 95% contours, created using the 'move' R package. (D) The dBBMM contours plotted onto an NDVI raster calculated from Landsat 8 surface reflectance and spectral indices data.

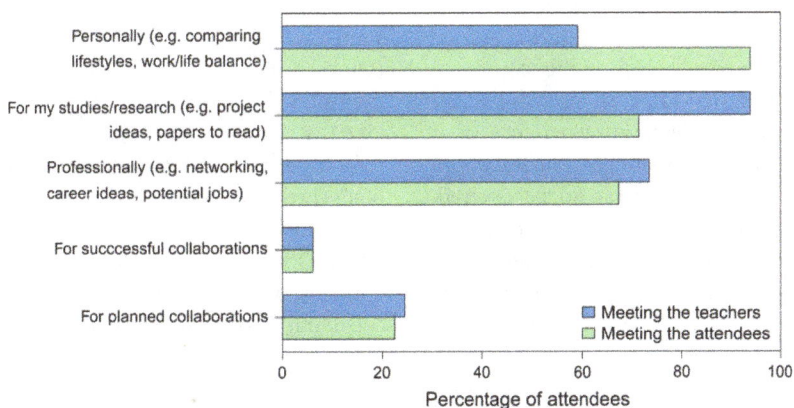

Figure 3. The percentage of the 49 attendees who found that meeting the teachers and other attendees was beneficial for each reason given in a multiple choice question.

Guidance for course organizers

Our open question asking 'What were the course's main benefits?' revealed numerous preferences of course attendees, and they reflect particular highlights as the question required free style answers. Small numbers of responses containing matching opinions are expected and opinions may be shared by participants who did not specifically expressed them. Twelve attendees praised the teaching and six particularly valued a group of specialists with diverse experience. Three participants appreciated the use of real-world data in lectures, and seven mentioned that practical examples were particularly useful, with three who would have liked more of this element. Learning to apply statistical methods to spatial datasets was also suggested. Participants want to finish a course with the ability to carry out a complete workflow, including data collection, processing, error checking, visualization, analysis and publication quality figures.

Course intensity

When asked what the main drawbacks of the course were, the most common response stated by 14 participants from all courses was that the course was too intensive or too short to cover the material. Attendees needed more time to go through practical exercises and use their own data, especially given the computational run time of some techniques. Planning and conducting courses for heterogeneous groups ranging from master's students to postdoctoral researchers is a challenging task. This leads to diverging requirements; our survey found that 16% of the participants strongly agreed and 12% agreed with the statement 'The course was too advanced', while 33% disagreed and 8% strongly disagreed. This is especially relevant for coding, which can be a major obstacle. A solution suggested in four responses is to provide more preparatory material before the course. We expand that to include example data formats for students to match their own data to in advance. Additionally, more explicit guidelines would allow those considering attending a course to better assess whether their skill level would allow them to get the most out of the course, and, if required, obtain more experience before the course. A small number of students per tutor is also important when individuals have differing levels of experience, as it allows for more questions. Running longer courses would reduce intensity and allow more material to be covered, but we recognize that longer courses are more costly and logistically challenging. Intensive courses may therefore be unavoidable given the constraints and complexity of the field. Having separate courses for different types of users could also reduce these problems.

Use of participants' own data

While simple datasets are required for teaching new methods, our survey showed that 94% of participants want to also have sessions for investigating their own data. Although this takes up time that could have been used to deliver more teaching material, attendees found working with their own data during the course was extremely productive as experts were available to solve problems quickly.

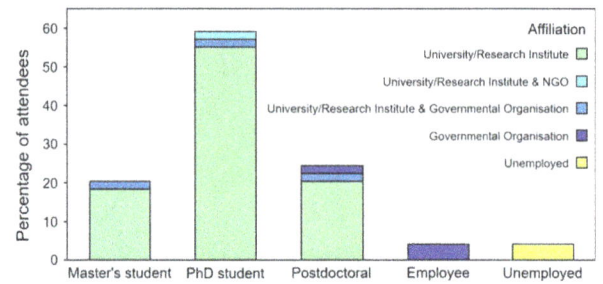

Figure 4. The position and affiliations of the 49 survey participants at the time of the course that they attended.

AniMove course participants had 5 days working with their own data, and many found this very constructive. Attendees may go to a course specifically to learn how to handle their own data, rather than to learn general concepts and techniques, which is likely as most attendees were postgraduate students and postdoctoral researchers (Fig. 4). It may also be easier to understand concepts and their applicability in relation to familiar data.

Software

Spatial analysis and visualization requires specialist software, and while many participants have experience with commercial software through institutional licenses, open source software is on the rise in ecology and conservation (Pettorelli et al. 2014a). Open source software is key to maximizing the future application of taught material by ensuring that students have unlimited access after the course. All attendees wanted to learn methods using open source software, such as R (R Core Team 2016) and QGIS (QGIS 2016), with only 18% wanting to also learn commercial software, such as ArcGIS (ESRI, Redlands, CA, USA). R in particular, provides a range of packages for remote sensing and movement analysis, such as 'RStoolbox' (Leutner and Horning 2016), 'move' (Kranstauber and Smolla 2016), and the 'adehabitat' packages (Calenge 2006). Open source also promotes the development of new methods, aids research outside of academia and facilitates collaboration (Rocchini and Neteler 2012; Neteler et al. 2012).

A major challenge is the use of different operating systems and software versions, as even minor differences can cause errors and disrupt practical sessions. To minimize disruption, courses could provide more installation and update information in advance, including the required R packages. Course organizers could reduce time spent troubleshooting by using a pre-configured virtual environment containing all the required course material, software and R packages (e.g. http://live.osgeo.org/en/index.html). The adapted virtual machine could then be imported and used seamlessly on any platform a participant may bring (e.g. www.virtualbox.org). However, we recognize that new software or operating systems might increase the complexity for organizers, and it is important to ensure that participants can repeat taught techniques after the course.

Recommendation summary

The most worthwhile training incorporates the coverage of theory, multiple analytical techniques, practical coding methods, data management and time to work with participants' own data. Logistical and financial constraints on course length mean that these courses become very intensive, so organizers could optimize time in the course by providing more preparatory material, guidelines on the preferred level of coding skills required, installation instructions and having small numbers of students per teacher.

Reaching a wider diversity of participants

Our ability to monitor animal movements and environmental change is improving as technological advances facilitate the collection of large high-resolution datasets (He et al. 2015; Ropert-Coudert and Wilson 2005; Hampton et al. 2013). Generating these datasets is becoming easier and cheaper, but analytical skills are lagging behind (Pettorelli et al. 2014a). We call for more training in this field and for the training to reach a wider diversity of participants, particularly ecologists and conservationists outside of academia, and researchers from developing countries. A course with a diverse group of attendees and tutors of different positions, affiliations, nationalities, research themes and experience was seen as highly beneficial in five responses to the open question on the main course benefits. For one course, a poor ratio of 0 female to 8 male teachers was also mentioned. Women are underrepresented in science, and even though biological science has a more equal ratio up until postgraduate level, there is still a long way to go (Blickenstaff 2005; Cronin and Roger 1999). Given the importance of diversity, we encourage appointing a more balanced ratio of female and male teachers.

Attendee affiliation

Training ecologists and conservationists contributes to Target 19 of the Aichi Biodiversity Targets by enhancing implementation through capacity building (O'Connor et al. 2015), but training in remote sensing for animal movement is reaching few practitioners outside academia. We found that 90% of course participants were affiliated with universities or research institutes, and only 9% of those were jointly affiliated with NGOs or governmental organizations (Fig. 4). Many knowledge gaps are filled through collaborations between remote sensing experts and ecologists or conservationists (Pettorelli et al. 2014b). But to keep up with the rate of animal movement data collection, practitioners will need the analytical tools to complete projects autonomously. Training targeted specifically at conservation practitioners may be the best way to achieve wider participation (Buchanan et al. 2015; Wegmann et al. 2016).

Course costs and travel

Financial cost and travel are key factors preventing the attendance of conservation practitioners and students; six participants mentioned course cost as a main drawback in response to the open question. People were willing to spend a median of 600 USD (mean of 708 USD) on a 5-day course, but travel and accommodation can be prohibitively expensive, due to the small number of courses worldwide. This issue is particularly important for attendees from developing countries, where funding is limited and conservation need is great (Waldron et al. 2013; Bruner et al. 2004). Two responses mentioned that serious jet lag made the course more difficult, which can only be solved by having courses within more time zones. Furthermore, learning in a non-native language was challenging for some, and despite an international student base, to the best of our knowledge, there are no similar courses taught in a language other than English. As such, more courses and more funding options for attending existing courses is key to maximizing the potential of remote sensing in ecology and conservation. We urge postgraduate supervisors, non-governmental organizations and employers to consider this type of training in budgets. We also encourage research council, institutes and universities to provide awards or scholarships for course attendance and organization.

Summary and Conclusions

Remote sensing is becoming an increasingly important tool for ecology and conservation. However, completing projects that span the disciplines of remote sensing and animal movement is challenging due to large datasets, lack of standard analytical techniques, limited coverage in standard curriculums and small communities within institutions. Training can bridge the gap by bringing newcomers up to date and forming a more cohesive network. We show that courses are highly valued by previous attendees and provide subject-specific knowledge, practical and coding skills, as well as opportunities for networking and collaboration, clarifying research ideas and gaining generic research and communication skills. We stress the importance of interdisciplinary training in increasing the application of remote sensing in conservation and ecology.

Here, we provide recommendations for three groups: those considering attending a course, course organizers, and those who may allocate funding for students or employees to attend courses. Firstly, we advise attendees to learn about general coding and statistics in advance, as courses in this field tend to be intensive. Secondly, we recommended that course organizers: use real-world examples in lectures; include many practical sessions; incorporate time for participants to investigate their own data; provide more preparatory material; and teach with only open source software. We also suggest that courses are taught by a diverse group of teachers, with different expertise and a more equal gender balance. Finally, we advocate more funding for underrepresented participants to attend existing courses and for new courses to be developed.

Acknowledgments

We thank all the course attendees who responded to our survey, especially Emma Fuller and Anna Kusler for answering follow-up questions. We thank the course organizers who shared the survey link: Martin Wegmann and Kate Christen of AniMove; Thomas Lund of CAn-Move, Lund University; Gabriele Cozzi of University of Zurich; and Oliver Hooker of PR Statistics Limited.

References

Blickenstaff, J. C. 2005. Women and science careers: leaky pipeline or gender filter? *Gend. Educ.* **17**, 369–386. doi:10.1080/09540250500145072.

Bruner, A. G., R. E. Gullison, and A. Balmford. 2004. Financial costs and shortfalls of managing and expanding protected-area systems in developing countries. *Bioscience* **54**, 1119–1126. doi:10.1641/0006-3568(2004) 054[1119:FCASOM] 2.0.CO;2.

Buchanan, G. M., A. B. Brink, A. K. Leidner, R. Rose, and M. Wegmann. 2015. Advancing terrestrial conservation through remote sensing. *Ecol. Inform.* **30**, 318–321. doi:10.1016/ j.ecoinf.2015.05.005.

Calenge, C. 2006. The package 'adehabitat' for the R Software: a tool for the analysis of space and habitat use by animals. *Ecol. Model.* **197**, 516–519. doi:10.1016/ j.ecolmodel.2006.03.017.

Cech, T. R., and G. M. Rubin. 2004. Nurturing interdisciplinary research. *Nat. Struct. Mol. Biol.* **11**, 1166–1169. doi:10.1038/nsmb1204-1166.

Cronin, C., and A. Roger. 1999. Theorizing progress: women in science, engineering, and technology in higher education. *J. Res. Sci. Teach.* **36**, 637–661. doi:10.1002/(SICI)1098-2736 (199908)36:6 < 637:AID-TEA4 > 3.0.CO;2-9.

Dray, S., M. Royer-carenzi, and C. Calenge. 2010. The exploratory analysis of autocorrelation in animal-movement studies. *Ecol. Res.* **25**, 673–681. doi:10.1007/s11284-010-0701-7.

Hampton, S. E., C. A. Strasser, J. J. Tewksbury, W. K. Gram, A. E. Budden, A. L. Batcheller, et al. 2013. Big data and the future of ecology. *Front. Ecol. Environ.* **11**, 165–62. doi:10.1890/120103.

He, K. S., B. A. Bradley, A. F. Cord, D. Rocchini, M. Tuanmu, S. Schmidtlein, et al. 2015. Will remote sensing shape the next generation of species distribution models? *Remote Sens. Ecol. Conserv.* **1**, 4–18. doi:10.1002/rse2.7.

Karnieli, A., Y. Bayarjargal, M. Bayasgalan, B. Mandakh, Ch Dugarjav, J. Burgheimer, et al. 2013. Do vegetation indices provide a reliable indication of vegetation degradation? A case study in the mongolian pastures. *Int. J. Remote Sens.* **34**, 6243–6262. doi:10.1080/01431161.2013.793865.

Kerr, J. T., and M. Ostrovsky. 2003. From space to species: ecological applications for remote sensing. *Trends Ecol. Evol.* **18**, 299–305. doi:10.1016/S0169-5347(03)00071-5.

Kranstauber, B., and M. Smolla. 2016. *Move: visualizing and analyzing animal track data. R Package version 1.6.541.* Available at: https://cran.r-project.org/package=move (accessed 8 February 2016).

Kranstauber, B., R. Kays, S. D. Lapoint, M. Wikelski, and K. Safi. 2012. A dynamic Brownian bridge movement model to estimate utilization distributions for heterogeneous animal movement. *J. Anim. Ecol.* **81**, 738–746. doi:0.1111/j.1365-2656.2012.01955.x.

Legendre, P. 1993. Spatial autocorrelation: trouble or new paradigm? *Ecology* **74**, 1659–1673. doi:10.2307/1939924.

Leutner, B., and N. Horning. 2016. *RStoolbox: tools for remote sensing data analysis. R Package version 0.1.4.* Available at: https://cran.r-project.org/package=RStoolbox (accessed 28 January 2016).

Mace, G. M., and A. Purvis. 2008. Evolutionary biology and practical conservation: bridging a widening gap. *Mol. Ecol.* **17**, 9–19. doi:10.1111/j.1365-294X.2007.03455.x.

Nagendra, H. 2001. Using remote sensing to assess biodiversity. *Int. J. Remote Sens.* **22**, 2377–2400. doi:10.1080/ 01431160117096.

Nathan, R. 2008. An emerging movement ecology paradigm. *Proc. Natl. Acad. Sci. USA* **105**, 19050–19051. doi:10.1073/ pnas.0808918105.

Nathan, R., W. M. Getz, E. Revilla, M. Holyoak, R. Kadmon, D. Saltz, et al. 2008. A movement ecology paradigm for unifying organismal movement research. *Proc. Natl. Acad. Sci. USA* **105**, 19052–19059. doi:10.1073/ pnas.0800375105.

Neteler, M., M. H. Bowman, M. Landa, and M. Metz. 2012. GRASS GIS: a multi-purpose open source GIS. *Environ. Model. Softw.* **31**, 124–130. doi:10.1016/j.envsoft.2011. 11.014.

O'Connor, B., C. Secades, J. Penner, R. Sonnenschein, A. Skidmore, N. D. Burgess, et al. 2015. Earth observation as a tool for tracking progress towards the Aichi biodiversity targets. *Remote Sens. Ecol. Conserv.* **2**, 19–28. doi:10.1002/rse2.4.

Pettorelli, N., W. F. Laurance, T. G. O. Brien, M. Wegmann, H. Nagendra, and W. Turner. 2014a. Satellite remote sensing for applied ecologists: opportunities and challenges. *J. Appl. Ecol.* **51**, 839–848. doi:10.1111/1365-2664.12261.

Pettorelli, N., K. Safi, and W. Turner. 2014b. Satellite remote sensing, biodiversity research and conservation of the future. *Philos. Trans. R. Soc. Lond. B Biol. Sci.* **369**, 20130190. doi:10.1098/rstb.2013.0190.

Pettorelli, N., M. Wegmann, A. Skidmore, S. Mucher, T. P. Dawson, M. Fernandez, et al. 2016. Framing the concept of satellite remote sensing essential biodiversity variables: challenges and future directions. *Remote Sens. Ecol. Conserv.* **1–10**, 1–10. doi:10.1002/rse2.15.

QGIS. 2016. *A free and open source geographic information system.* Available at: http://www.qgis.org/ (accessed 1 July 2016).

R Core Team. 2016. *R: a language and environment for statistical computing.* R Foundation for Statistical Computing, Vienna, Austria. Available at: https://www.r-project.org/ (accessed 1 July 2016).

Rocchini, D., and M. Neteler. 2012. Let the four freedoms paradigm apply to ecology. *Trends Ecol. Evol.* **27**:310–311. doi:10.1016/j.tree.2012.03.009.

Ropert-Coudert, Y., and R. P. Wilson. 2005. Trends and perspectives in animal-attached remote sensing. *Front. Ecol. Environ.* **3**, 437–444. doi:10.2307/3868660.

Rose, R. A., D. Byler, J. R. Eastman, E. Fleishman, G. Geller, S. Goetz, et al. 2014. Ten ways remote sensing can contribute to conservation. *Conserv. Biol.* **29**, 350–359. doi:10.1111/cobi.12397.

Roughgarden, J., S. W. Running, and P. A. Matson. 1991. What does remote sensing do for ecology? *Ecology* **72**, 1918–1922. doi:10.2307/1941546.

Scales, K. L., P. I. Miller, C. B. Embling, S. N. Ingram, E. Pirotta, and S. C. Votier. 2014. Mesoscale fronts as foraging habitats: composite front mapping reveals oceanographic drivers of habitat use for a pelagic seabird. *J. R. Soc. Interface* **11**, 20140679. doi:10.1098/rsif.2014.0679.

Skidmore, A. K., N. Pettorelli, N. C. Coops, G. N. Geller, M. Hansen, R. Lucas, et al. 2015. Agree on biodiversity metrics to track from space. *Nature* **523**, 403–405. doi:10.1038/523403a.

Turner, W., S. Spector, N. Gardiner, M. Fladeland, E. Sterling, and M. Steininger. 2003. Remote sensing for biodiversity science and conservation. *Trends Ecol. Evol.* **18**, 306–314. doi:10.1016/S0169-5347(03)00070-3.

Turner, W., C. Rondinini, N. Pettorelli, B. Mora, A. K. Leidner, Z. Szantoi, et al. 2015. Free and open-access satellite data are key to biodiversity conservation. *Biol. Conserv.* **182**, 173–176. doi:10.1016/j.biocon.2014.11.048.

Waldron, A., A. O. Mooers, D. C. Miller, N. Nibbelink, D. Redding, T. S. Kuhn, et al. 2013. Targeting global conservation funding to limit immediate biodiversity declines. *Proc. Natl. Acad. Sci. USA* **110**, 12144–12148. doi:10.5061/dryad.p69t1.

Wegmann, M., B. Leutner, and S. Dech. 2016. *Remote sensing and GIS for ecologists: using open source software.* Pelagic Publishing Ltd, Exeter.

Wikelski, M., R. W. Kays, N. J. Kasdin, K. Thorup, J. A. Smith, and G. W. Swenson. 2007. Going wild: what a global small-animal tracking system could do for experimental biologists. *J. Exp. Biol.* **210**, 181–186. doi:10.1242/jeb.02629.

Wilson, R. P., E. L. C. Shepard, and N. Liebsch. 2008. Prying into the intimate details of animal lives: use of a daily diary on animals. *Endanger. Species Res.* **4**, 123–137. doi:10.3354/esr00064.

5

Investigating animal activity patterns and temporal niche partitioning using camera-trap data: challenges and opportunities

Sandra Frey[1], Jason T. Fisher[1,2], A. Cole Burton[3] & John P. Volpe[1]

[1]School of Environmental Studies, University of Victoria, 3800 Finnerty Rd., Victoria, BC V8W 2Y2, Canada
[2]Ecosystem Management Unit, InnoTech Alberta, 3-4476 Markham St., Victoria, BC V8Z 7X8, Canada
[3]Department of Forest Resources Management, University of British Columbia, Vancouver, BC V6T 1Z4, Canada

Keywords
Activity patterns, camera trapping, competition, niche partitioning, species coexistence, species interactions

Correspondence
Sandra Frey, School of Environmental Studies, University of Victoria, 3800 Finnerty Rd.,Victoria, British Columbia V8W 2Y2, Canada. E-mail: safrey@uvic.ca

Funding Information
This work was supported by Innotech Alberta Grant 18788023, the University of Victoria and a NSERC Canada Graduate Scholarship to SF.

Editor: Marcus Rowcliffe

Abstract

Time-stamped camera data are increasingly used to study temporal patterns in species and community ecology, including species' activity patterns and niche partitioning. Given the importance of niche partitioning for facilitating coexistence between sympatric species, understanding how emerging environmental stressors – climate and landscape change, biodiversity loss and concomitant changes to community composition – affect temporal niche partitioning is of immediate importance for advancing ecological theory and informing management decisions. A large variety of analytical approaches have been applied to camera-trap data to ask key questions about species activity patterns and temporal overlap among heterospecifics. Despite the many advances for describing and quantifying these temporal patterns, few studies have explicitly tested how interacting biotic and abiotic variables influence species' activity and capacity to segregate along the temporal niche axis. To address this gap, we suggest coordinated distributed experiments to capture sufficient camera-trap data across a range of anthropogenic stressors and community compositions. This will facilitate a standardized approach to assessing the impacts of multiple variables on species' behaviours and interactions. Ultimately, further integration of spatial and temporal analyses of camera-trap data is critical for improving our understanding of how anthropogenic activities and landscape changes are altering competitive interactions and the dynamics of animal communities.

Introduction

Global biodiversity declines are being driven by the direct and indirect effects of anthropogenic disturbances (Cardinale et al. 2012; Hooper et al. 2012). Although these direct effects manifest in obvious ways through habitat loss and wildlife population declines, more subtle are the myriad indirect and cascading effects of human-driven disturbances, including altered species behaviours and interspecific interactions. A better understanding of these indirect impacts is needed to inform effective conservation planning. Recent technological and statistical advances in the application of camera trapping suggest that this emerging methodology may help provide such understanding.

Camera trapping is widely used in ecology and conservation for investigating species' distributions, estimating population densities and inventorying biodiversity (O'Connell et al. 2011; Burton et al. 2015; Steenweg et al. 2017). While camera-trap studies have typically focused on the spatial and numerical aspects of species and population ecology (e.g. Karanth and Nichols 1998; Linkie et al. 2007; Tobler et al. 2008), they have less often examined species' behaviours and interactions and their associated consequences for community structure.

Only recently have researchers focused attention on the finer scaled temporal data provided by time-stamped camera-trap images (e.g. Ridout and Linkie 2009; Rowcliffe et al. 2014), which detail the timing of wildlife occurrences across points in space. While such temporal data present analytical challenges, they are critical for developing a more complete understanding of population and community dynamics in the face of global change.

Temporal camera-trap data offer the opportunity to address unresolved questions regarding species ecology and community interactions, such as variation in activity patterns and partitioning along the temporal niche axis. These temporal insights are not only valuable from an ecological perspective, but they also provide insight into human-driven changes to species behaviours and interactions, and the resulting impacts on niche partitioning and community structure. The increase in camera-trap studies focused on temporal analyses is beginning to generate new ecological and applied insights, but a synthesis of recent approaches and trends is lacking. In this review, we pursue this synthesis through exploring several principal questions and analytical approaches for investigating temporal data collected by wildlife cameras. These questions reflect common themes we observed in the literature, and associated methods for analysing temporal data in the context of species' behaviour and interactions. Based on an *ad hoc* review, we provide a synthetic overview of frequently cited and more recent papers, building on notable past reviews (Bridges and Noss 2011) by adding more recent advances in approaches and thought. We review the theoretical basis for activity patterns and temporal niche partitioning, summarize current approaches, assess current limitations to more complete analyses and highlight significant advances in gaining a fuller understanding of species and community ecology. Ultimately, species' interactions and community dynamics can only be fully resolved by combining spatial and temporal data, therefore we also discuss new directions where combined spatiotemporal aspects of species niche partitioning and responses to environmental stimuli can be explored.

Exploring Time as a Niche Axis

Temporal dynamics are integral to niche theory (Hutchinson 1957, 1959; MacArthur and Levins 1967), including species autecology and community assembly, diel activity patterns and temporal niche partitioning among sympatric heterospecifics. Animal activity – quantifying how species distribute their activity over the day – is an important dimension of animal behaviour; how

species use time as a resource provides valuable information about their ecological niche (Schoener 1974). Extending to the community level, understanding how sympatric species partition time provides insight into the mechanisms facilitating stable coexistence (Carothers and Jaksić 1984; Kronfeld-Schor and Dayan 2003). Numerous studies employing camera-trap data have observed temporal niche partitioning as an important strategy for enabling the coexistence of ecologically similar species (e.g. Di Bitetti et al. 2010; Monterroso et al. 2014; Sunarto et al. 2015).

As diel activities are adapted to local conditions (Halle 2000), the influence of abiotic and biotic variables on activity patterns and temporal niche partitioning is a primary question for both ecological research and biodiversity conservation. Already there is mounting evidence from camera-trap studies that human-driven landscape and community impacts – including land-use change (Ramesh and Downs 2013), human activity (Wang et al. 2015; Ngoprasert et al. 2017), hunting (Di Bitetti et al. 2008), predator control (Brook et al. 2012) and presence of invasive competitors or predators (Gerber et al. 2012; Zapata-Ríos and Branch 2016) – may alter species' activity patterns and competitive or predatory interactions through altered temporal niche partitioning. Therefore, effective conservation decisions must also consider how environmental stressors and shifts in community composition may impact sympatric species' ability to segregate not just spatially, but also temporally.

The circular distribution of temporal data comes with its own set of analytical challenges, and very large sample sizes are required to explore fine-scale temporal responses across spatial gradients. Recent statistical and software developments have made important strides in tackling the challenges of temporal camera-trap data analysis (e.g. Ridout and Linkie 2009; Oliveira-Santos et al. 2013), thereby facilitating characterization of activity patterns and temporal niche overlap. Nevertheless, modelling the degree to which external variables (habitat characteristics, community structure, disturbance variables, etc.) cumulatively influence species' activity patterns and temporal niche partitioning continues to present considerable challenges. To date, few researchers have attempted such multivariate analyses with temporal data (e.g. Norris et al. 2010; Wang et al. 2015). Even more challenging is combining both spatial and temporal species' distributions to gain a fuller resolution of the underlying dynamics structuring interspecific interactions and community-level responses. Tackling this challenge starts with analysing the activity patterns of single species, and builds iteratively towards more complex multispecies and multivariable models.

Current Approaches to the Analysis of Activity Patterns

Activity data reflect an important dimension of animal behaviour and ecology, as they provide relevant information on species' natural history and ecological niche. Temporal data extracted from time-stamped wildlife images have provided some of the first analyses of diel (or circadian) activity of populations and species (e.g. Gerber et al. 2012; Bu et al. 2016).

Early camera-trap studies derived descriptive inferences from tabulated records or graphical displays of activity over discrete time periods of the diel cycle (e.g. van Schaik and Griffiths 1996; Lizcano and Cavelier 2000; Jácomo et al. 2004). This allowed assignment of taxa to general behavioural groups (e.g. diurnal, nocturnal) and better describe the temporal aspects of species' ecological niches. More recently, graphical displays of diel activity patterns use nonparametric kernel density estimates (e.g. Ridout and Linkie 2009; Linkie and Ridout 2011; Farris et al. 2015) to view species' activity as a continuous distribution over the wrapped 24-h cycle. Kernel density functions yield a continuous measure of the density of data points across their scale (Worton 1989), treating the estimates as a random sample from an underlying continuous distribution instead of grouping them into discrete time categories. Meredith and Ridout's (2014) R package 'Overlap' produces kernel density curves of species activity patterns from camera-trap data, with a similar function offered by the R package 'Circular' (Agostinelli and Lund 2013). Such graphical displays of activity patterns reflect aspects of temporal variability in species activity over the diel cycle, including basic behavioural categorizations (e.g. diurnality vs. nocturnality) and periods of peak activity. This approach dramatically improved the level of insight gained without any further investment in data acquisition, and thus represents significantly improved return on investment of camera-trap arrays.

Quantitatively investigating activity patterns comes with various challenges. Time is a wrapped distribution with an arbitrary zero point, thus classical statistical methods cannot be applied (Zar 2010). To solve this, circular statistics use trigonometric functions to derive descriptive statistics of temporal data, including mean time of activity (the mean vector), circular median, standard deviation and variance, as well as other dispersal estimates such as concentration (Batschelet 1981). Various software packages offer functions for deriving the statistical parameters on circular data, including ORIANA (Kovach 2011) and the R packages 'CircStats' (Lund and Agostinelli 2007) and 'Circular' (Agostinelli and Lund 2013). However, multimodal distributions indicating multiple peaks of activity (e.g. a crepuscular species showing activity peaks at dawn and dusk) do not yield intuitive statistical estimates of centrality (Batschelet 1981). As bimodal activity patterns are widespread (Aschoff 1966), the derived mean vector may fall between the two activity modes. Although studies have reported the mean vector to quantify species' mean activity time (e.g. Di Bitetti et al. 2010; Norris et al. 2010; Ramesh et al. 2012), this should be done with great caution to ensure the derived mean vector reflects a biologically accurate and meaningful value.

Oliveira-Santos et al. (2013) proposed conditional circular kernel density functions to characterize 'activity range' and 'activity core' from time of detection camera data. Following an approach similar to telemetry-based home range contours, they created density functions yielding 95% isopleths representing the time interval in which 95% of the animal activity occurs – an ecologically relevant activity range that eliminates outlying periods of activity produced by the statistical smoothing process. More conservatively, the 50% isopleths can be used to determine during which time interval(s) core activity is focused. This approach allows for a more quantitative analysis of temporal data, delimiting hours of peak activity to characterize specific aspects of species' circadian activities. Rowcliffe et al. (2014) also applied kernel density functions to camera data in developing an analysis to quantify the overall proportion of time that an animal spends active (i.e. activity level). The R package 'activity' (Rowcliffe 2016) fits circular distributions to temporal camera-trap data to create activity schedules and calculate species' activity level, thereby facilitating inquiry into animal energetics, predation risk and foraging effort, although key assumptions for deriving this metric may not be met in certain populations (Rowcliffe et al. 2014).

Species' activity patterns may also be characterized according to selection for certain time periods by discretizing the 24-h diel cycle into categories such as dawn, day, dusk and night. Chi-square tests determine if species' activity patterns are non-random (e.g. Bu et al. 2016). Resource selection functions (Manly et al. 2002) have also been used to determine how species distribute activity over various time periods given their availability (e.g. Gerber et al. 2012; Bu et al. 2016), which provides an approach to ascribing behavioural categorizations to species' activity patterns (e.g. diurnal, nocturnal or crepuscular). Species can also be assigned into such categorizations using niche selectivity indices, such as Ivlev's Electivity Index (Ivlev 1961) or its derived Jacobs Selectivity Index (Jacobs 1974). Using a novel approach to investigating how species selectively use different time periods, Farris et al. (2015) used hierarchical Bayesian Poisson analysis by modelling photographic rate (capture events/available hours) for each time category.

Camera-trap studies using such descriptive and quantitative approaches have produced considerable insight into the activity patterns of a wide range of species from diverse systems. These have included carnivore guilds (Di Bitetti et al. 2010; Monterroso et al. 2014), ungulates (Ferreguetti et al. 2015), rodents (Meek et al. 2012), primates (Gerber et al. 2012), birds (Srbek-Araujo et al. 2012) and various other mammals (Oliveira-Santos et al. 2008; Galetti et al. 2015). Interestingly, some conclusions from camera-trap research on species activity patterns have challenged previous conclusions regarding species-specific temporal activity (Bischof et al. 2014), which may arise from past sampling constraints that did not allow non-invasive, 24-h sampling. However, we are aware of no studies that have directly compared animal activity patterns generated via camera-trap data with more complete descriptions of activity derived from high-frequency GPS telemetry relocations. It is possible that activity data collected by camera traps may contain biases related to temporal variability of detectability caused by temperature, humidity or other factors suppressing detectability, but these remain untested to the best of our knowledge.

Despite the potential limitations of sampling species' activity patterns using camera-trap data, many emerging advances in documenting these patterns have been developed. The logical first step is comparing these activity patterns among sympatric species to ask how species divide the temporal niche axis.

Analyses of Temporal Niche Partitioning

Perhaps the ecologically most interesting question asked of species activity data is how sympatric species partition their activities to promote stable coexistence. MacArthur and Levins' (1967) limiting similarity theory predicts that no two species can coexist in time and space; thus, sympatry demands species divide their resources to avoid extinction by competition (Fig. 1). Time can be considered as a resource as it is 'consumed' analogously to other resources with limited availability (Halle 2000). Although not previously emphasized as an important mechanism for reducing competition, partitioning time of activity may be one of the most relevant strategies for the coexistence of species (Schoener 1974). Understanding how ecologically similar species coexist is not just a key question in ecology, but also crucial for understanding community diversity.

Early investigations of temporal niche partitioning relied on qualitative analyses of histograms. Researchers later began using linear frequency statistical procedures with the 24-h cycle categorized in contingency tables

(Jácomo et al. 2004; Lucherini et al. 2009; Gerber et al. 2012). Measures of niche similarity and overlap – such as Renkonen's similarity index and Pianka's measure of niche overlap (Krebs 1998) – evaluate differential use and partitioning of time as a resource (e.g. Lucherini et al. 2009; Hofmann et al. 2016), although these require discretization of data into arbitrary bin sizes.

Software packages which fit nonparametric circular density functions to camera-trap data allow researchers to analyse activity through a circular inferential statistical approach. A descriptive measure of the degree of similarity between two kernel density curves can be calculated following Ridout and Linkie's (2009) innovative coefficient of overlap, which fits camera-trap data to a kernel density function and then estimates a symmetrical overlapping coefficient between species using a total variation distance function (Fig. 2). This coefficient of overlap (Δ), whose precision can be estimated via bootstrapping and ranges from 0 (*no overlap*) to 1 (*complete overlap*), has often been used to investigate potential competitive and interaction possibilities between species (e.g. Linkie and Ridout 2011; Farris et al. 2015; Cusack et al. 2017). As Δ is a relative measure, interspecific differences in activity patterns may also be tested for statistical significance. The nonparametric circular Mardia–Watson–Wheeler (MWW) statistical test (Batschelet 1981) and Watson U^2 test (Zar 2010) have both been used to determine if two or more circular distributions vary significantly. Meredith and Ridout's (2014) 'Overlap' package remains a popular tool for presenting the overlap of two activity curves visually and estimating Δ, despite the biases introduced by the smoothing process when applying kernel density functions to temporal data and deriving an estimation of Δ (as discussed by Ridout and Linkie 2009).

Exploring temporal niche partitioning with camera traps has highlighted the prevalence and importance of segregation along the temporal axis for enabling coexistence within diverse assemblages of sympatric species. For example, Bischof et al. (2014) concluded that the elusive Altai mountain weasel *Mustela altaica* compensates for spatial overlap with intraguild predators by adopting an inverse activity pattern to its sympatric dominant predators while still maintaining spatial access to prey. Ferreguetti et al. (2015) concluded that two sympatric deer species may mitigate competition for similar space and food resources through differences in their activity patterns. Di Bitetti et al.'s (2010) analysis of Neotropical felid species activity patterns observed diurnal, nocturnal and cathemeral behaviours among species. Morphologically similar species had the most contrasting activity patterns, suggesting that the ability of species to segregate temporal activities may explain the lack of

Figure 1. Sympatric species must partition time or space to co-exist. These four species (clockwise: grizzly bear *Ursus arctos*, wolverine *Gulo gulo*, mule deer *Odocoileus hemionus*, moose *Alces alces*) were detected at the same camera-trap location. Spatiotemporal partitioning reduces competition and the potential for agonistic encounters.

character displacement seen in certain assemblages (Di Bitetti et al. 2010). Similarly, Sunarto et al. (2015) observed that within a tropical community of felids, those species with the most similar body size or with similarly sized prey had the lowest temporal overlap. Monterroso et al. (2014) observed a negative correlation between mean pairwise temporal overlap and species richness (number of species with at least 10 detections) across a mesocarnivore community. They suggest that temporal niche partitioning may be influenced by community diversity and likely plays an important role in facilitating stable coexistence in mesocarnivore guilds showing high diversity.

With statistical techniques to quantify temporal niche partitioning using camera data quickly developing, it is increasingly possible to ask questions about the factors that affect partitioning, including anthropogenic pressures induced by landscape and climate change.

Investigating Changes to Species Activity Patterns and Niche Partitioning

Animal activity patterns evolve via processes of natural selection (Kronfeld-Schor and Dayan 2003), such as historic co-evolutionary competitive interactions ('the ghost of competition past', Connell 1980), but behavioural plasticity may allow flexible changes to activity patterns in response to environmental stimuli (Halle 2000). Environmental cues such as predation risk, resource availability and the potential for agonistic encounters with dominant competitors influence behavioural decisions that alter a species' activity (Halle 2000). Activity during suboptimal times of higher predation risk, increased energy demand or lower prey availability may incur fitness costs. Comparing activity patterns in response to external stimuli provides insight into the degree of plasticity in species

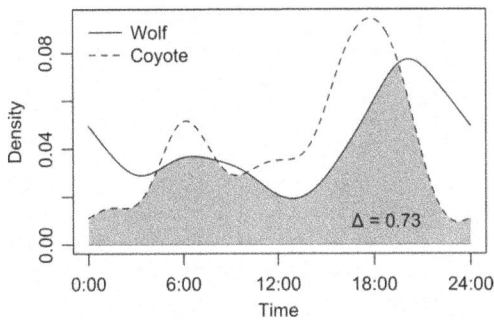

Figure 2. An example of the characterization of diel activity patterns from camera-trap data. Kernel density functions were used to depict grey wolf *Canis lupus* and coyote *Canis latrans* activity sampled via camera trapping during October–March 2006 to 2008, in the Willmore Wilderness Area, Alberta, Canada. The overlap coefficient (Δ) is the area under the minimum of the two density estimates (denoted in grey).

activity schedules and into the extent to which various environmental factors may alter an animal's activity pattern.

Changes to species' activity patterns may lead to altered temporal niche partitioning between species, with potential repercussions to species interactions such as intraguild competition and predator–prey dynamics. Indirect effects of anthropogenic stressors such as climate and landscape change could increase temporal overlap between species, augmenting interspecific conflict and exploitation of prey, or conversely, releasing species from predation or competitive pressure with reduced overlap. However, very few studies have empirically quantified how external factors may influence temporal niche partitioning (but see Wang et al. 2015).

Investigations of altered activity patterns, as with simpler investigations of animal activity, have typically involved descriptive comparisons of activity distributions from graphical displays, but also paired with simple statistical tests to determine whether two or more circular distributions differ significantly. Largely, these data have come from time-stamped wildlife images collected via camera trapping (but see Suselbeek et al. 2014). Generally, authors have divided the camera-trap data into two or three treatment groups based on abiotic or biotic factors such as season, lunar phase, presence/absence of predators or competitors, human activity or landscape change. Significant differences between activity times may be quantified statistically through a chi-squared contingency table of frequency of photographic records (e.g. Jácomo et al. 2004), but again this requires categorization of the temporal data into discrete time bins. The aforementioned MWW and Watson U^2 tests have also both been used to determine if activity distributions between populations vary significantly. For example, Di Bitetti

et al. (2009) observed that pampas foxes *Lycalopex gymnocercus* showed significantly different activity patterns in areas where the competitively dominant crab-eating fox occurred. Likewise, statistical comparisons of activity records between two colour morphs of oncilla *Leopardus tigrinus* revealed significant intraspecific differences in diel activity patterns (Graipel et al. 2014).

Intraspecific comparisons between study systems or treatment groups have also been performed using Ridout and Linkie's Δ (e.g. Monterroso et al. 2014; Wang et al. 2015). For example, Monterroso et al. (2014) observed a considerable degree of plasticity in European mesocarnivore nocturnal activity times between seasons and sites based on mean Δ values. By overlaying intraspecific activity curves of predators experiencing high versus low levels of human disturbance, Wang et al. (2015) demonstrated the timing and direction of activity shifts between two treatment groups. Activity overlap may also be quantified at conditional isopleths to determine whether overlap is more concentrated in the activity cores of the species (Oliveira-Santos et al. 2013). Rheingantz et al. (2016) observed very low activity overlap at 95% and 50% conditional isopleths between the two studied otter populations (45.6% and 14.1% respectively), suggesting a high level of plasticity in activity patterns; this was hypothesized to be a product of human activity or shifts in prey availability.

To date, the majority of studies evaluating the impact of external variables on species activity patterns have analysed the effect of a single variable at a time. Comparative tests do not allow for modelling multiple explanatory variables, potentially missing cumulative effects of multiple stressors, and interaction terms. Moreover, differences arising between treatment groups may potentially manifest in response to confounding (or collinear) variables. Alternative options include angular–linear correlations, as done by Hofmann et al. (2016) in comparing peccary activity time in relation to air temperature. Using an information-theoretic analysis of species activity, Norris et al. (2010) used linear mixed effects models to evaluate how abiotic conditions and human disturbance influenced activity pattern of three Amazonian terrestrial mammals. They observed that the time since isolation of forest patches had the strongest influence on agouti activity timing (Norris et al. 2010). However, care should be taken to ensure that the linear (as opposed to circular) scale used to define activity patterns upholds biological relevance; as mentioned, there is little biological difference but marked statistical difference between 2355 h and 0005 h on the linear scale.

A simple test for evaluating the impact of abiotic or biotic variables on temporal niche partitioning between sympatric species could involve directly comparing the bootstrapped mean overlap coefficient and 95% confidence intervals between species pairs across two or more treatment groups. Despite the relative simplicity and potential insights that could be gained from such a comparison, we are aware of no studies that have examined this direct comparison of interspecific temporal overlap across experimental treatments.

One noteworthy study by Wang et al. (2015) evaluated the influence of external variables on temporal niche partitioning in areas of ex-urban development near the Santa Cruz Mountains of California. Using an information-theoretic approach, these authors modelled Δ between mesocarnivore species pairs as a response to landscape development, human activity and forest cover. Wang et al.'s (2015) approach represents one of the few studies that simultaneously models the effect of multiple variables on species' activities and partitioning along the temporal axis. However, such fine-scale inferential analysis requires large amounts of data and a robust sampling design for capturing the effect of multiple explanatory variables across a spatial gradient. Many studies of species activity patterns and temporal niche partitioning are performed as secondary investigations, repurposing camera-trap data collected primarily for analysing spatial patterns or other responses (e.g. Di Bitetti et al. 2006; Sunarto et al. 2015; Ikeda et al. 2016). For all the reasons detailed above, spatially focused study designs with sample sizes only sufficient to confidently yield spatial and numerical responses may not be adequate to extend insight to complex and fine-scale investigation of species' activity patterns and temporal niche partitioning.

In summary, scientists have only begun to delve into discovering how animals spend their days, how species divide up time among them and how our marked impacts on landscapes, climates and biotic communities change these temporal processes. Moreover, although it is tacitly understood that space and time are inextricably linked, their integration in this context remains to be explored.

Future Directions: Analysing Spatiotemporal Species Interactions

With an increasing number of statistical approaches, and emerging studies of species behaviours and partitioning along both the spatial and temporal niche dimensions, our understanding of species interactions across time and space is mounting. However, this subfield is still relatively young, and most studies use opportunistic, not purpose-designed, data. There are many interacting ecological processes and cumulative effects of anthropogenic impacts yet to disentangle. This is a key future area of research, as the indirect effects of environmental stressors on species

activity and interactions may be as important as the direct effects (Strauss 1991; Schoener 1993; Abrams 1995).

The opportunity to parse the relative influences of space and time in species sympatric coexistence is an intriguing prospect. The competitive interactions shaping community structure likely manifest as both spatial and temporal patterns, but few studies to date have directly assessed such spatiotemporal interactions (but see Lewis et al. 2015; Swanson et al. 2016; Cusack et al. 2017; Karanth et al. 2017). Based on a comparison of approaches, Cusack et al. (2017) suggested that approaches using the combined spatial and temporal data generated by camera traps yield better insight into the associative patterns between sympatric species.

A second key opportunity is in using environmental stressors as 'treatments' in large-scale experiments designed specifically to understand the factors affecting species activity and interactions. Very little is known about how natural and anthropogenic changes to landscapes and biotic communities influence competitive interactions in animal populations. As it stands, it is difficult to predict how climate change, landscape change and anthropogenic changes to community composition may impact the competitive interactions and behavioural adaptations integral to maintaining biodiversity and ecosystem stability. Altered spatiotemporal interactions between sympatric species in communities could have rippling effects throughout the entire ecosystem (Crooks and Soulé 1999). With anthropogenic landscape changes projected to continue globally (Theobald 2005; Seto et al. 2011; Maxwell et al. 2016), focusing research efforts on understanding species spatiotemporal responses to those impacts is vital to sound conservation and management decisions. However, these questions are exceedingly difficult to answer within a single landscape.

Camera-trap surveys are invaluable for tracking direct effects of anthropogenic change on species distributions and abundances. However, the indirect effects of human influence, mediated by interactions among species in shifting communities, reside at the frontier of our knowledge of wildlife responses in the Anthropocene. With the growth of camera-trap networks deployed across multiple landscapes (Ahumada et al. 2013; Burton et al. 2015; McShea et al. 2016), hopefully growing into a global biodiversity network (Steenweg et al. 2017), network nodes deployed as coordinated distributed experiments (*sensu* Fraser et al. 2013) may help tease apart the effects of landscape and climate change on species interactions in complex environments. This research coordination and accompanying sampling designs remain the greatest opportunity for this emerging field of research. Fully capitalizing on the multi-scale spatial and temporal data produced by these networks may represent one of our best chances of advancing our ecological discoveries and meeting the pressing demands of biodiversity conservation.

Acknowledgments

This work was supported by Innotech Alberta Grant 18788023, the University of Victoria and a NSERC Canada Graduate Scholarship to SF. The paper was significantly improved by helpful revisions offered by P. Jansen and J. Cusack. The authors have no conflict of interest to declare.

References

Abrams, P. A. 1995. Implications of dynamically variable traits for identifying, classifying, and measuring direct and indirect effects in ecological communities. *Am. Nat.* **146**, 112–134.

Agostinelli, C., and U. Lund. 2013. R package 'circular': Circular Statistics (version 0.4-7). Available at: https://r-forge. r-project. org/projects/circular.

Ahumada, J. A., J. Hurtado, and D. Lizcano. 2013. Monitoring the status and trends of tropical forest terrestrial vertebrate communities from camera trap data: a tool for conservation. *PLoS ONE* **8**, e73707.

Aschoff, J. 1966. Circadian activity pattern with two peaks. *Ecology* **47**, 657–662.

Batschelet, E. 1981. *Circular statistics in biology*. Academic Press, New York.

Bischof, R., H. Ali, M. Kabir, S. Hameed, and M. A. Nawaz. 2014. Being the underdog: an elusive small carnivore uses space with prey and time without enemies. *J. Zool.* **293**(1), 40–48.

Bridges, A. S., and A. J. Noss. 2011. Behavior and activity patterns. Pp. 57–69. In *Camera Traps in Animal* Springer Japan, Japan.

Brook, L. A., C. N. Johnson, and E. G. Ritchie. 2012. Effects of predator control on behaviour of an apex predator and indirect consequences for mesopredator suppression. *J. Appl. Ecol.* **49**, 1278–1286.

Bu, H., F. Wang, W. J. McShea, Z. Lu, D. Wang, and S. Li. 2016. Spatial co-occurrence and activity patterns of mesocarnivores in the temperate forests of Southwest China. *PLoS ONE* **11**, e0164271.

Burton, A. C., E. Neilson, D. Moreira, A. Ladle, R. Steenweg, J. T. Fisher, et al. 2015. Wildlife camera trapping: a review and recommendations for linking surveys to ecological processes. *J. Appl. Ecol.* **52**, 675–685.

Cardinale, B. J., J. E. Duffy, A. Gonzalez, D. U. Hooper, C. Perrings, P. Venail, et al. 2012. Biodiversity loss and its impact on humanity. *Nature* **486**, 59–67.

Carothers, J. H., and F. M. Jaksic. 1984. Time as a niche difference: the role of interference competition. *Oikos* **42**, 403–406.

Connell, J. H. 1980. Diversity and the coevolution of competitors, or the ghost of competition past. *Oikos* **35**, 131–138.

Crooks, K. R., and M. E. Soulé. 1999. Mesopredator release and avifaunal extinctions in a fragmented system. *Nature* **400**, 563–566.

Cusack, J. J., A. J. Dickman, M. Kalyahe, J. M. Rowcliffe, C. Carbone, D. W. Macdonald, et al. 2017. Revealing kleptoparasitic and predatory tendencies in an African mammal community using camera traps: a comparison of spatiotemporal approaches. *Oikos* **126**, 812–822.

Di Bitetti, M. S., A. Paviolo, and C. De Angelo. 2006. Density, habitat use and activity patterns of ocelots (*Leopardus pardalis*) in the Atlantic Forest of Misiones, Argentina. *J. Zool.* **270**, 153–163.

Di Bitetti, M. S., A. Paviolo, C. A. Ferrari, C. De Angelo, and Y. Di Blanco. 2008. Differential responses to hunting in two sympatric species of brocket deer (*Mazama americana* and *M. nana*). *Biotropica* **40**, 636–645.

Di Bitetti, M. S., Y. E. Di Blanco, J. A. Pereira, A. Paviolo, and I. J. Pírez. 2009. Time partitioning favors the coexistence of sympatric crab-eating foxes (*Cerdocyon thous*) and pampas foxes (*Lycalopex gymnocercus*). *J. Mammal.* **90**, 479–490.

Di Bitetti, M. S., C. D. De Angelo, Y. E. Di Blanco, and A. Paviolo. 2010. Niche partitioning and species coexistence in a Neotropical felid assemblage. *Acta Oecologica* **36**, 403–412.

Farris, Z. J., B. D. Gerber, S. Karpanty, A. Murphy, V. Andrianjakarivelo, F. Ratelolahy, et al. 2015. When carnivores roam: temporal patterns and overlap among Madagascar's native and exotic carnivores. *J. Zool.* **296**, 45–57.

Ferreguetti, Á. C., W. M. Tomás, and H. G. Bergallo. 2015. Density, occupancy, and activity pattern of two sympatric deer (Mazama) in the Atlantic Forest, Brazil. *J. Mammal.* **96**, 1245–1254.

Fraser, L. H., H. A. Henry, C. N. Carlyle, S. R. White, C. Beierkuhnlein, J. F. Cahill, et al. 2013. Coordinated distributed experiments: an emerging tool for testing global hypotheses in ecology and environmental science. *Front. Ecol. Environ.* **11**, 147–155.

Galetti, M., H. Camargo, T. Siqueira, A. Keuroghlian, C. I. Donatti, M. L. S. Jorge, et al. 2015. Diet overlap and foraging activity between feral pigs and native peccaries in the Pantanal. *PLoS ONE* **10**, e0141459.

Gerber, B. D., S. M. Karpanty, and J. Randrianantenaina. 2012. Activity patterns of carnivores in the rain forests of Madagascar: implications for species coexistence. *J. Mammal.* **93**, 667–676.

Graipel, M. E., L. G. R. Oliveira-Santos, F. V. B. Goulart, M. A. Tortato, P. R. M. Miller, and N. C. Cáceres. 2014. The role of melanism in oncillas on the temporal segregation of nocturnal activity. *Braz. J. Biol.* **74**, S142–S145.

Halle, S., 2000. Ecological relevance of daily activity patterns. Pp. 67–90. In *Activity Patterns in Small Mammals*. Springer, Berlin Heidelberg.

Hofmann, G. S., I. P. Coelho, V. A. G. Bastazini, J. L. P. Cordeiro, and L. F. B. de Oliveira. 2016. Implications of climatic seasonality on activity patterns and resource use by sympatric peccaries in northern Pantanal. *Int. J. Biometeorol.* **60**, 421–433.

Hooper, D. U., E. C. Adair, B. J. Cardinale, J. E. Byrnes, B. A. Hungate, K. L. Matulich, et al. 2012. A global synthesis reveals biodiversity loss as a major driver of ecosystem change. *Nature* **486**, 105–108.

Hutchinson, G. E. 1957. The multivariate niche. In cold Spring Harbor Symp *Quant. Biol.* **22**, 415–421.

Hutchinson, G. E. 1959. Homage to Santa Rosalia or why are there so many kinds of animals? *Am. Nat.* **93**, 145–159.

Ikeda, T., K. Uchida, Y. Matsuura, H. Takahashi, T. Yoshida, K. Kaji, et al. 2016. Seasonal and diel activity patterns of eight sympatric mammals in Northern Japan revealed by an intensive camera-trap survey. *PLoS ONE* **11**, e0163602.

Ivlev, V. S. 1961. *Experimental ecology of the feeding of fishes.* Yale University Press, New Haven.

Jacobs, J. 1974. Quantitative measurement of food selection. *Oecologia* **14**, 413–417.

Jácomo, A. T. A., L. Silveira, and J. A. F. Diniz-Filho. 2004. Niche separation between the maned wolf (*Chrysocyon brachyurus*), the crab-eating fox (*Dusicyon thous*) and the hoary fox (*Dusicyon vetulus*) in central Brazil. *J. Zool.* **262**, 99–106.

Karanth, K. U., and J. D. Nichols. 1998. Estimation of tiger densities in India using photographic captures and recaptures. *Ecology* **79**, 2852–2862.

Karanth, K. U., A. Srivathsa, D. Vasudev, M. Puri, R. Parameshwaran, and N. S. Kumar. 2017. Spatio-temporal interactions facilitate large carnivore sympatry across a resource gradient. *Proc. R. Soc. B.* **284**, 20161860. https://doi.org/10.1098/rspb.2016.1860.

Kovach, W. L. 2011. Oriana–circular statistics for windows, ver. 4. Kovach Computing Services, Pentraeth.

Krebs, C. J. 1998. Niche measures and resource preferences. Pp. 455–495 *in* C. J. Krebs, ed. *Ecological methodology.* Benjamin Cummings, Menlo Park, CA.

Kronfeld-Schor, N., and T. Dayan. 2003. Partitioning of time as an ecological resource. *Annu. Rev. Ecol., Evol. Syst.* **34**, 153–181.

Lewis, J. S., L. L. Bailey, S. Vande Woude, and K. R. Crooks. 2015. Interspecific interactions between wild felids vary across scales and levels of urbanization. *Ecol. Evol.* **5**, 5946–5961.

Linkie, M., and M. S. Ridout. 2011. Assessing tiger–prey interactions in Sumatran rainforests. *J. Zool.* **284**, 224–229.

Linkie, M., Y. Dinata, A. Nugroho, and I. A. Haidir. 2007. Estimating occupancy of a data deficient mammalian species living in tropical rainforests: sun bears in the Kerinci Seblat region, Sumatra. *Biol. Cons.* **137**, 20–27.

Lizcano, D. J., and J. Cavelier. 2000. Daily and seasonal activity of the mountain tapir (*Tapirus pinchaque*) in the Central Andes of Colombia. *J. Zool.* **252**, 429–435.

Lucherini, M., J. I. Reppucci, R. S. Walker, M. L. Villalba, A. Wurstten, G. Gallardo, et al. 2009. Activity pattern

segregation of carnivores in the high Andes. *J. Mammal.* **90**, 1404–1409.

Lund, U., and C. Agostinelli. 2007. Circstats: circular statistics, from "topics in circular statistics" (2001). S-plus original by Lund, U. R port by Agostinelli, C. R package version 0.2-3.

Macarthur, R., and R. Levins. 1967. The limiting similarity, convergence, and divergence of coexisting species. *Am. Nat.* **101**, 377–385.

Manly, B. F. L., L. McDonald, D. Thomas, T. L. McDonald, and W. P. Erickson. 2002. *Resource selection by animals: statistical design and analysis for field studies,* 2nd ed.. Kluwer Academic Publishers, Dordrecht.

Maxwell, S. L., R. A. Fuller, T. M. Brooks, and J. E. Watson. 2016. Biodiversity: the ravages of guns, nets and bulldozers. *Nature* **536**, 143–145.

McShea, W. J., T. Forrester, R. Costello, Z. He, and R. Kays. 2016. Volunteer-run cameras as distributed sensors for macrosystem mammal research. *Landscape Ecol.* **31**, 55–66.

Meek, P. D., F. Zewe, and G. Falzon. 2012. Temporal activity patterns of the swamp rat (*Rattus lutreolus*) and other rodents in north-eastern New South Wales, Australia. *Aust. Mammal.* **34**, 223–233.

Meredith, M., and M. Ridout. 2014. overlap: Estimates of coefficient of overlapping for animal activity patterns. R package version 0.2.4. Available at: http://CRAN.R-project.org/package=overlap

Monterroso, P., P. C. Alves, and P. Ferreras. 2014. Plasticity in circadian activity patterns of mesocarnivores in Southwestern Europe: implications for species coexistence. *Behav. Ecol. Sociobiol.* **68**, 1403–1417.

Ngoprasert, D., A. J. Lynam, and G. A. Gale. 2017. Effects of temporary closure of a national park on leopard movement and behaviour in tropical Asia. *Mamm. Biol.-Zeitschrift für Säugetierkunde* **82**, 65–73.

Norris, D., F. Michalski, and C. A. Peres. 2010. Habitat patch size modulates terrestrial mammal activity patterns in Amazonian forest fragments. *J. Mammal.* **91**, 551–560.

O'Connell, A. F., J. D. Nichols, and K. U. Karanth. 2011. Pp. 253–263. *Camera traps in animal ecology: methods and analyses.* Springer, New York.

Oliveira-Santos, L. G. R., M. A. Tortato, and M. E. Graipel. 2008. Activity pattern of Atlantic Forest small arboreal mammals as revealed by camera traps. *J. Trop. Ecol.* **24**, 563–567.

Oliveira-Santos, L. G. R., C. A. Zucco, and C. Agostinelli. 2013. Using conditional circular kernel density functions to test hypotheses on animal circadian activity. *Anim. Behav.* **85**, 269–280.

Ramesh, T., and C. T. Downs. 2013. Impact of farmland use on population density and activity patterns of serval in South Africa. *J. Mammal.* **94**, 1460–1470.

Ramesh, T., R. Kalle, K. Sankar, and Q. Qureshi. 2012. Spatio-temporal partitioning among large carnivores in relation to major prey species in Western Ghats. *J. Zool.* **287**, 269–275.

Rheingantz, M. L., C. Leuchtenberger, C. A. Zucco, and F. A. Fernandez. 2016. Differences in activity patterns of the Neotropical otter Lontra longicaudis between rivers of two Brazilian ecoregions. *J. Trop. Ecol.* **32**, 170–174.

Ridout, M. S., and M. Linkie. 2009. Estimating overlap of daily activity patterns from camera trap data. *J. Agric. Biol. Environ. Stat.* **14**, 322–337.

Rowcliffe, J. M.. 2016. activity: Animal Activity Statistics. R package version 1.1. Available at: http://CRAN.R-project.org/package=activity

Rowcliffe, J. M., R. Kays, B. Kranstauber, C. Carbone, and P. A. Jansen. 2014. Quantifying levels of animal activity using camera trap data. *Methods Ecol. Evol.* **5**, 1170–1179.

van Schaik, C. P., and M. Griffiths. 1996. Activity periods of Indonesian rain forest mammals. *Biotropica* **000**, 105–112.

Schoener, T. W. 1974. Resource partitioning in ecological communities. *Science* **185**, 27–39.

Schoener, T. W.. 1993. On the relative importance of direct versus indirect effects in ecological communities. Pp. 365–411. In *Mutualism and community organization: behavioral, theoretical and food web approaches*. Oxford University Press, Oxford, UK.

Seto, K. C., M. Fragkias, B. Güneralp, and M. K. Reilly. 2011. A meta-analysis of global urban land expansion. *PLoS ONE* **6**, e23777.

Srbek-Araujo, A. C., L. F. Silveira, and A. G. Chiarello. 2012. The red-billed curassow (*Crax blumenbachii*): social organization, and daily activity patterns. *The Wilson J. Ornithol.* **124**, 321–327.

Steenweg, R., M. Hebblewhite, R. Kays, J. Ahumada, J. T. Fisher, C. Burton, et al. 2017. Scaling-up camera traps: Monitoring the planet's biodiversity with networks of remote sensors. *Front. Ecol. Environ.* **15**, 26–34.

Strauss, S. Y. 1991. Indirect effects in community ecology: their definition, study and importance. *Trends Ecol. Evol.* **6**, 206–210.

Sunarto, S., M. J. Kelly, K. Parakkasi, and M. B. Hutajulu. 2015. Cat coexistence in central Sumatra: ecological characteristics, spatial and temporal overlap, and implications for management. *J. Zool.* **296**, 104–115.

Suselbeek, L., W. J. Emsens, B. T. Hirsch, R. Kays, J. M. Rowcliffe, V. Zamora-Gutierrez, et al. 2014. Food acquisition and predator avoidance in a Neotropical rodent. *Anim. Behav.* **88**, 41–48.

Swanson, A., T. Arnold, M. Kosmala, J. Forester, and C. Packer. 2016. In the absence of a "landscape of fear": how lions, hyenas, and cheetahs coexist. *Ecol. Evol.* **6**, 8534–855.

Theobald, D. M. 2005. Landscape patterns of exurban growth in the USA from 1980 to. *Ecol. Soc.* **10**, 32.

Tobler, M. W., S. E. Carrillo-Percastegui, R. Leite Pitman, R. Mares, and G. Powell. 2008. An evaluation of camera traps for inventorying large-and medium-sized terrestrial rainforest mammals. *Anim. Conserv.* **11**, 169–178.

Wang, Y., M. L. Allen, and C. C. Wilmers. 2015. Mesopredator spatial and temporal responses to large predators and human development in the Santa Cruz Mountains of California. *Biol. Cons.* **190**, 23–33.

Worton, B. J. 1989. Kernel methods for estimating the utilization distribution in home-range studies. *Ecology* **70**, 164–168.

Zapata-Ríos, G., and L. C. Branch. 2016. Altered activity patterns and reduced abundance of native mammals in sites with feral dogs in the high Andes. *Biol. Cons.* **193**, 9–16.

Zar, J. H. 2010. *Biostatistical analysis*, 5th ed.. Pearson Prentice Hall, New Jersey.

Will remote sensing shape the next generation of species distribution models?

Kate S. He[1], Bethany A. Bradley[2], Anna F. Cord[3], Duccio Rocchini[4], Mao-Ning Tuanmu[5], Sebastian Schmidtlein[6], Woody Turner[7], Martin Wegmann[8,9] & Nathalie Pettorelli[10]

[1]Department of Biological Sciences, Murray State University, Murray, Kentucky 42071, USA
[2]Department of Environmental Conservation, University of Massachusetts, Amherst, Massachusetts 01003, USA
[3]Department of Computational Landscape Ecology, Helmholtz Centre for Environmental Research – UFZ, Permoserstraße 15, 04318, Leipzig, Germany
[4]GIS and Remote Sensing Unit, Department of Biodiversity and Molecular Ecology, Research and Innovation Center, Fondazione Edmund Mach, Via E. Mach 1, S. Michele all'Adige, Trento 38010, Italy
[5]Department of Ecology and Evolutionary Biology, Yale University, New Haven, Connecticut 06520, USA
[6]Institute of Geography and Geoecology, Karlsruhe Institute of Technology (KIT), 76131 Karlsruhe Germany
[7]Earth Science Division, NASA, Washington, District of Columbia, USA
[8]Department of Remote Sensing, University of Wuerzburg, 97074, Wurzburg, Germany
[9]CEOS Biodiversity at German Remote Sensing Data Centre, German Aerospace Centre DLR, 82234 Wessling, Germany
[10]Institute of Zoology, Zoological Society of London, Regent's Park, London, NW1 4RY, UK

Keywords

Ecological niche modeling, habitat suitability modeling, hyperspectral and multispectral data, LiDAR and RADAR metrics, predictor and response variables, spatial and temporal resolution

Correspondence

Kate S. He, Department of Biological Sciences, Murray State University, Murray, KY 42071. E-mail: xhe@murraystate.edu

Funding Information

No funding information provided.

Editor: Harini Nagendra
Associate Editor: Ned Horning

Abstract

Two prominent limitations of species distribution models (SDMs) are spatial biases in existing occurrence data and a lack of spatially explicit predictor variables to fully capture habitat characteristics of species. Can existing and emerging remote sensing technologies meet these challenges and improve future SDMs? We believe so. Novel products derived from multispectral and hyperspectral sensors, as well as future Light Detection and Ranging (LiDAR) and RADAR missions, may play a key role in improving model performance. In this perspective piece, we demonstrate how modern sensors onboard satellites, planes and unmanned aerial vehicles are revolutionizing the way we can detect and monitor both plant and animal species in terrestrial and aquatic ecosystems as well as allowing the emergence of novel predictor variables appropriate for species distribution modeling. We hope this interdisciplinary perspective will motivate ecologists, remote sensing experts and modelers to work together for developing a more refined SDM framework in the near future.

Introduction

Over the past two decades, a tremendous amount of work has been undertaken to map species' distributions and use the collected information to identify suitable habitats (Austin, 2002; Araújo et al. 2005; Franklin 2010). An array of sophisticated modeling tools are available to ecologists interested in predicting species occurrence (Elith and Leathwick 2009; Kissling et al. 2012) and species distribution models (SDMs) are now commonly used for pursuing diverse research endeavors, such as testing ecological theories (e.g. Petitpierre et al. 2012); predicting species range dynamics in response to environmental change (e.g. Schurr et al. 2012; Fordham et al. 2013, 2014; Dolos et al. 2015);

assessing invasion risks of introduced species (e.g. Bradley et al. 2009); and facilitating the design and selection of nature reserves (e.g. Kremen et al. 2008).

In most of the SDMs published in the last decade, the response variable (species occurrence data) is derived from herbaria or atlases, whereas predictor variables are mostly derived from spatially interpolated data (e.g. climate variables of climate research unit (CRU), New et al. 2002 and Worldclim, Hijmans et al. 2005), or categorical data (e.g. land cover and vegetation type). Occurrence data derived from remote sensing technology have started to be used in SDM studies (e.g. Bradley and Mustard 2006; Andrew and Ustin 2009), yet the utilities of remotely derived occurrence or abundance data remain largely unexplored. Environmental predictor variables derived from remote sensing data are more common in SDMs; this is particularly true when thinking of topographical information derived from the Shuttle Radar Topography Mission (SRTM) (see examples in Franklin 2010) and land cover maps (e.g. Pearson et al. 2004; Thuiller et al. 2004; Luoto et al. 2007; Newton-Cross et al. 2007; Morán-Ordóñez et al. 2012; Rickbeil et al. 2014). Continuous remotely sensed metrics as predictors of habitat condition, such as the normalized different vegetation index (NDVI) and leaf area index (LAI), both effective proxies for vegetation productivity (Zimmermann et al. 2007; Buermann et al. 2008; Pettorelli 2013), are still relatively under used. Yet, these and other remotely sensed products are becoming increasingly available for ecological analyses. We believe that continuous remote sensing metrics have become an integral part of SDM studies and will contribute significant amount of spatially explicit data for multi-scales and multi-taxa distribution models given recent development in remote sensing technologies and products.

Here, we describe examples of response and predictor variables derived from remote sensing that could provide novel information for species distribution modeling. We focus our attention on spaceborne and airborne systems, targeting both passive and active sensors. Passive sensors considered in this work range from panchromatic (e.g. high-resolution aerial photography with a single grayscale spectral band) to multispectral (e.g. moderate resolution sensors like Landsat collecting information in 4–11 bands) and hyperspectral (e.g. airborne high to moderate resolution data from AVIRIS with over one hundred narrow spectral bands). Active sensors include laser-light remote sensing Light Detection and Ranging (LiDAR) and microwave RADARs. Specific information on these sensors and others is provided in Table 1. We demonstrate how remotely derived variables have helped improve our understanding of species distribution over the past decade with a few case studies, while pointing out the uncertainty and constraints related to the use of remote sensing variables in SDMs. Lastly, we discuss how new technologies and products may shape the next generations of SDMs (NG-SDMs).

Remote Sensing of Species Distributions: The Response Variable

In SDMs, presence data are the most common response variable, with presence/absence or abundance data only occasionally available (Elith and Leathwick 2009). Occurrence records are generally derived from herbarium and museum collections, national atlases, large-scale field surveys, regional checklists, expert range maps and collections from citizen science groups (Jetz et al. 2012). However, these data can be associated with a variety of limitations, including sampling biases, inaccuracies in geo-referencing and taxonomy (Dickinson et al. 2010). Species occurrence data such as presence and absence records from museum and herbarium collections and field sampling can indeed be quite biased. This can sometimes be traced back to the distribution of collection sites, with some sites being under-sampled due to accessibility and other logistics issues. Reliable species absence data can be even more problematic to acquire since some species can be present in the considered site, but undetected. As demonstrated below, these limitations can be overcome in certain cases by using remotely derived species occurrence records.

Plants

Remote detection of plant species is most likely to be viable if the target plant species has a unique growth form or phenology. Many ecologists are familiar with global or national land cover classifications derived from satellite reflectance data (e.g. Friedl et al. 2002). Even with a few spectral bands, it is possible to separate functional types of vegetation (i.e. grasslands, forests, deserts, salt marshes, etc.) across broad spatial extents (He et al. 2009). A similar approach could enable species-level detection in cases where the target plant is the dominant form or a homogenous stand. For example dominant tree species in shrublands or grasslands have been identified based on unique vegetation index time series signatures (extracted from MODIS; Morisette et al. 2006) as well as through object-based identification of tree crowns given high enough spatial resolution (based on aerial photos; Weisberg et al. 2007). In a perennial shrubland, invasive annual grasses were detectable using Landsat imagery (Peterson 2005).

In addition to identifying distinct plant functional types through growth form, multispectral remote sensing can be used to identify plants with unique phenologies. This approach has been used most often to identify invasive plants (Bradley 2014). For example inter-annual variability in phenology has been used to identify annual grasses

Table 1. Specifics on sensors and missions with platform types, spatial and temporal resolutions and swath width information provided.

Platform and sensor	Spatial resolution (pan)	Spatial resolution (multi)	Spatial resolution (thermal)	Swath width	Revisiting time (theoretical maximum)	Temporal availability
Passive sensors						
Multispectral						
Worldview (-3)	0.31 m	1.24 m		13 km	1–4 days	2009(14)-present
Quickbird	0.61 m	2.4 m		17 km	1–3 days	2001-present
Pleiades	0.7 m	2 m		20–120 km	Daily	2000-present
Ikonos	0.82 m	3.2 m		11 km	1–3 days	1999-present
TopSAT	2.8 m	5.6 m		10–15 km	Daily	2005-present
RapidEye		6.5 m		77 km	5.5 days	2008-present
RapidEye+	<1 m	Similar to RapidEye				Launch 2019
SPOT (5)	5 m	10–20 m		60 km	2–3 days	1986 (2002)-present
SENTINEL 2A		10–60 m		290 km	5 days	Launch in 2015
ASTER		15–30 m	90 m	60 km	16 days	2000-present
CBERS	20 m	20–260 m	80 m	120–890 km	26 days	1999-present
Landsat TM 4/5	30 m	30 m		185 km	16 days	1982-present
Landsat 7/8	15 m	30 m	60/100 m	185 km	16 days	1999/2013-present
MODIS		250–1000 m	1000 m	2330 km	1–2 days	2000-present
AVHRR		1090 m	1090 m	2600 km	Daily	1978-present
Hyperspectral						
Hyperion		30 m		7.7 km	Tasked	2000-present
HyMap	Spatial resolution depending on flight altitude (c. 3–20 m), availability on request, up to several hundred bands					
HySpex						
AVIRIS						
EnMAP	Hyperspectral space-borne mission by DLR, c. 30 m spatial resolution, launch planned before c. 2020					
OMI	Hyperspectral space-borne mission for atmospheric parameters by NIVR and FMI, planned with 13–24 km spatial resolution					

Platform and sensor	Spatial resolution	Revisiting time (theoretical maximum)	Temporal availability
Active sensors			
COSMO-Skymed	1–15 m	1–15 days	2007-present
TerraSAR-X	1–18 m	2.5 days	2007-present
Tandem-X	1–18 m	2.5 days	2010-present
RADARSAT-2	3–100 m	24 days	2007-present
Sentinel 1	5–40 m	c. 12 days	2014-present
ENVISAT	30–1000 m	35 days	2002–2012
ICESAT 1 (2)	2003–2009 (launch 2017, Advanced Topographic Laser Altimeter System)		
LiDAR	Airborne, availability on request		
GEDI	Space-borne LiDAR, planned		
TRMM	5 km	16 times per day	1997-present
GPM	5 km	c. 3 hours	2014-present
SMAP	1–3 km	2–3 days	2015-present
SRTM	Global DEM, 30–90 m spatial resolution		
ASTER-DEM	Global DEM, 30 m spatial resolution		
TerraSAR-X/Tandem-X	Global DEM, 12 m spatial resolution, forthcoming		

in desert ecosystems, including cheatgrass (*Bromus tectorum*) (Bradley and Mustard 2005) and Lehmann lovegrass (*Eragrostis lehmanniana*) (Huang and Geiger 2008). Early growth and late senescence has been used to map dominant forest understory species including two bamboo species (*Bashania faberi* and *Fargesia robusta*) (Tuanmu et al. 2010) and honeysuckle (*Lonicera maackii*) (e.g. Wilfong et al. 2009).

The previous examples of broad-scale plant detection rely on unique functional or phenological properties. But, detection of plant species is also possible thanks to the higher thematic details provided by hyperspectral data. With over a hundred spectral bands being monitored, hyperspectral sensors can detect subtle differences in reflectance resulting from unique plant chemistries. This could help reduce misidentification and taxonomic biases found in field surveys. Numerous case studies of successful plant species detection using hyperspectral information can be found for exotic and invasive plants (Huang and Asner 2009; He et al. 2011). For example Andrew and Ustin (2008) used HyMap to identify unique white flowers of invasive pepperweed (*Lepidium latifolium*) near Sacramento, California. Similarly, Mitchell and Glenn (2009) also used HyMap to identify the unique yellow bracts of invasive leafy spurge (*Euphorbia esula*) in south-east Idaho. In Hawaii, a combination of differences in pigmentation and leaf water content enabled the detection of non-native trees using AVIRIS (Asner et al. 2008a). Other tree species were also successfully mapped with hyperspectral data within the tropics and subtropics (Clark et al. 2005; Carlson et al. 2007; Lucas et al. 2008; Yang et al. 2009; Féret and Asner 2013) as well as in temperate forest ecosystems (Fassnacht et al. 2014). Given sufficient expertise, effective classification algorithms and available data, many more plant species could be detectable using hyperspectral data.

LiDAR coupled with multispectral or hyperspectral data has also been used for identifying tree species (Jones et al. 2010; Heinzel and Koch 2011; Dalponte et al. 2012; Alonzo et al. 2014; Ghosh et al. 2014). This approach takes the advantage of using complementary information gathered from spectral reflectance and vertical structure of target species. Using a multi-sensor system (hyperspectral AISA, multispectral GeoEye-1, and high point density LiDAR), Dalponte et al. (2012) identified eight tree species in the Southern Alps with accuracies ranging from 76.5 to 93.2%. Similar conclusions were also made when mapping eleven tree species in coastal south-western Canada thanks to a combination of hyperspectral imagery and LiDAR data (Jones et al. 2010). In Hawaii, Asner et al. (2008b) employed a hybrid airborne system, combining the Carnegie Airborne Observatory small-footprint LiDAR system with AVIRIS to map the three-dimensional spectral and structural properties of three highly invasive trees. In this particular study, the authors separated the tree species based on their unique biophysical properties with a multi-stage spectral mixture analysis.

Animals

Tracking the presence of animal species using satellite remote sensing is feasible given fine enough pixel resolution and large enough animals under an unobstructed view. For example Yang et al. (2014) used both expert interpretation and an automated object-based classification to estimate populations of zebra (*Equus quagga burchellii*) and wildebeest (*Connochaetes taurinus*) and investigate their migration patterns in the open savannah of the Maasai Mara National Reserve, Kenya, thanks to very high-resolution GeoEye-1 satellite images (0.5 m resolution). Fretwell et al. (2014) identified 55 southern right whales (*Eubalaena australis*) in a breeding ground off the coast of Argentina based on brighter reflectance from WordView-2 (50 cm resolution). Similar approaches have been used to identify polar bears (*Ursus maritimus*) (Stapleton et al. 2014), walrus (*Odobenus rosmarus*) (Platonov et al. 2013) and emperor penguins (*Aptenodytes forsteri*) (Fretwell et al. 2012).

Spaceborne and airborne remote sensing can be very effective in supplementing species occurrence data (presence-absence, presence-only and point events), but getting very high-resolution remote sensing imagery is still costly in general even though no-cost imagery and open-source software for imagery processing are an increasingly common practice worldwide (Wegmann et al. 2015). At times, these high costs can be reduced by employing light-weight unmanned aerial vehicles (UAVs; Anderson and Gaston 2013).

For example UAVs mounted with off the shelf cameras and GPS were used to count marine mammals (dugong, *Dugong dugon*) in western Australia (Hodgson et al. 2013), along with a variety of other marine species. In a terrestrial case study, UAV images were used to identify orangutan (*Pongo abelii*) and elephant (*Elephas maximus*) in Sumatra, Indonesia (Koh and Wich 2012).

High spatial resolution remote sensing of terrestrial and marine animals is an excellent tool for measuring populations and identifying important habitat (e.g. stop-overs on migratory routes, breeding grounds). Thus far, most animal detection studies have focused on a small area due to the reliance on very high spatial resolution data. However, increasingly available high-resolution imagery and inexpensive UAVs coupled with object-based identification of animals might enable much broader scale identification of animal occurrence. The use of UAVs can also be a particularly cost-efficient way to collect input data for model calibration and validation.

Similarly to species occurrence data collected from field surveys, remotely derived response variables come with uncertainty and errors. These errors are typically introduced during data acquisition and processing, and through associated analytical algorithms. Studies have used various metrics to estimate classifiers' performance, ideally based on independent validation data, such as the Cohen's kappa statistics, confusion matrix, F-scores, overall accuracy and the receiver operating characteristic (ROC) curve (obtained by plotting fraction of true positive against fraction of false positive). The typically acceptable accuracies range from 60 to 90% for plant species and error rates for validation or training data are not reported in most studies (He et al. 2011). Furthermore, false-positive and false-negative detections can be identified by using plot-scale surveys guided by remotely sensed data (Asner et al. 2008c). When modeling species potential ranges with presence-only data, omission errors (identifying a species as absent when it is actually present) can lead to underestimates of potential range. In contrast, commission errors (identifying a species as present when it is actually absent) can lead to overestimates of potential range. The relative amount of omission versus commission error in the initial remote sensing map can inform interpretation of a resulting SDM output.

Regardless of the relative error rates, remote sensing of plants and animals provides both presence and absence data, which is more informative for SDMs than presence-only data, a relative measure that only represents a partial estimate of species occurrence (Fithian and Hastie 2013). Recent developments in SDMs have shown that a Poisson process model can combine presence-absence and presence-only data for correcting sampling bias in SDMs (Fithian et al. 2015). Furthermore, treating presence-only data as point events and estimating the intensity of the spatial location of presence points with a point process modeling framework, may also reduce uncertainty in SDMs (Renner et al. 2015). In the latter case, remote sensing can effectively provide point event data at various spatial scales to complement survey data. Most importantly, remote sensing may provide an estimate of absolute population density rather than relative density and this can be achieved with LiDAR and RADAR systems, particularly for tree species. Lastly, we want to point out that response variables derived from remote sensing can be updated every year or at desired time span, which allows a more dynamic approach to understanding habitat suitability and species range expansion or contraction. As time series data become more readily available, phenological events in plants and animals can be tracked and linked to fine-scale species distribution studies.

Remote Sensing of Environmental Conditions: The Predictor Variable

Good distribution models require spatial predictor variables that are ecologically relevant (Franklin 1995) for the modeled organisms. In some cases, remote sensing metrics can be challenging to translate into meaningful ecological entities, particularly those that provide indirect measures of ecosystem processes (e.g. surface roughness from RADAR; Buermann et al. 2008) making it unclear why to consider them in SDMs in the first place.

Abiotic predictor variables

Climate data have been commonly used to predict potential species distributions across broad spatial scales (e.g. Franklin 2010). CRU (New et al. 2002), WorldClim (Hijmans et al. 2005), CliMond (Kriticos et al. 2012) and PRISM (US only; Daly et al. 2002), are all examples of spatially explicit datasets of climatic conditions. These datasets encompass information on modeled temperature, precipitation, solar radiation and soil moisture (along with several derived bioclimatic combinations of temperature and precipitation), which are based on interpolations from global weather stations. However, interpolations are only as good as the underlying data, and uneven geographical coverage leads to high model uncertainty, especially in developing countries where few weather stations are in place (Daly 2006; Bedia et al. 2013; Waltari et al. 2014). When uncertainty in spatial climate variables is not accounted for, coefficient estimates tend to be biased which lead to poor performances of SDMs as shown with recent simulations (Stoklosa et al. 2015).

Remotely sensed climate data are continuously observed without interpolation and geographical biases. Therefore, satellite-based temperature, precipitation and radiation measurements could improve climate predictor variables. For example land surface temperature (LST) is measured globally four times per day by the MODIS Terra and Aqua satellites (Wan et al. 2004; Sims et al. 2008) and a derived product at 250 m spatial resolution is freely available at http://gis.cri.fmach.it/eurolst for Europe (Metz et al. 2014). MODIS LST data are increasingly being used in SDMs to understand and predict ecological processes (Buermann et al. 2008; Bisrat et al. 2012; Neteler et al. 2013; Pau et al. 2013; Still et al. 2014). Recently, efforts have been made to use LST to facilitate interpolation of weather station data as weather station data have a long temporal span, which cannot be fully covered by remote sensing data (Parmentier et al. 2015). In addition, the global UV-B radiation dataset from NASA Aura-OMI (Beckmann et al. 2014), designed for macroecological studies, offers exciting opportunities for both correlative and process-based species distribution modeling.

Precipitation estimates from satellite are available historically from TRMM (Tropical Rainfall Measuring Mission; Huffman et al. 2007) at a 4 km spatial resolution covering the tropical region ($20°N–20°S$) and extended to $50°N–50°S$ (Wang et al. 2014). New rainfall products are just becoming available from the Global Precipitation Measurement (GPM) mission, which has replaced TRMM (TRMM data collection stopped in 2105). With the recent launch of NASA's Soil Moisture Active Passive (SMAP) Mission, high-resolution soil moisture data (3 and 9 km) with global coverage will also soon be available. Finally, an analysis of global cloud cover from MODIS can serve as a proxy for average precipitation (A. Wilson and W. Jetz, pers.comm. 2015). Satellite measurements of temperature, precipitation and soil moisture may thus soon provide better wall to wall estimates of climatic conditions than weather station interpolations and are becoming increasingly accessible to ecologists for building SDMs with less uncertainty.

Topographic features of land surface derived from SRTM digital elevation data (the DEM products) and GDEM (Global Digital Elevation Map) from ASTER are already commonly used as predictor variables in SDMs (Franklin 2009). For finer-scale studies of local and micro-topography, airborne LiDAR and stereographic DEMs from WorldView-2 are both options. Very high-resolution topographic data derived from LiDAR have been incorporated in SDMs while assessing habitat suitability of eleven at-risk plant species in Hawaii (Questad et al. 2014) and for assessing diversity and composition of a temperate montane forest in Germany (Leutner et al. 2012). However, both datasets are costly and LiDAR data are currently limited in temporal coverage and spatial extent. An emerging alternative source of very high-resolution DEMs are UAVs (Anderson and Gaston 2013), which are rapidly becoming more reliable, lightweight and cost effective for LiDAR instrumentation (Watts et al. 2012).

In the marine realm, sea-surface temperature derived from Aqua MODIS (https://podaac.jpl.nasa.gov/SeaSurfaceTemperature), with global resolutions as fine as 1 × 1 km, has been one of the most influential predictors in SDMs for identifying productivity hotspots and seascape modeling (Louzao et al. 2011; Ramírez et al. 2014). Furthermore, the Bio-ORACLE (ocean rasters for analyses of climate and environment at http://www.bio-oracle.ugent.be), a marine counterpart of the WorldClim database has been developed, consisting of 23 environmental rasters, derived from both satellite-based and *in situ* data for modeling the distribution of shallow water marine species at a global scale (Tyberghein et al. 2012). A comprehensive review on using remotely derived variables to inform marine habitat mapping and monitoring can be found in Kachelriess et al. (2014).

Biotic predictor variables

Vegetation characteristics can be important predictors of species' habitat, acting as a proxy for sources of food availability or shelter. Many studies have used remotely sensed variables to model habitat suitability for animals, in particular using satellite-derived land cover classifications (Leyequien et al. 2007) as well as continuous metrics of vegetation productivity such as NDVI (Pettorelli

et al. 2011). For example NDVI data from MODIS was used as a predictor of food availability in a model of vervet monkey (*Chlorocebus pygerythrus*) habitat in Africa (Willems et al. 2009). Similarly, researchers used NDVI data from AVHRR to assess habitat availability of the Iberian mole (*Talpa occidentalis*) along a biogeographic gradient in Spain (Suárez-Seoane et al. 2014). Although being increasingly used (Bradley et al. 2012; Cord et al. 2013), incorporating remotely sensed metrics of vegetation into models of habitat suitability requires a more careful approach, particularly when it comes to plants. Bradley et al. (2012) indeed caution that the use of NDVI metrics in plant models could create biases in cases where they measure properties of the target species directly.

Vegetation structure derived from RADAR or LiDAR could also be an important predictor of habitat (Vierling et al. 2008). For example Buermann et al. (2008) used RADAR data from QuikSCAT as a proxy for Amazonian forest canopy roughness and found that it improved habitat suitability models for several species of birds. Similarly, Farrell et al. (2013) concluded that models incorporating LiDAR-derived metrics, such as tree height, improved model predictions of bird habitat in Texas.

In addition, vegetation phenology derived from satellite time series can provide important information about timing of biological events (Morisette et al. 2009) and serve as a proxy for habitat. For example NDVI-based estimate of the length of summer was an important predictor of moose (*Alces alces*) body weight, and therefore habitat quality, in Norway (Herfindal et al. 2006). Tuanmu et al. (2011) suggested that multi-year phenology metrics derived from MODIS can reduce model complexity and multicollinearity among predictor variables and thus improve model transferability (i.e. the ability of a model developed in one time period/area to be applied to a different time period/area) for giant panda (*Ailuropoda melanoleuca*) habitat change in China. Furthermore, remotely sensed seasonal variation in vegetated land surfaces can be influential predictor variables when modeling species distribution and habitat suitability (Osborne and Suárez-Seoane 2007).

Multi-year NDVI and its predicted values under climate change scenarios have been used to assess likely impacts of environmental change on future species distributions and extinction risks, which are a major motivation for SDM research. Singh and Milner-Gulland (2011) for example used 25 years of temperature and NDVI data to identify the changing drivers of migratory saiga (*Saiga tatarica*) distribution in Central Kazakhstan under a range of scenarios, including changes in temperature and productivity. In this study, projected NDVI values were proven as one of the critical predictors in modeling future saiga distribution and changes in population density.

Thus, if remotely sensed predictors, such as NDVI in this case, improve SDMs, then predicted extinction risks from environmental change are going to become more reliable.

Information on biotic conditions can also be derived from UAVs. For example Koh and Wich (2012) note that imagery from a UAV in Sumatra, Indonesia could detect evidence of small-scale human disturbance, including logging and local oil palm plantations. Similarly, Getzin et al. (2014) used a UAV to identify small forest gaps in Germany, which could be an important predictor of early successional species occurrence.

For a few satellite missions (e.g. Landsat), data archives are now decades long, enabling the tracking of temporal changes in ecosystems. While land use/land cover change is a long recognized discipline (e.g. Meyer and Turner 1994), including change metrics as predictors in SDMs is exceedingly rare. Yet, temporal trends in NDVI (e.g. Verbyla 2008) and other satellite measurements may be key indicators of ecological changes likely to influence the distribution of species. With MODIS archives reaching 15 years and Landsat 40, plus the recently launched Sentinel 1 and 2 missions, the combination of higher spatial and longer-term temporal analyses is increasingly possible.

Future Opportunities

New space missions and sensor networks

Remote sensing products provide dynamic information that is increasingly relevant to the fields of ecology and conservation biology (Pettorelli et al. 2014a; Turner 2014). In recent years, the potential of remote sensing to support ecological research has been boosted by the prospects of new technological developments and new space missions, including a number of very high spatial and spectral resolution passive optical satellites as well as active optical (LiDAR) and RADAR imaging systems equipped with state-of-the-art technology (Pettorelli et al. 2014b).

New optical satellite missions include the European Sentinel-2 satellites, the Pleiades of CNES, the TopSat (UK), CBERS (China and Brazil) and the Resourcesat series (India) along with a host of private sector missions seeking to offer high spatial resolution imagery of the sunlit Earth essentially everywhere at all times. Recently launched and planned RADAR and LiDAR missions include Sentinel 1 with C-Band, the TerraSAR-X/Tan-DEM-X mission, the RadarSat program of the Canadian Space Agency, the RADAR mission by the JAXA space agency in Japan and the NASA Global Ecosystem Dynamics Investigation (GEDI) LiDAR planned for the International Space Station (Koch 2010). Detailed information on sensors and missions is listed in Table 1. In addition, there are continued new developments in low-cost, light-weight, and long-duration UAVs (Lucieer et al. 2014). New missions and new sensors will allow mapping and monitoring of global ecosystems at an unprecedented level of detail (sub-meter spatial resolution and 3-D profiles are now possible), potentially providing invaluable data for improving the predictive power of SDMs.

Novel predictor variables bring new possibilities

Biophysical, biochemical and physiological predictors derived from modern remote sensing have huge potential when it comes to improving the predictive power of SDMs. Advances over the more widely used NDVI include LAI3g (third generation of LAI data with best-quality and significantly improved post-processing algorithms) and fraction of absorbed photosynthetically active radiation (fPAR, especially fPAR3g), both of which are available from MODIS data. In addition to detecting plant pigmentation, hyperspectral data can be used to measure leaf water content and leaf nitrogen content, along with other unique chemical signals. Active LiDAR and RADAR can estimate canopy/tree height and stem density, canopy moisture and 3-D habitat structural profiles (a vertical description of the habitat, such as the position of leaves, branches and ground) (Simonson et al. 2014). These and other potential predictor variables are outlined in Table 2.

With new high-resolution sensors, remotely sensed data could add insight into the spatial patterns of plant interactions at local to landscape scales. These sorts of biotic interactions are absent from SDMs due to lack of data (Kissling et al. 2012). But, hyperspectral data can be used to classify vegetation communities into plant functional types based on optical reflectance values (Ustin and Gamon 2010), creating very high-resolution maps of plant assemblages, which can provide information on interactions among species in terms of competition for water and light. Hyperspectral data have also been used to map plant communities based on competitive, stress tolerant, ruderal strategy (Schmidtlein et al. 2012) where C strategists are highly competitive, S strategists are stress tolerant and R strategists are ruderals with rapid growth and short life spans (sensu Grime 1974, 1977). At broader scales, remote measurements including fPAR, NDVI, LAI and tree/canopy height can combine to estimate overall ecological diversity (van Ewijk et al. 2014), a proxy for competition. These sorts of remote sensing products could enable assessments of intensity and spatial location of biotic interactions across thousands of hectares, much larger than current plot studies.

Table 2. Remotely derived predictor variables with sources and case studies provided.

Predictor variables	Source	Examples
Abiotic predictors		
Land cover	MODIS, Landsat, Landsat ETM+	Pearson et al. (2004); Thuiller et al. (2004); Luoto et al. (2007); Newton-Cross et al. (2007); Morán-Ordóñez et al. (2012); Rickbeil et al. (2014); Tuanmu and Jetz (2014)
Topographic features/ elevation	SRTM, (DEM products), LiDAR, WorldView-2, ASTER, GTOPO30, GMTED2010, UAVs	Buermann et al. (2008); Franklin (2010);Pradervand et al. (2014); Questad et al. (2014); van Ewijk et al. (2014)
Land surface temperature (LST)	Landsat-8, MODIS	Cord and Rödder (2011); Bisrat et al. (2012); Neteler et al. (2013); Pau et al. (2013); Still et al. (2014)
Sea-surface temperature (SST)	Aqua MODIS	Louzao et al. (2011); Ramírez et al. (2014); Rickbeil et al. (2014)
Precipitation	TRMM, GPM, MODIS cloud cover	Saatchi et al. (2008); Rovero et al. (2014); Seiler et al. (2014)
Soil moisture	NASA SMAP	Not in SDM literature yet
Biotic predictors		
Normalized difference vegetation index (NDVI)	AVHRR, Landsat, MODIS, QuickBird	Morisette et al. (2006); Zimmermann et al. (2007); Prates-Clark et al. (2008); Pettorelli et al. (2011); Feilhauer et al. (2012); Hall et al. (2012); van Ewijk et al. (2014)
Vegetation phenology	MODIS, Landsat	Bradley and Mustard (2005); Morisette et al. (2009); Wilfong et al. (2009); Tuanmu et al. (2010)
Leaf area index (LAI)	MODIS	Buermann et al. (2008); Prates-Clark et al. (2008); Saatchi et al. (2008)
Fraction of absorbed photosynthetically active radiation (fPAR)	MODIS, Landsat	Bisrat et al. (2012); Fitterer et al. (2012); Rickbeil et al. (2014); Gould et al. (2015)
Canopy/tree height	LiDAR and RADAR	Goetz et al. (2007); Swatantran et al. 2012; Tattoni et al. (2012); Farrell et al. (2013); Alonzo et al. (2014); Ficetola et al. (2014); Simonson et al. (2014); van Ewijk et al. (2014)
Stem density	LiDAR and RADAR	Swatantran et al. (2012)
Canopy moisture	Hyperspectral sensors, QSCAT	Buermann et al. (2008); Prates-Clark et al. (2008)
Canopy roughness	QSCAT	Saatchi et al. (2008)
3-D habitat structural profile	LiDAR and RADAR	Bergen et al. (2009); Goetz et al. (2010); Simonson et al. (2014)
Leaf water content	Hyperspectral sensors	Not in SDM literature yet
Leaf nitrogen content	Hyperspectral sensors	Not in SDM literature yet
Spectral heterogeneity/ functional types	Hyperspectral sensors, Landsat	Morán-Ordóñez et al. (2012); Schmidtlein et al. (2012); Henderson et al. (2014); Pottier et al. (2014)
Spatial heterogeneity of vegetation	MODIS, Landsat	Lahoz-Monfort et al. (2010); Culbert et al. (2012); Tuanmu and Jetz (2015)

The NG-SDMs

Given the opportunities provided by remote sensing presented above, along with data collected from well-designed experiments, field plots and *in situ* sensors, the NG-SDMs could develop rapidly, building upon recent development in SDMs (Lurgi et al. 2015; Renner et al. 2015). NG-SDMs could be integrative models that (1) operate at different areas along the correlative-process model continuum (sensu Dormann et al. 2012). The majority of current SDMs either fall at one extreme end of the continuum for bring correlative (with explanatory variables which may or may not be casual factors for species occurrence) or fit at the other extreme end of the continuum for being process-based (with clearly defined ecological meaningful parameters); (2) form a hierarchically nested predictive framework, allowing for assessing species distribution at multiple biological levels and spatial scales; and (3) explicitly consider biotic interactions and variation in demographic rates with a process-oriented approach driven by underlying mechanisms (Schurr et al. 2012; Kissling 2013; Wisz et al. 2013; Renner et al. 2015). We provide a comparative modeling framework between the current SDMs and the NG-SDMs proposed in this perspective in Figure 1.

The concrete contributions to the development of NG-SDMs from remote sensing could include a variety of ecologically meaningful predictors. First, ecophysiologically relevant variables, such as remotely derived earth surface temperature, precipitation and MODIS phenologi-

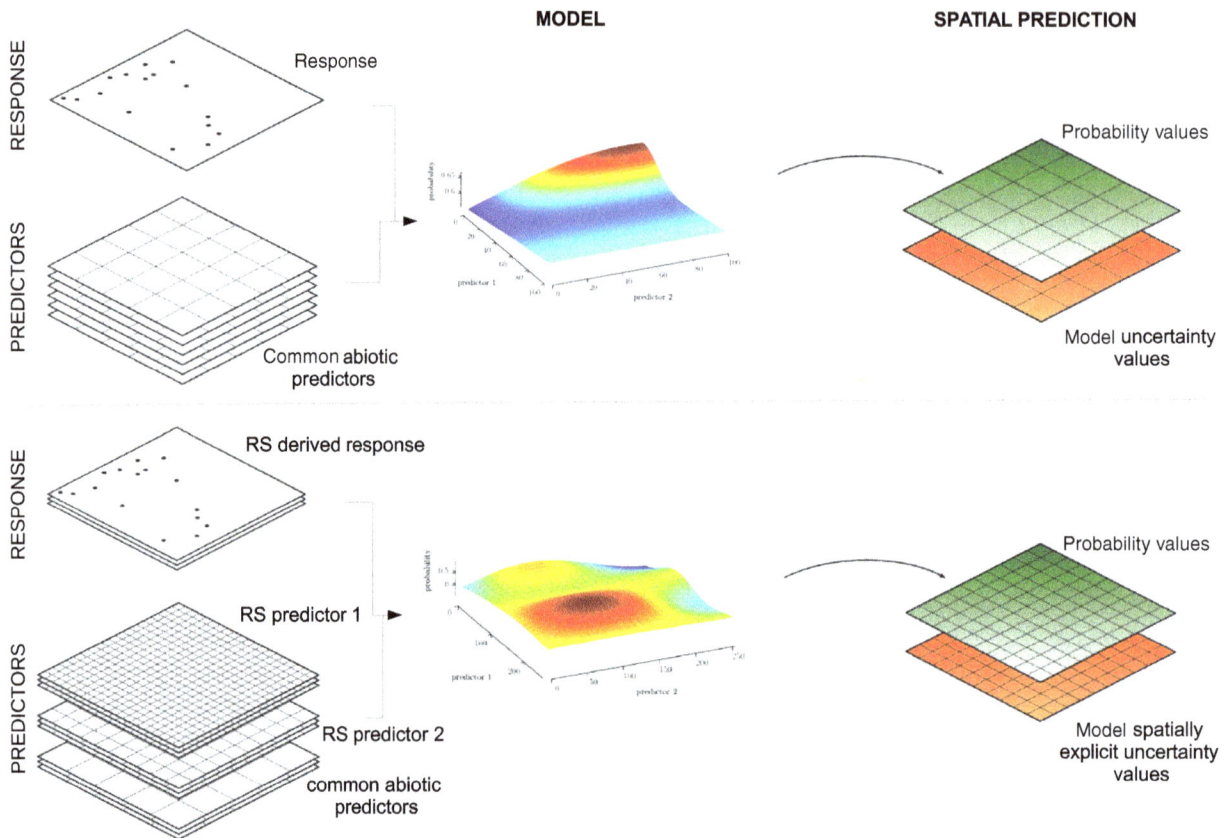

Figure 1. A comparative modeling framework of the current SDMs (above) and the NG-SDMs (below), showing remotely derived response variable and multi-scale predictor variables, including spatially explicit uncertainty of predictor variables. In classical SDMs, uncertainty is often not reported in a spatially explicit manner and one layer per predictor is used. In contrast, NG-SDMs can have a stack of images organized systematically by scales in time to capture each predictor, thus resulting predictions with high accuracy. NG-SDMs, next generation species distribution models.

cal metrics as discussed in previous sections, can be the basis for mapping a species' tolerance to abiotic constraints. These variables are mostly suited for broad-scale models developed at the correlative end of the continuum for both plant and animal species. Second, demographic parameters capturing differences in life histories, such as a population's growth rates obtained from LiDAR (e.g. tree height, stem density for both canopy and sub-canopy layers at different points in time) and species' biophysical traits derived from hyperspectral sensors (e.g. leaf water and nitrogen contents, pigment characteristics and other biochemistry traits), provide opportunities for developing process-based models at the local scale. Third, biotic predictors, including plant functional types, fPAR (an indirect proxy for light competition and growth) and 3-D habitat structure (capable of depicting reliance among species), can be related to biotic interactions at both local and broader scales. The response variables in NG-SDMs will be multi-level in nature, and could include the pres-

ence/absence of a single taxon; species fitness metrics; trait diversity information; and occurrence or abundance of aggregated taxa, functional groups and community assemblages. Being integrative models as suggested by Lurgi et al. (2015), the NG-SDMs can handle a wide range of data types and resolutions, and model uncertainty, while being capable of revealing the underlying causal factors of shaping species distribution and abundance.

Finally, we also want to stress the limitations and challenges of remote sensing in NG-SDMs. First of all, not all plant and animal species can be detected by remote sensors. Understory species and species with lesser distinctive spectral features are difficult to detect. To this end, sensor networks and data fusion (optical/RADAR/LiDAR) may play a key role in tracking species distribution (Koch 2010; Dalponte et al. 2012). Second, there is always a trade-off between spatial and temporal resolutions and a trade-off between spatial and spectral resolution. For example high temporal resolution data usually have low spatial resolution (such as time series of multispectral sensors, Landsat or MODIS). Low spatial resolution can hardly discriminate objects on the ground resulting in lower classification accuracy. In general, finer spatial resolution increases classification accuracy, but at the same time, smaller pixels increase spectral variance resulting in decreased spectral separability of classes (Nagendra and Rocchini 2008). Third, remote sensing data are limited by the short time span of their availability and their contributions to modeling future projections of species distributions under climate change scenarios are limited at this stage. However, current archives of remote sensing data provide important baseline information such as changes in plant physiology and phenology for future climate change studies. New sensors with high temporal resolutions will become an integral part of monitoring instrument for tracking and predicting future species distribution under global climate change. Fourth, using species distributions that have been derived from remote sensing as responses in image-based SDMs bears the risk of circularity. Even if we have an independent response and aim at using remote sensing as predictor we should consider that the response may have had an influence on reflectance. Fifth, to fully utilize remote sensing data, one needs expertise in data processing and software development. In recent years, this has been facilitated by open-source algorithms and software as well as powerful computing capacities. Lastly, accessibility to free remote sensing data with global coverage can be challenging and this is particularly true in developing countries where data processing, storage and sharing are still hampered by information technology and archiving capability (Pettorelli et al. 2014b).

To overcome these limitations and constraints, we call for (1) the creation of sensor networks and improved interoperability between remotely sensed information and *in situ* biological data collections, as *in situ* data provide powerful information for accurate imagery interpretation; (2) development of ecologically meaningful predictors and application of cross-scale approaches; and (3) targeted coordination of field campaigns and the acquisition of remote sensing data.

Conclusions

Remote sensing has been one of the most powerful approaches to provide observations of key species distribution patterns in terms of reduced time and costs. Novel analytical techniques, increasing computational capacity, enhanced sensor fusion and networking capability as well as free access to satellite data (Turner 2014) have greatly promoted the use of remote sensing in species distribu-

tion modeling and provide the opportunity to develop a novel modeling framework as we propose here, the NG-SDMs. This modeling framework will bring new possibilities for hypothesis testing and further exploration of generalized patterns of biodiversity and underlying environmental drivers in both terrestrial and aquatic ecosystems. We hope that this interdisciplinary perspective will stimulate more discussions on species distribution modeling and motivate ecologists, remote sensing experts and modelers to work together for developing a more refined modeling framework in the near future.

Acknowledgments

This paper emerged from a Plenary Symposium 'Tracking Changes from Space: Advances of Remote Sensing in Biogeography' at the 7th Biennial Conference of the International Biogeography Society (IBS), held in Bayreuth, Germany from 8 to 12 January 2015. The symposium organizers, Kate S. He, Anna F. Cord and Mao-Ning Tuanmu, thank the IBS for making the symposium possible. We are also very fortunate to have had very thorough and insightful reviewers; the current paper has benefited greatly from their comments and suggestions.

Conflict of Interest

None declared.

References

Alonzo, M., B. Bookhagen, and D. A. Roberts. 2014. Urban tree species mapping using hyperspectral and LiDAR data fusion. Remote Sens. Environ. 148:70–83.

Anderson, K., and K. J. Gaston. 2013. Lightweight unmanned aerial vehicles will revolutionize spatial ecology. Front. Ecol. Environ. 11:138–146.

Andrew, M. E., and S. L. Ustin. 2008. The role of environmental context in mapping invasive plants with hyperspectral image data. Remote Sens. Environ. 112:4301–4317.

Andrew, M. E., and S. L. Ustin. 2009. Habitat suitability modelling of an invasive plant with advanced remote sensing data. Divers. Distrib. 15:627–640.

Araújo, M. B., R. G. Pearson, W. Thuiller, and M. Erhard. 2005. Validation of species–climate impact models under climate change. Glob. Change Biol. 11:1504–1513.

Asner, G. P., M. O. Jones, R. E. Martin, D. E. Knapp, and R. F. Hughes. 2008a. Remote sensing of native and invasive species in Hawaiian forests. Remote Sens. Environ. 112:1912–1926.

Asner, G. P., D. E. Knapp, T. Kennedy-Bowdoin, M. O. Jones, R. E. Martin, J. Boardman, et al. 2008b. Invasive species detection in Hawaiian rainforests using airborne imaging spectroscopy and LiDAR. Remote Sens. Environ. 112:1942–1955.

Asner, G. P., R. F. Hughes, P. M. Vitousek, E. David, D. E. Knapp, T. Kennedy-Bowdoin, et al. 2008c. Invasive plants transform the three-dimensional structure of rain forests. Proc. Natl. Acad. Sci. USA 105:4519–4523.

Austin, M.P. 2002. Spatial prediction of species distribution: an interface between ecological theory and statistical modelling. Ecol. Model. 157:101–118.

Beckmann, M., T. Václavík, A. M. Manceur, L. Šprtová, H. von Wehrden, E. Welk, et al. 2014. glUV: a global UV-B radiation dataset for macroecological studies. Methods Ecol. Evol. 5:372–383.

Bedia, J., S. Herrera, and J. M. Gutierrez. 2013. Dangers of using global bioclimatic datasets for ecological niche modeling. Limitations for future climate projections. Global Planet. Change 107:1–12.

Bergen, K., S. Goetz, R. Dubayah, G. Henebry, M. Imhoff, R. Nelson, et al. 2009. Remote sensing of vegetation 3D structure for biodiversity and habitat: review and implications for LiDAR and RADAR spaceborne missions. J. Geophys. Res. 114:GE00E06.

Bisrat, S. A., M. A. White, K. H. Beard, and D. Richard Cutler. 2012. Predicting the distribution potential of an invasive frog using remotely sensed data in Hawaii. Divers. Distrib. 18:648–660.

Bradley, B. A. 2014. Remote detection of invasive plants: a review of spectral, textural and phenological approaches. Biol. Invasions 16:1411–1425.

Bradley, B. A., and J. F. Mustard. 2005. Identifying land cover variability distinct from land cover change: cheatgrass in the Great Basin. Remote Sens. Environ. 94:204–213.

Bradley, B. A., and J. F. Mustard. 2006. Characterizing the landscape dynamics of an invasive plant and risk of invasion using remote sensing. Ecol. Appl. 16:1132–1147.

Bradley, B. A., M. Oppenheimer, and D. S. Wilcove. 2009. Climate change and plant invasion: restoration opportunities ahead? Glob. Change Biol. 15:1511–1521.

Bradley, B. A., A. D. Olsson, O. Wang, B. G. Dickson, L. Pelech, S. E. Sesnie, et al. 2012. Species detection vs. habitat suitability: are we biasing habitat suitability models with remotely sensed data? Ecol. Model. 244:57–64.

Buermann, W., S. Saatchi, T. Smith, B. Zutta, J. Chaves, B. Mila, et al. 2008. Predicting species distributions across the Amazonian and Andean regions using remote sensing data. J. Biogeogr. 35:1160–1176.

Carlson, K. M., G. P. Asner, R. F. Hughes, R. Ostertag, and R. E. Martin. 2007. Hyperspectral remote sensing of canopy biodiversity in Hawaiian lowland rainforests. Ecosystems 10:526–549.

Clark, M., D. A. Roberts, and D. B. Clark. 2005. Hyperspectral discrimination of tropical rain forest tree species at leaf to crown scales. Remote Sens. Environ. 96:497–508.

Cord, A. F., and D. Rödder. 2011. Inclusion of habitat availability in species distribution models through multi-temporal remote-sensing data? Ecol. Appl. 21:3285–3298.

Cord, A. F., R. K. Meentemeyer, P. J. Leitão, and T. Václavík. 2013. Modelling species distributions with remote sensing data: bridging disciplinary perspectives. J. Biogeogr. 40:2226–2227.

Culbert, P. D., V. C. Radeloff, V. St-Louis, C. H. Flather, C. D. Rittenhouse, T. P. Albright, et al. 2012. Modeling broad-scale patterns of avian species richness across the Midwestern United States with measures of satellite image texture. Remote Sens. Environ. 118:140–150.

Dalponte, M., L. Bruzzone, and D. Gianelle. 2012. Tree species classification in the Southern Alps based on the fusion of very high geometrical resolution multispectral/hyperspectral images and LiDAR data. Remote Sens. Environ. 123:258–270.

Daly, C. 2006. Guidelines for assessing the suitability of spatial climate data sets. Int. J. Climatol. 26:707–721.

Daly, C., W. P. Gibson, G. H. Taylor, G. L. Johnson, and P. Pasteris. 2002. A knowledge-based approach to the statistical mapping of climate. Clim. Res. 22:99–113.

Dickinson, J. L., B. Zuckerberg, D. N. Bonter, A. F. Dickinson, and L. Janis. 2010. Citizen science as an ecological research tool: challenges and benefits. Annu. Rev. Ecol. Evol. Syst. 41:149–172.

Dolos, K., A. Bauer, and S. Albrecht. 2015. Site suitability for tree species: is there a positive relation between a tree species' occurrence and its growth? Eur. J. Forest Res. 134:609–621.

Dormann, C. F., S. J. Schymanski, J. Cabral, I. Chuine, C. Graham, F. Hartig, et al. 2012. Correlation and process in species distribution models: bridging a dichotomy. J. Biogeogr. 39:2119–2131.

Elith, J., and J. R. Leathwick. 2009. Species distribution models: ecological explanation and prediction across space and time. Annu. Rev. Ecol. Evol. Syst. 40:677–697.

van Ewijk, K. Y., C. F. Randin, P. M. Treitz, and N. A. Scott. 2014. Predicting fine-scale tree species abundance patterns using biotic variables derived from LiDAR and high spatial resolution imagery. Remote Sens. Environ. 150:120–131.

Farrell, S. L., B. A. Collier, K. L. Skow, A. M. Long, A. J. Campomizzi, M. L. Morrison, et al. 2013. Using LiDAR-derived vegetation metrics for high-resolution, species distribution models for conservation planning. Ecosphere 4:42. doi: 10.1890/ES12-000352.1

Fassnacht, F. E., C. Neumann, M. Förster, H. Buddenbaum, A. Ghosh, A. Clasen, et al. 2014. Comparison of feature reduction algorithms for classifying tree species with hyperspectral data on three central European test sites. IEEE J. Sel. Top. Appl. Earth Obs. Remote Sens. 7:2547–2561.

Feilhauer, H., K. S. He, and D. Rocchini. 2012. Modeling species distribution using niche-based proxies derived from composite bioclimatic variables and MODIS NDVI. Remote Sens. 4:2057–2075.

Féret, J., and G. P. Asner. 2013. Tree species discrimination in tropical forests using airborne imaging spectroscopy. IEEE Trans. Geosci. Remote Sens. 51:73–84.

Ficetola, G. F., A. Bonardi, C. A. Mücher, N. L. M. Gilissen, and E. Padoa-Schioppa. 2014. How many predictors in species distribution models at the landscape scale? Land use versus LiDAR-derived canopy height. Int. J. Geogr. Inf. Sci. 28:1723–1739.

Fithian, W., and T. Hastie. 2013. Finite-sample equivalence in statistical models for presence-only data. Ann. Appl. Stat. 7:1917–1939.

Fithian, W., J. Elith, T. Hastie, and D. A. Keith. 2015. Bias correction in species distribution models: pooling survey and collection data for multiple species. Methods Ecol. Evol. 6:424–438.

Fitterer, J. L., T. A. Nelson, N. C. Coops, and M. A. Wulder. 2012. Modelling the ecosystem indicators of British Columbia using Earth observation data and terrain indices. Ecol. Ind. 20:151–162.

Fordham, D. A., H. R. Akcakaya, B. W. Brook, A. Rodrıguez, P. C. Alves, E. Civantos, et al. 2013. Adapted conservation measures are required to save the Iberian lynx in a changing climate. Nat. Clim. Chang. 3:899–903.

Fordham, D. A., K. T. Shoemaker, N. H. Schumaker, H. R. Akcakaya, N. Clisby, and B. W. Brook. 2014. How interactions between animal movement and landscape processes modify local range dynamics and extinction risk. Biol. Lett. 10:20140198.

Franklin, J. 1995. Predictive vegetation mapping: geographic modelling of biospatial patterns in relation to environmental gradients. Prog. Phys. Geogr. 19:474–499.

Franklin, J. 2009. Mapping species distributions: spatial inference and prediction. Cambridge University Press, Cambridge.

Franklin, J. 2010. Moving beyond static species distribution models in support of conservation biogeography. Divers. Distrib. 16:321–330.

Fretwell, P. T., M. A. LaRue, P. Morin, G. L. Kooyman, B. Wienecke, N. Racliffe, et al. 2012. An emperor penguin population estimate: the first global, synoptic survey of a species from space. PLoS One 7:e33751.

Fretwell, P. T., I. J. Staniland, and J. Forcada. 2014. Whales from space: counting southern right whales by satellite. PLoS One 9:e88655.

Friedl, M. A., D. K. McIver, J. C. F. Hodges, X. Y. Zhang, D. Muchoney, A. H. Strahler, et al. 2002. Global land cover mapping from MODIS: algorithms and early results. Remote Sens. Environ. 83:287–302.

Getzin, S., R. S. Nuske, and K. Wiegand. 2014. Using unmanned aerial vehicles (UAV) to quantify spatial gap patterns in forests. Remote Sens. 6:6988–7004.

Ghosh, A., F. E. Fassnacht, P. K. Joshi, and B. Koch. 2014. A framework for mapping tree species combining hyperspectral and LiDAR data: role of selected classifiers and sensor across three spatial scales. Int. J. Appl. Earth Obs. Geoinf. 26:49–63.

Goetz, S. J., D. Steinberg, R. Dubayah, and B. Blair. 2007. Laser remote sensing of canopy habitat heterogeneity as a predictor of bird species richness in an eastern temperate forest, USA. Remote Sens. Environ. 108:254–263.

Goetz, S. J., D. Steinberg, M. G. Betts, R. T. Holmes, P. J. Doran, R. Dubayah, et al. 2010. Lidar remote sensing variables predict breeding habitat of a Neotropical migrant bird. Ecology 91:1569–1576.

Gould, S. F., S. Hugh, L. L. Porfirio, and B. Mackey. 2015. Ecosystem greenspots pass the first test. Landscape Ecol. 30:141–151.

Grime, J. P. 1974. Vegetation classification by reference to strategies. Nature 250:26–31.

Grime, J. P. 1977. Evidence for the existence of three primary strategies in plants and its relevance to ecological and evolutionary theory. Am. Nat. 111:1169–1194.

Hall, K., T. Reitalu, M. T. Sykes, and H. C. Prentice. 2012. Spectral heterogeneity of QuickBird satellite data is related to fine-scale plant species spatial turnover in semi-natural grasslands. Appl. Veg. Sci. 15:145–157.

He, K. S., J. Zhang, and R. Zhang. 2009. Linking variability in species composition and MODIS NDVI based on beta diversity measurements. Acta Oecol. 35:14–21.

He, K. S., D. Rocchini, M. Neteler, and H. Nagendra. 2011. Benefits of hyperspectral remote sensing for tracking plant invasions. Divers. Distrib. 17:381–392.

Heinzel, J., and B. Koch. 2011. Exploring full-waveform LiDAR parameters for tree species classification. Int. J. Appl. Earth Obs. Geoinf. 2011:152–160.

Henderson, E. B., J. L. Ohmann, M. J. Gregory, H. M. Roberts, and H. Zald. 2014. Species distribution modelling for plant communities: stacked single species or multivariate modelling approaches? Appl. Veg. Sci. 17:516–527.

Herfindal, I., E. J. Solberg, B. E. Saether, K. A. Hogda, and R. Andersen. 2006. Environmental phenology and geographical gradients in moose body mass. Oecologia 150:213–224.

Hijmans, R. J., S. E. Cameron, J. L. Parra, P. G. Jones, and A. Jarvis. 2005. Very high resolution interpolated climate surfaces for global land areas. Int. J. Climatol. 25:1965–1978.

Hodgson, A., N. Kelly, and D. Peel. 2013. Unmanned aerial vehicles (UAVs) for surveying marine fauna: a dugong case study. PLoS One 8:e79556.

Huang, C. Y., and G. P. Asner. 2009. Applications of remote sensing to alien invasive plant studies. Sensors 9:4869–4889.

Huang, C. Y., and E. L. Geiger. 2008. Climate anomalies provide opportunities for large-scale mapping of non-native plant abundance in desert grasslands. Divers. Distrib. 14:875–884.

Huffman, G. J., D. T. Bolvin, E. J. Nelkin, D. B. Wolff, R. F. Adler, G. Gu, et al. 2007. The TRMM multisatellite precipitation analysis (TMPA): quasi-global, multiyear, combined-sensor precipitation estimates at fine scales. J. Hydrometeorol. 8:38–55.

Jetz, W., J. M. McPherson, and R. P. Guralnick. 2012. Integrating biodiversity distribution knowledge: toward a global map of life. Trends Ecol. Evol. 27:151–159.

Jones, T. G., N. C. Coops, and T. Sharma. 2010. Assessing the utility of airborne hyperspectral and LiDAR data for species distribution mapping in the coastal Pacific Northwest, Canada. Remote Sens. Environ. 114:2841–2852.

Kachelriess, D., M. Wegmann, M. Gollock, and N. Pettorelli. 2014. The application of remote sensing for marine protected area management. Ecol. Ind. 36:169–177.

Kissling, W. D. 2013. Estimating extinction risk under climate change: next-generation models simultaneously incorporate demography, dispersal, and biotic interactions. Front. Biogeogr. 5:163–165.

Kissling, W. D., C. F. Dormann, J. Groeneveld, T. Hickler, I. Kühn, G. J. McInerny, et al. 2012. Towards novel approaches to modelling biotic interactions in multispecies assemblages at large spatial extents. J. Biogeogr. 39:2163–2178.

Koch, B. 2010. Status and future of laser scanning, synthetic aperture RADAR and hyperspectral remote sensing data for forest biomass assessment. ISPRS J. Photogramm. Remote Sens. 65:581–590.

Koh, L. P., and S. A. Wich. 2012. Dawn of drone ecology: low-cost autonomous aerial vehicles for conservation. Trop. Conserv. Sci. 5:121–132.

Kremen, C., A. Cameron, A. Moilanen, S. J. Phillips, C. D. Thomas, H. Beentje, et al. 2008. Aligning conservation priorities across taxa in Madagascar with high-resolution planning tools. Science 320:222–226.

Kriticos, D. J., B. L. Webber, A. Leriche, N. Ota, I. Macadam, J. Bathols, et al. 2012. CliMond: global high resolution historical and future scenario climate surfaces for bioclimatic modelling. Methods Ecol. Evol. 3:53–64.

Lahoz-Monfort, J. J., G. Guillera-Arroita, E. J. Milner-Gulland, R. P. Young, and E. Nicholson. 2010. Satellite imagery as a single source of predictor variables for habitat suitability modelling: how Landsat can inform the conservation of a critically endangered lemur. J. Appl. Ecol. 47:1094–1102.

Leutner, B. F., B. Reineking, J. Müller, M. Bachmann, C. Beierkuhnlein, S. Dech, et al. 2012. Modelling forest α-diversity and floristic composition – on the added value of LiDAR plus hyperspectral remote sensing. Remote Sens. 4:2818–2845.

Leyequien, E., J. Verrelst, M. Slot, G. Schaepman-Strub, I. M. A. Heitkonig, and A. Skidmore. 2007. Capturing the fugitive: applying remote sensing to terrestrial animal distribution and diversity. Int. J. Appl. Earth Obs. Geoinf. 9:1–20.

Louzao, M., D. Pinaud, C. Peron, K. Delord, T. Wiegand, and H. Weimerskirch. 2011. Conserving pelagic habitats:

seascape modelling of an oceanic top predator. J. Appl. Ecol. 48:121–131.

Lucas, R. M., A. C. Lee, and P. J. Bunting. 2008. Retrieving forest biomass through integration of CASI and LiDAR data. Int. J. Remote Sens. 29:1553–1577.

Lucieer, A., Z. Malenovsky, T. Yeness, and L. Wallace. 2014. HyperUAS – imaging spectroscopy from a multirotor unmanned aircraft system. J. Field Robotics 31:571–590.

Luoto, M., R. Virkkala, and R. K. Heikkinen. 2007. The role of land cover in bioclimatic models depends on spatial resolution. Glob. Ecol. Biogeogr. 16:34–42.

Lurgi, M., B. W. Brook, F. Saltre, and D. A. Fordham. 2015. Modelling range dynamics under global change: which framework and why? Methods Ecol. Evol. 6:247–256.

Metz, M., D. Rocchini, and M. Neteler. 2014. Surface temperatures at continental scale: tracking changes with remote sensing at unprecedented detail. Remote Sens. 6:3822–3840.

Meyer, W. B., and B. L. Turner, eds. 1994. Changes in land use and land cover: a global perspective Vol. 4. Cambridge University Press, Cambridge, UK.

Mitchell, J. J., and N. F. Glenn. 2009. Subpixel abundance estimates in mixture-turned matched filtering classifications of leafy spurge (*Euphorbia esula* L.). Int. J. Remote Sens. 30:6099–6119.

Morán-Ordóñez, A., S. Suárez-Seoane, J. Elith, L. Calvo, and E. de Luis. 2012. Satellite surface reflectance improves habitat distribution mapping: a case study on heath and shrub formations in the Cantabrian Mountains (NW Spain). Divers. Distrib. 18:588–602.

Morisette, J. T., C. S. Jarnevich, A. Ullah, W. J. Cai, J. A. Pedelty, J. E. Gentle, et al. 2006. A tamarisk habitat suitability map for the continental United States. Front. Ecol. Environ. 4:11–17.

Morisette, J. T., A. D. Richardson, A. K. Knapp, J. I. Fisher, E. A. Graham, J. Abatzoglou, et al. 2009. Tracking the rhythm of the seasons in the face of global change: phenological research in the 21st century. Front. Ecol. Environ. 7:253–260.

Nagendra, H., and D. Rocchini. 2008. High resolution satellite imagery for tropical biodiversity studies: the devil is in the detail. Biodivers. Conserv. 17:3431–3442.

Neteler, M., M. Metz, D. Rocchini, A. Rizzoli, E. Flacio, L. Engeler, et al. 2013. Is Switzerland suitable for the invasion of *Aedes albopictus*? PLoS One 8:e82090.

New, M., D. Lister, M. Hulme, and I. Makin. 2002. A high-resolution data set of surface climate over global land areas. Clim. Res. 21:1–25.

Newton-Cross, G., P. C. L. White, and S. Harris. 2007. Modelling the distribution of badgers *Meles meles*: comparing predictions from field-based and remotely derived habitat data. Mamm. Rev. 37:54–70.

Osborne, P., and S. Suárez-Seoane. 2007. Identifying core areas in a species' range using temporal suitability analysis: an example using little bustards *Tetrax tetrax* L. in Spain. Biodivers. Conserv. 16:3505–3518.

Parmentier, B., B. J. McGill, A. M. Wilson, J. Regetz, W. Jetz, R. Guralnick, et al. 2015. Using multi-timescale methods and satellite-derived land surface temperature for the interpolation of daily maximum air temperature in Oregon. Int. J. Climatol., doi: 10.1002/joc.4251

Pau, S. P., E. J. Edwards, and C. J. Still. 2013. Improving our understanding of environmental controls on the distribution of C3 and C4 grasses. Glob. Change Biol. 19:184–196.

Pearson, R. G., T. E. Dawson, and C. Liu. 2004. Modelling species distributions in Britain: a hierarchical integration of climate and land-cover data. Ecography 27:285–298.

Peterson, E. B. 2005. Estimating cover of an invasive grass (*Bromus tectorum*) using tobit regression and phenology derived from two dates of Landsat ETM plus data. Int. J. Remote Sens. 26:2491–2507.

Petitpierre, B., C. Kueffer, O. Broennimann, C. Randin, C. Daehler, and A. Guisan. 2012. Climatic niche shifts are rare among terrestrial plant invaders. Science 335:1344–1348.

Pettorelli, N. 2013. The normalised difference vegetation index. Oxford University Press, Oxford, UK.

Pettorelli, N., S. Ryan, T. Mueller, N. Bunnefeld, B. Jedrzejewska, M. Lima, et al. 2011. The normalized difference vegetation index (NDVI): unforeseen successes in animal ecology. Clim. Res. 46:15–27.

Pettorelli, N., B. Laurance, T. O'Brien, M. Wegmann, H. Nagendra, and W. Turner. 2014a. Satellite remote sensing for applied ecologists: opportunities and challenges. J. Appl. Ecol. 51:839–848.

Pettorelli, N., K. Safi, and W. Turner. 2014b. Satellite remote sensing, biodiversity research and conservation of the future. Philos. Trans. R. Soc. B 369:20130190.

Platonov, N. G., I. N. Mordvintsev, and V. V. Rozhnov. 2013. The possibility of using high resolution satellite imagery for detection of marine mammals. Biol. Bull. 40:197–205.

Pottier, J., Z. Malenovský, A. Psomas, L. Homolová, M.E. Schaepman, P. Choler, W. Thuiller, A. Guisan, and N. E. Zimmermann. 2014. Modelling plant species distribution in alpine grasslands using airborne imaging spectroscopy. Biol. Lett. 10:20140347.

Pradervand, J.-N., A. Dubuis, L. Pellissier, A. Guisan, and C. Randin. 2014. Very high resolution environmental predictors in species distribution models: moving beyond topography? Prog. Phys. Geogr. 38:79–96.

Prates-Clark, C. D. C., S. Saatchi, and D. Agosti. 2008. Predicting geographical distribution models of high-value timber trees in the Amazon Basin using remotely sensed data. Ecol. Model. 211:309–323.

Questad, E. J., J. R. Kellner, K. Kinney, S. Cordell, G. P. Asner, J. Thaxton, et al. 2014. Mapping habitat suitability for at-risk plant species and its implications for restoration and reintroduction. Ecol. Appl. 24:385–395.

Ramírez, F., I. Afán, K. A. Hobson, M. Bertellotti, G. Blanco, and M. G. Forero. 2014. Natural and anthropogenic factors affecting the feeding ecology of a top marine predator, the Magellanic penguin. Ecosphere 5:art38.

Renner, I. W., J. Elith, A. Baddeley, W. Fithian, T. Hastie, S. J. Phillips, et al. 2015. Point process models for presence-only analysis. Methods Ecol. Evol. 6:366–379.

Rickbeil, G. J. M., N. C. Coops, M. C. Drever, and T. A. Nelson. 2014. Assessing coastal species distribution models through the integration of terrestrial, oceanic and atmospheric data. J. Biogeogr. 41:1614–1625.

Rovero, F., M. Menegon, J. Fjeldså, L. Collett, N. Doggart, C. Leonard, et al. 2014. Targeted vertebrate surveys enhance the faunal importance and improve explanatory models within the Eastern Arc Mountains of Kenya and Tanzania. Divers. Distrib. 20:1438–1449.

Saatchi, S., W. Buermann, H. ter Steege, S. Mori, and T. Smith. 2008. Modelling distribution of Amazonian tree species and diversity using remote sensing measurements. Remote Sens. Environ. 112:2000–2017.

Schmidtlein, S., H. Feihauer, and H. Bruelheide. 2012. Mapping plant strategy types using remote sensing. J. Veg. Sci. 23:395–405.

Schurr, F. M., J. Pagel, J. S. Cabral, J. Groeneveld, O. Bykova, R. B. O'Hara, et al. 2012. How to understand species' niches and range dynamics: a demographic research agenda for biogeography. J. Biogeogr. 39:2146–2162.

Seiler, C., R. W. A. Hutjes, B. Kruijt, J. Quispe, S. Añez, V. K. Arora, et al. 2014. Modeling forest dynamics along climate gradients in Bolivia. J. Geophys. Res. Biogeosci. 119:758–775.

Simonson, W. D., H. D. Allen, and D. A. Coomes. 2014. Applications of airborne LiDAR for the assessment of animal species diversity. Methods Ecol. Evol. 5:719–729.

Sims, D. A., A. F. Rahman, V. D. Cordova, B. Z. El-Masri, D. D. Baldocchi, P. V. Bolstad, et al. 2008. A new model of gross primary productivity for North American ecosystems based solely on the enhanced vegetation index and land surface temperature from MODIS. Remote Sens. Environ. 112:1633–1646.

Singh, N. J., and E. J. Milner-Gulland. 2011. Conserving a moving target: planning protection for a migratory species as its distribution changes. J. Appl. Ecol. 48:35–46.

Stapleton, S., M. LaRue, N. Lecomte, S. Atkinson, D. Garshelis, C. Porter, et al. 2014. Polar bears from space: assessing satellite imagery as a tool to track Arctic wildlife. PLoS One 9:e101513.

Still, C. J., S. Pau, and E. J. Edwards. 2014. Land surface skin temperature captures thermal environments of C3 and C4 grasses. Glob. Ecol. Biogeogr. 23:286–296.

Stoklosa, J., C. Daly, S. D. Foster, M. B. Ashcroft, and D. I. Warton. 2015. A climate of uncertainty: accounting for error in climate variables for species distribution models. Methods Ecol. Evol. 6:412–423.

Suárez-Seoane, S., E. Virgós, O. Terroba, X. Pardavila, and J. M. Barea-Azcón. 2014. Scaling of species distribution models across spatial resolutions and extents along a biogeographic gradient. The case of the Iberian mole *Talpa occidentalis*. Ecography 37:279–292.

Swatantran, A., R. Dubayah, S. Goetz, M. Hofton, M. G. Betts, M. Sun, et al. 2012. Mapping migratory bird prevalence using remote sensing data fusion. PLoS One 7:e28922.

Tattoni, C., F. Rizzolli, and P. Pedrini. 2012. Can LiDAR data improve bird habitat suitability models? Ecol. Model. 245:103–110.

Thuiller, W., M. B. Araújo, and S. Lavorel. 2004. Do we need land-cover data to model species distributions in Europe? J. Biogeogr. 31:353–361.

Tuanmu, M.-N., and W. Jetz. 2014. A global 1-km consensus land-cover product for biodiversity and ecosystem modelling. Glob. Ecol. Biogeogr. 23:1031–1045.

Tuanmu, M.-N., and W. Jetz, 2015. A global, remote sensing-based characterization of terrestrial habitat heterogeneity for biodiversity and ecosystem modeling. Glob. Ecol. and Biogeogr, doi: 10.1111/geb.12365.

Tuanmu, M.-N., A. Viña, S. Bearer, W. Xu, Z. Ouyang, H. Zhang, et al. 2010. Mapping understory vegetation using phenological characteristics derived from remotely sensed data. Remote Sens. Environ. 114:1833–1844.

Tuanmu, M.-N., A. Viña, G. J. Roloff, W. Liu, Z. Ouyang, H. Zhang, et al. 2011. Temporal transferability of wildlife habitat models: implications for habitat monitoring. J. Biogeogr. 38:1510–1523.

Turner, W. 2014. Sensing biodiversity. Science 346:301–302.

Tyberghein, L., H. Verbruggen, K. Pauly, C. Troupin, F. Mineur, and O. De Clerck. 2012. Bio-ORACLE: a global environmental dataset for marine species distribution modelling. Glob. Ecol. Biogeogr. 21:272–281.

Ustin, S. L., and J. A. Gamon. 2010. Remote sensing of plant functional types. New Phytol. 186:795–816.

Verbyla, D. 2008. The greening and browning of Alaska based on 1982–2003 satellite data. Glob. Ecol. Biogeogr. 17:547–555.

Vierling, K. T., L. A. Vierling, W. A. Gould, S. Martinuzzi, and R. M. Clawges. 2008. LiDAR: shedding new light on habitat characterization and modeling. Front. Ecol. Environ. 6:90–98.

Waltari, E., R. Schroeder, K. McDonald, R. P. Anderson, and A. Carnaval. 2014. Bioclimatic variables derived from remote sensing: assessment and application for species distribution modelling. Methods Ecol. Evol. 5:1033–1042.

Wan, Z., Y. Zhang, Q. Zhang, and Z. L. Li. 2004. Quality assessment and validation of the MODIS global land surface temperature. Int. J. Remote Sens. 25:261–274.

Wang, J. J., R. F. Adler, G .J. Huffman, and D. Bolvin. 2014. An updated TRMM composite climatology of tropical rainfall and its validation. J. Clim. 27:273–284.

Watts, A. C., V. G. Ambrosia, and E. A. Hinkley. 2012. Unmanned aircraft systems in remote sensing and scientific research: classification and considerations of use. Remote Sens. 4:1671–1692.

Wegmann, M., B. Leutner, and S. Dech. 2015. Remote sensing and GIS for ecologists: using open source software. Pelagic Publishing, Exeter, UK.

Weisberg, P. J., E. Lingua, and R. B. Pillai. 2007. Spatial patterns of pinyon-juniper woodland expansion in central Nevada. Rangeland Ecol. Manag. 60:115–124.

Wilfong, B. N., D. L. Gorchov, and M. C. Henry. 2009. Detecting an invasive shrub in deciduous forest understories using remote sensing. Weed Sci. 57:512–520.

Willems, E. P., R. A. Barton, and R. A. Hill. 2009. Remotely sensed productivity, regional home range selection, and local range use by an omnivorous primate. Behav. Ecol. 20:985–992.

Wisz, M. S., J. Pottier, W. D. Kissling, L. Pellissier, J. Lenoir, C. F. Damgaard, et al. 2013. The role of biotic interactions in shaping distributions and realised assemblages of species: implications for species distribution modelling. Biol. Rev. 88:15–30.

Yang, C., J. H. Everitt, R. S. Fletcher, R. R. Jensen, and P. W. Mausel. 2009. Evaluating AISA+hyperspectral imagery for mapping black mangrove along the south Texas Gulf Coast. Photogramm. Eng. Remote Sensing 75:425–435.

Yang, Z., T. Wang, A. K. Skidmore, J. de Leeuw, M. Y. Said, and J. Freer. 2014. Spotting East African mammals in open savannah from space. PLoS One 9:e115989.

Zimmermann, N. E., T. C. Edwards, G. G. Moisen, T. S. Frescino, and J. A. Blackard. 2007. Remote sensing-based predictors improve distribution models of rare, early successional and broadleaf tree species in Utah. J. Appl. Ecol. 44:1057–1067.

7

Remote sensing training in African conservation

Helen Margaret de Klerk[1] & Graeme Buchanan[2]

[1]Department of Geography and Environmental Studies, Stellenbosch University, P. Bag X1 Matieland, Stellenbosch 7602, South Africa
[2]Conservation Science, RSPB Scotland Headquarters, 2 Lochside View, Edinburgh Park, Edinburgh EH12 9DH, United Kingdom

Keywords
Academic programs, Africa, conservation implementation, human capacity, remote sensing, training

Correspondence
Helen Margaret de Klerk, Department of Geography and Environmental Studies, Stellenbosch University, P. Bag X1 Matieland, Stellenbosch 7602, South Africa.
E-mail: hdeklerk@sun.ac.za

Editor: Harini Nagendra
Associate editor: Martin Wegmann

Abstract

The potential of remote sensing (RS) to assist with conservation planning, implementation and monitoring is well described, and particularly relevant in African areas that are inaccessible due to terrain, finances or politics. We provide an African perspective on remote sensing (RS) training for conservation and ecology over the last decade through investigating (1) recent use of RS in African conservation literature, (2) use of RS in African conservation agencies, (3) RS training by African institutions and (4) RS capacity development by ad hoc events. Africa does not produce most of the research using RS in conservation and ecological studies conducted on Africa, with authors with correspondence addresses in the USA predominating (33% of a bibliometric analysis), although South Africa-based authors constituted 20% (with an increase between 2000 and 2015), Kenya 6% and Tanzania and Ethiopia 4% each. Ideally research should be conducted close to the point of use to ensure relevance and data residence in the country concerned. This is a point for attention, possibly through international funding to increase the capacity of African academic institutions to conduct research using RS to answer conservation questions. Part of this will need to include attention on data and software costs, internet speeds and human capacity. Data costs have been alleviated by free Landsat and MODIS data, and the Copernicus programs, but there is need for higher resolution imagery to be freely available for certain conservation projects. Open Source software may well offer a long-term solution to software costs. This would require that teaching is realigned to employer requirements, which are shifting in many countries and agencies from proprietary software to Open Source due to licensing costs. Low internet connectivity in many areas of Africa might limit the uptake of new data processing options that require connectivity, although over time these tools may become available to more users. However, human capacity is developing. Of the 72 academic institutions surveyed, a number of conservation programs supplied either tailored RS teaching or used 'service modules' to provide RS skills to young graduating conservation professionals, showing a recognition of the importance of RS in conservation in Africa. This study highlights the success of capacity development in Africa, and the increasing use of remote sensing for conservation in Africa.

Introduction

It is well recognized in the literature that remote sensing (RS) has potential to provide much data and knowledge to aid conservation managers and decision makers (Buchanan et al. 2009; Turner et al. 2003; Turner 2014; O'Connor et al. 2015). There has been a long history of the application of remote sensing in Africa. One of the first publications to recommend the use of RS in conservation in Africa is Wicht's (1945) use of aerial photography for mapping vegetation in South Africa. Wicht saw the value of remote sensing images to provide back-drop information on infrastructure (railways and fire breaks) and natural processes (e.g. fire and invasive alien plant infestations). Further applications have seen remote sensing being used to map and monitor land cover

(Stuckenberg et al. 2012; Verhulp and Van Niekerk 2016) and threats to Important Bird Areas (Buchanan et al. 2009; Tracewski et al. 2016), habitat availability for specific species (Buchanan et al. 2011; Piel et al. 2015), potential conflicts between wildlife and agriculture (Wallin et al. 1992) and habitat degradation (Lück-Vogel et al. 2013). Despite the breath and history of applications of remote sensing to conservation, many have suggested that the remote sensing community is not meeting the needs of the conservation community and that there is a need for a better dialog between the two (Rose et al. 2015; Pettorelli et al. 2014). Topics identified by conservationists in which development is needed to improve the uptake and use of remote sensing in conservation is diverse, but a recurring theme is education. The development of remote sensing capacity and knowledge within the conservation community was identified as one of the possible solutions to the disconnect between the two fields (Turner 2014). Many of the assessments of needs from the conservation community have been undertaken in and primarily focused upon users based in North America and Europe (e.g. Rose et al. 2015). Although many of the participants have experience in working outside these areas, there has been no dedicated assessment of the state of use of remote sensing, and user needs in Africa. This is despite this continent being biodiversity rich, and a region to which remote sensing could make a large difference for conservation monitoring.

Here, we describe the use of remote sensing in African conservation through (1) a bibliometric analysis of peer-reviewed published research literature over the last 15 years; (2) describe the current use of remote sensing data in conservation through interviews with conservation field staff and conservation specialist technical staff in various conservation organizations and (3) describe the processes in place to educate the conservation community of Africa in the use of RS by surveying various academic conservation programs. In doing so we hope to provide a record

of where the current state of play lies, and also identify where processes need to be improved. We hope that the study will feed into the improved provision of remote sensing training for African conservationists, tailored to the needs of this community.

The recent uses of remote sensing in African conservation in scientific literature

To describe the recent history of scientific output of the application of remote sensing in conservation and ecology research in Africa, a bibliometric analysis was conducted in Web of Science using the following combination of keywords and criteria (adapted from de Araujo Barbosa et al. 2015): (1) the literature should include the following combination of keywords: remote sensing OR earth observation OR Landsat OR Lidar OR MODIS OR SPOT OR Radar, AND conservation OR ecology, AND each of the African country names (see Appendix S1 for list of individual African countries listed); (2) only scientific peer-reviewed journals, including reviews, were considered; (3) articles written in English, or other languages with English bibliometric information, were considered and (4) only articles published between 2000 and 2015 were analyzed. We used the output of these searches to describe trends in the number of publications produced over years and according to author affiliation. For the six African countries who had published the highest number of articles, we were able to look at trends in their publication rate over time.

We identified 580 studies which met our search criteria. The number of publications that integrate remote sensing and conservation in Africa has increased fivefold from 2000 to 2015 (Fig. 1). Around 34% of the research is led by researchers whose correspondence address was in the USA. About 21% of the literature was led by authors whose address was in South Africa, whereas authors in

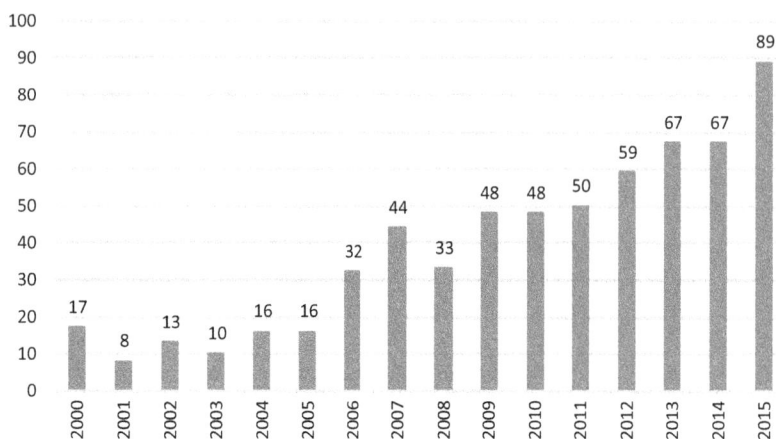

Figure 1. Number of peer-reviewed publications integrating remote sensing and conservation in Africa (meeting all the search criteria) per year from 2000 to 2015.

Kenya produced 6%. Authors in Ethiopia and Tanzania each produced 4% (Fig. 2). Together authors with addresses in African countries accounted for 51% of records (Table 1, Fig. 3). A steady growth of publications on conservation and RS is seen in South Africa, but not in Kenya, Ethiopia or Tanzania (Fig. 4).

Based on cases where the source of data was given in the abstract of the article, Landsat data were the most widely used data, accounting for around 50% of all studies. MODIS data in one form or another came second, at around 20% of studies (some studies could occur in both categories). Thus, these two free platforms were used in the majority of studies. This no doubt highlights the importance of free data (see, e.g. Turner 2014). Active data (radar and LiDAR) featured to a much lower extent with just 22 and 17 studies. These data types are newer, and until recently not widely or freely available. Sentinel 1, as part of Copernicus program, is making radar data widely available for free, so we might see an increase in the use of these technologies in the future. The cloud penetrating abilities of radar might make it particularly useful in areas of Arica which have heavy cloud cover, and for which optical data are of little value.

Figure 5 provides a broad illustration of the topics occurring within the keywords of the articles we considered. After conservation (one of the search terms), vegetation and forest appeared to be prevalent words. Management and National Park were also apparently frequently occurring words. This might indicate that studies were being undertaken with a view to informing management.

Use of remote sensing in conservation organizations in Africa

To determine the current use of RS (data and tools) in conservation to support decision making and environmental management, we used semi-structured interviews to determine the current use of RS in four conservation organizations in Africa (see Appendix S2). The focus was on conservation field staff and the use of RS products or tools to aid their daily core functions, and conservation specialist technical staff who produce data, information products and tools to assist conservation field staff and decision makers at all levels (those reviewing development applications; regional development and conservation plans and national policies). We present results in a discursive rather than quantitative way. Surveys indicated that remote sensing data were initially used as backdrop images for mapping relevant features by the majority of respondents. In particular, the applications were for digitization of land cover features such as wetlands, vegetation, fire scars and patches of invasive alien plants, and infrastructure, roads, trails and fire breaks. These initial steps were generally taken around the late 1990's/early 2000's. Such data capture was often based on hard copy aerial photographs. One notable example of where these proved useful in generating more interest was in CapeNature where the distribution of digital images (with ArcView) leads to requests for and the subsequent acquisition of an organization-wide Landsat mosaic (15 m pan sharpened). As field staff gained confidence with the use of digital tools, they required finer spatial (and temporal) resolution. A national initiative that provided SPOT (2.5 m) country-wide annually from 2005 provided a significant boost to the spatial and temporal frequency with which important features could be mapped. The use of visual interpretation remains a large focus of remote sensing data analysis and the majority of respondents noted that online data sources (Google Earth™) were very valuable in their work, and continue to be used.

The interviewees' perceptions of the barriers inhibiting a greater application of satellite remote sensing in conservation in Africa were, in order of importance, data costs, specialized software costs and the costs of training and

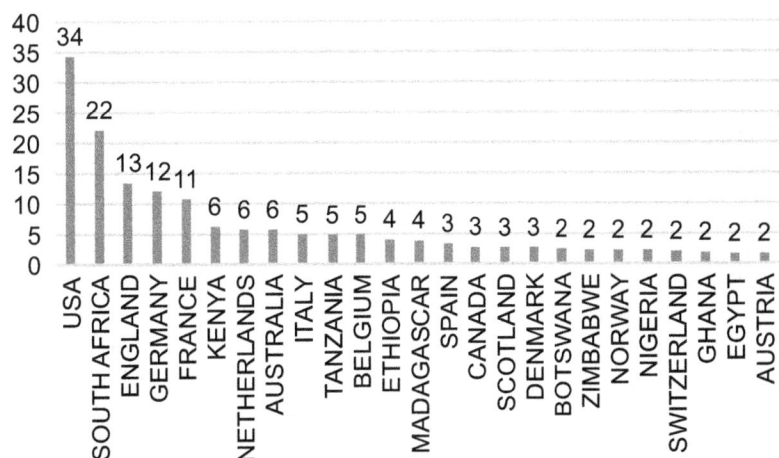

Figure 2. Percentage of peer-reviewed publications integrating remote sensing and conservation in Africa (meeting all the search criteria) per country of authors 2000 to 2015 (countries with two or more percent shown).

Table 1. Number and percentage of records with authors in different countries who have published articles meeting the search criteria.

Countries	No. of records	Percentage of records
USA	189	34
South Africa	122	22
England	74	13
Germany	66	12
France	59	11
Kenya	34	6
Australia	31	6
Netherlands	31	6
Italy	27	5
Belgium	26	5
Tanzania	26	5
Ethiopia	22	4
Madagascar	21	4
Spain	18	3
Canada	15	3
Denmark	14	3
Scotland	14	3
Botswana	13	2
Nigeria	12	2
Norway	12	2
Zimbabwe	12	2
Switzerland	11	2
Hungary	10	2
Austria	9	2
Egypt	9	2
Cameroon	8	1
Democratic Republic Congo	8	1
Namibia	8	1
Uganda	8	1
Wales	8	1
Burkina Faso	7	1
Benin	6	1
Finland	6	1
Morocco	6	1
Portugal	6	1
Senegal	6	1
Sweden	6	1
Algeria	5	1
Japan	5	1
Indonesia	4	1
Mozambique	4	1
Peoples R China	4	1
Poland	4	1
Sudan	4	1
Tunisia	4	1
Zambia	4	1
Brazil	3	1
Czech Republic	3	1
India	3	1
New Caledonia	3	1
Panama	3	1
Papua N Guinea	3	1
Reunion	3	1
Angola	2	0

(Continued)

Table 1. Continued.

Countries	No. of records	Percentage of records
Argentina	2	0
Chile	2	0
Congo	2	0
Cote d'ivoire	2	0
Fr Polynesia	2	0
Israel	2	0
Jordan	2	0
Laos	2	0
Nepal	2	0
New Zealand	2	0
South Korea	2	0
Taiwan	2	0
Thailand	2	0
Bhutan	1	0
Burundi	1	0
Cambodia	1	0
Central Africa Republic	1	0
Chad	1	0
Comoros	1	0
Costa Rico	1	0
Cyprus	1	0
Ecuador	1	0
Fiji	1	0
Greece	1	0
Guinea Bissau	1	0
Iran	1	0
Ireland	1	0
Luxembourg	1	0
Malaysia	1	0
Mali	1	0
Mauritius	1	0
Mongol Peo Rep	1	0
Oman	1	0
Peru	1	0
Philippines	1	0
Qatar	1	0
Russia	1	0
Saudi Arabia	1	0
Seychelles	1	0
Singapore	1	0
Sri Lanka	1	0
Swaziland	1	0
Syria	1	0
Vietnam	1	0

skills development. IT infrastructure and access to internet is also a significant factor (see also Szantoi et al. 2016; Clerici et al. 2013). Many of the low-to-medium resolution datasets have been free for many years (e.g. MODIS and Spot Vegetation and Proba V which replaced SPOT Vegetation). In 2007, the medium-resolution Landsat data (30 m) became free, something welcomed by the conservation community (Turner 2014). The release of all Landsat data for free saw even greater uptake of these data

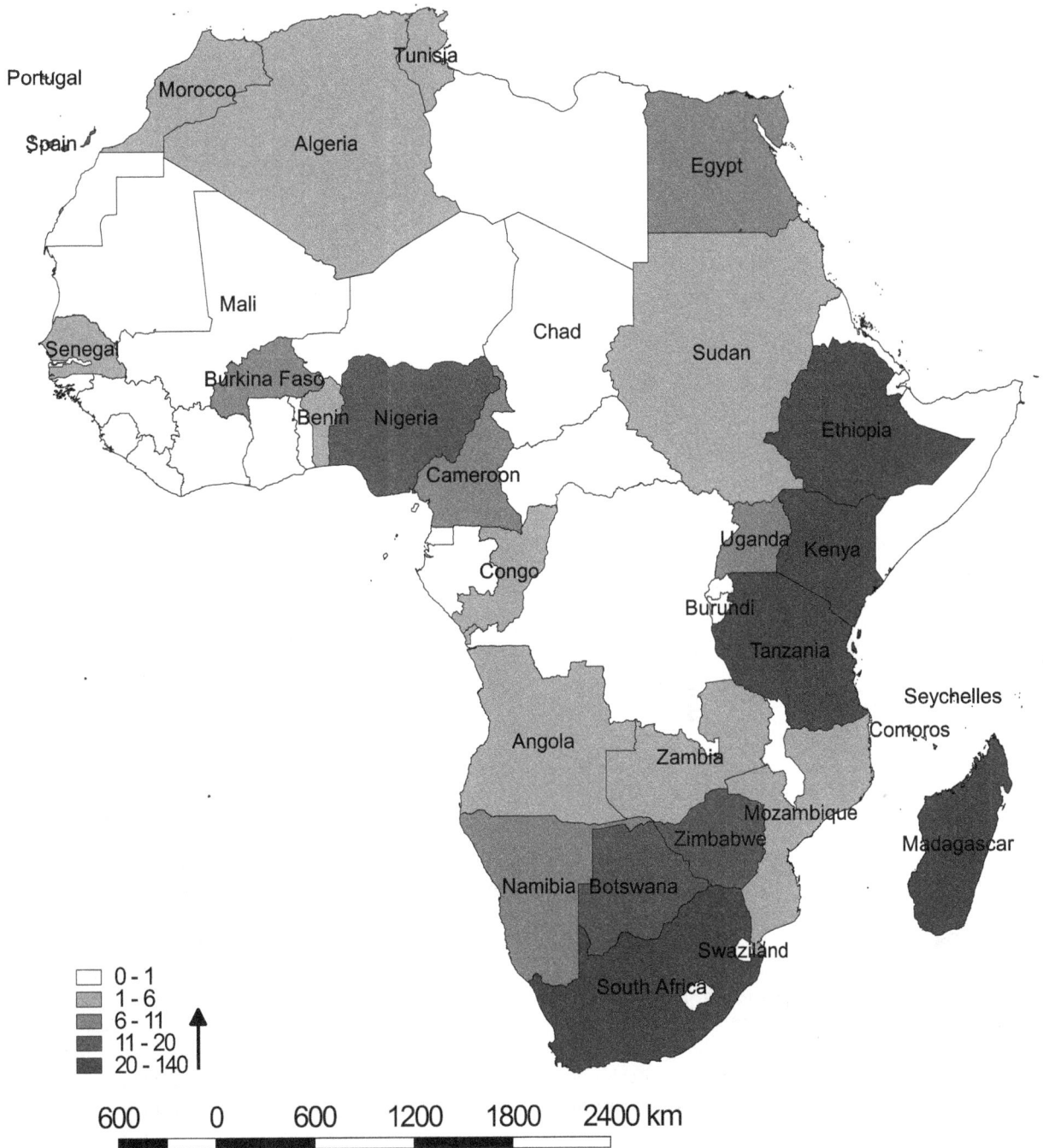

Figure 3. Number of peer-reviewed publications integrating remote sensing and conservation in Africa (meeting all the search criteria) produced by authors with addresses in Africa.

(see increase in papers post 2007 in Fig. 1). This allowed RS specialists to move on to image interpretation, including the evaluation of various fire products for use in specific vegetation contexts (de Klerk 2008; de Klerk et al. 2011), investigating the use of RS classification approaches (pixel-based and object-based) and

classification algorithms to solve various conservation questions, such as the mapping of low-density or isolated invasive alien plants and naturally fragmented and isolated environmental features (de Klerk et al. in press).

Interviewees indicated that there remained certain applications that require spatial resolutions of 1–5 m,

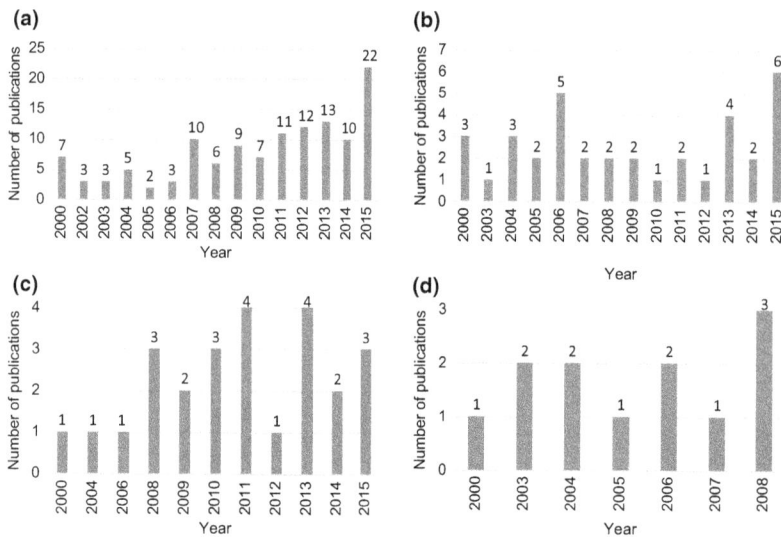

Figure 4. Trends in the number of peer-reviewed publications integrating remote sensing and conservation (that meet all the search criteria) with authors that have correspondence addresses in (A) South Africa (127 in total), (B) Kenya (35 in total), (C) Tanzania (26 in total) and (D) Ethiopia (24 in total).

Figure 5. The current use of remote sensing data in conservation organizations based on keywords.

such as mapping bracken fern invasion in Nyika Plateau, where SPOT 6 or even finer-scale World View imagery is preferred. These datasets remain unaffordable for African conservation agencies and can only be utilized through partnerships with European agencies or institutions. As such they are not suitable for operational monitoring on a regular basis. However, images down to approximately 10 m, including radar data will be made available through the Copernicus program, making progress toward the operationalization of these analyses.

Full and proper interrogation of images requires that first they are pre-processed. The survey respondents who were using these data in the early 2000's had to apply time-consuming pre-processing for atmospheric correction, radiometric correction, orthorectification and pan sharpening, reprojecting and mosaicing before they were able to use images or distribute them to their field-based colleagues. Data are now often supplied with these corrections having been applied (i.e. level 3 processing). These images can now be used off the shelf, reducing the amount of time dedicated to the pre-processing of

images, although this is still required for more advanced application of RS imagery for specific classification exercises, which then requires access to licenses of expensive software. The Biodiversity and Protected Areas Management Program's (BIOPAMA, http://rris.biopama.org/) Regional Reference Information System RRIS provides a solution to the preprocessing of Landsat imagery and outputs classified land parcels (segments) which is useful for landscape or habitat monitoring (Szantoi et al. 2016). Most of the advanced software is generally priced beyond the budget of African conservation organizations. The majority of those who responded noted that they used QGIS, a free GIS package with image processing capabilities. While GRASS (https://grass.osgeo.org/), Python (https://www.python.org) and R (https://www.r-project.org/) provide some future potential, the current workforce is not highly conversant with these Open Source software and programming languages. This situation could change as individuals are taught how to use these packages as increasing numbers of universities train students in R and other Open Source software. However, these skills are not being taught by all universities. Development of training material targeted at those who want to use these freeware packages, including code (e.g. Wegmann et al. 2016), is a route to develop these skills.

In terms of future data and products needs expressed by conservation agencies in Africa, the key needs are mapping the extent, and changes in the extent, of habitats (~land cover), assessing the effectiveness of conservation activities, mapping ecosystem services and developing capacity. We have already considered the latter above. BIOPAMA RRIS addresses changes in six to eight land class types (Szantoi et al. 2016). Mapping land cover encompasses many attributes, including mapping land cover disturbances, invasive species and agricultural encroachment and mapping the impacts of domestic grazing. These largely match previous assessments of the needs of the conservation community primarily based on surveys of conservation communities in North America and Europe (Rose et al. 2015; Green et al. 2011), as well as assessments that included the needs of users outside these areas (e.g. Buchanan et al. 2015).

RS training by African institutions

Of the 360 universities in Africa (Association of African Universities AAU; http://www.aau.org/membership/), we surveyed: (1) at least two to four institutions per country; (2) whose program details were available online (and working at the time of the study); (3) whose program information was available in English (we acknowledge that this is a limitation in Arabic, French and Portuguese speaking countries); (4) by preference including the larger institutions within a country, in the capital or major cities and (5) who offered programs that focused on conservation and environment, in terms of biodiversity (rather than conservation of soils and farming practices, such as agro-biodiversity conservation practices). A number of institutions surveyed did not offer conservation or environmental programs and focused on business and/or technology. Determination of the depth of the RS teaching in a program was difficult, but at the minimum we accepted the description of a remote sensing course that taught basic georeferencing and orthorectification; collection of control points and warping and conversion between vector and raster as acceptable (e.g. see Table 2). In some cases this was hard to define and we erred on the "generous" side where additional information warranted the decision. For instance, Egerton in Kenya offers four programs that are relevant to environmental management. The website (http://www.egerton.ac.ke/index.php/Faculty-of-Environment-and-Resources-Development/faculty-of-environment-and-resources-development.html) does not provide detailed course information for the RS and GIS components, but the lecturer responsible for RS and GIS into these programs, Dr. George Eshiamwata, provided more insights and his webpage indicates his RS and GIS skills. Where a university offers more than one conservation/environmental program, we have listed each separately, as some might include RS and GIS and others not.

Of the 72 university programs surveyed across Africa (Fig. 6), 47 either did not offer environmental or conservation programs or had no course material online to evaluate whether RS teaching was integrated into the program. Seventeen did integrate RS teaching into their environmental or conservation programs, while nine offering environmental or conservation programs did not include RS teaching. A number of those that did integrate RS into environmental or conservation programs, did so as contextual learning, that is, RS teaching was tailored to the conservation students. Examples include the University of the Western Cape's BSc Biodiversity and Conservation Biology and Kenyatta University's Environmental Studies (Resource Conservation). These are both undergraduate programs. Similar approaches are used in post-graduate programs, such as the MPhil Environmental Science at the University of Ghana and the MSc Conservation Biology at the University of Cape Town, FitzPatrick Institute. Frequently these RS and GIS courses are comprehensive, such as the example from Egerton in Kenya where the course "ENSC 404: Environmental Information System" offered in four of the environmental degrees covers "concepts and foundation of geo-informatics; remote sensing; photographic systems, thermal and multi-spectral scanning and image processing; components and applications of an environmental information system (EIS); characteristics of spatial data; models of spatial information; spatial relationships and algorithms; spatial analysis (such as route planning, map overlay, buffer zoning, etc.); database models for spatial data; errors in spatial data; sources of raster spatial data; sources of vector spatial data; ethical issues and spatial data; cartographic communication – the display of spatial data; coordinate systems and map projections; and Mobile EIS (location based services, combination with positioning, e.g. GPS, Galileo)" (Dr. George Eshiamwata pers. comm.). The University of Rwanda has even created a degree that acknowledges how crucial RS and GIS is to environmental management in its name; the "MSc in Geo-Information of Environmental and Sustainable Development" where "the module aims at equipping students with methods for formulating predictive models and applies them in environmental modelling using GIS and RS data and tools" (Dr. Emmanuel Havugimana pers. comm.). In other cases, descriptions are brief and hard to evaluate, such as that of Masters of Geography at the University of Liberia, which simply describes course GEOG 410 as "Air Photo Interpretation; Spatial Organization" (http://www.tlcafrica.com/lu/ul_course_master_list_geography.htm).

Other conservation programs utilized 'service' modules from Geomatics (or GeoInformatics) Departments within the same institution, such as the MSc Environmental Protection and Management at The Chinhoyi University of Technology, Malawi, The Polytechnic and Stellenbosch University's Department of Conservation and Ecology BSc in Conservation Ecology.

Encouragingly, there was evidence that integration of RS teaching into conservation programs can be achieved

Table 2. Example of academic programs for conservation and ecology students in Africa that include remote sensing teaching.

Academic program	Remote sensing teaching	GIS teaching	Description or comment	Source
University of the Western Cape: BSc Biodiversity and Conservation Biology	Y	Y	RS: Biodiversity Information Management:; Classification of satellite imagery (supervised, unsupervised and object-based approaches); General internet mapping; Geographical Information Systems (GIS) and remote sensing GIS: Mapping using a Global Positioning System (GPS) and analysis of these data; Digitizing using online resources; Geostatistics and spatial interpolations for modeling point data; Use of spatial data to develop species distribution data and to define meta-populations and identify species with conservation-critical distributions; Development of principles of a biodiversity/taxonomic data base.	https://www.uwc.ac.za/Faculties/NS/ Biodiversity_Conservation_Biology/ Pages/Programmes.aspx
Kenyatta University: Environmental Studies (Resource Conservation)	Y	Y	RS: Remote Sensing for Environmental Sciences; GIS: Natural resource mapping and cartography: Introduction to basic concepts and applications of geographic information systems. Spatial analysis systems; applications of GIS technology to natural resource systems; spatial analysis systems; applications of GIS on natural resource systems; spatial data entry; data compilation; and map output; basic concepts on cartography; Concepts and foundations; energy interaction in the atmosphere and with earth features; data acquisition; photographic and electronic data types; platforms and sensors; data processing; film spectral sensitivity and processing – black and white color films; color concepts; fundamentals of aerial photography, imagery interpretations and mapping; earth resource satellite and sensors (Landsat, SPOT, NOAA, Radarsat and others); applications in environmental monitoring and management	http://www.ku.ac.ke/schools/ environmental/index.php/progra mmes/undergraduate-programmes/ 91-programmes/308-bachelor-of-environmental-studies-resource-conservation
Egerton University: BSc in Natural Resources	Y	Y	One GIS and Remote sensing is a Unit = Concepts and foundation of geo-informatics; Remote sensing; Photographic systems, thermal and multi-spectral scanning and image processing; Components and applications of a Environmental information system (EIS); Characteristics of spatial data; Models of spatial information; Spatial relationships and algorithms; Spatial analysis (such as route planning, map overlay, buffer zoning, etc.); Database models for spatial data; Errors in spatial data; Sources of raster spatial data and introduction to	George Eshiamwata (gweshiamwata@gmail.com); http://www.egerton.ac.ke/index. php/Natural-Resources/dr-george-w-eshiamwata-phd.html; http:// www.egerton.ac.ke/index.php/ Faculty-of-Environment-and-Resources-Development/faculty-of-environment-and-resources-development.html

(Continued)

Table 2. Continued.

Academic program	Remote sensing teaching	GIS teaching	Description or comment	Source
			remote sensing; Sources of vector spatial data; Ethical issues and spatial data; Cartographic communication – the display of spatial data. Coordinate systems and map projections; Remote sensing, Geo-DBMS (spatial ADT's, spatial indexing, etc.), Mobile EIS (location-based services, combination with positioning, e.g. GPS, Galileo); Examination of remote sensing and EIS applications in agriculture; Conventions and policy issues; Computer models; Laboratory assignments.	

even in smaller institutions, such as the MSc Environmental Protection and Management at the Chinhoyi University of Technology, Malawi (see the course entitled "Geographical Information Systems & Remote Sensing" at http://www.poly.ac.mw/index.php/poly/msc_environ_protection). The delivery of this course by a small university should be seen as an indication of what can be achieved.

One lecturer mentioned that the costs of proprietary software place limitations on the number of students they can train. A number of institutions utilize or rely heavily upon Open Source software such as QGIS and GRASS. Open Source software is being taught in other institutions due to the growing demand of employers for trained professionals able to use Open Source software, as some government and conservation agencies are also battling with licensing costs of proprietary software. This is a trend seen in many parts of Africa.

RS capacity development by ad hoc events

Training of specialist RS and GIS staff in conservation organizations in Africa was initially through self-training, on-the-job learning and occasional workshops and course work. We are aware of a number of initiatives that have resulted in RS training being provided to the conservation community in Africa. These include the ESRI Society for Conservation GIS conferences, a series of training events by Conservation International for mapping forest loss, multiple courses by the European Commission's Joint Research Centre and the workshops run by the Cooperative Institute for Meteorological Satellite Studies (CIMSS; https://cimss.ssec.wisc.edu/; accessed 8 July 2016) and Space Science and Engineering Center (SSEC; http://www.ssec.wisc.edu/) at the University of

Wisconsin-Madison presented in southern Africa. These ad hoc training events have played an important role in training specialist RS staff, although younger employees have been through formal academic RS programs. SSEC has an active education and outreach policy and program with professional development programs offered around the world (see http://cimss.ssec.wisc.edu/rss/; accessed 8 July 2016). We are aware that one work unit in the European Commission's Joint Research Centre (EC-JRC) has run four courses alone in the last 2 years. These have attracted a range of attendees from academic, governmental and park management, and the courses focused on mapping land cover change using a free tool (Szantoi et al. 2016). This one unit has also delivered training in Monitoring for Environment and Security in Africa (MESA; http://rea.au.int/mesa/) and specific conservation training with the eStation (Clerici et al. 2013) in the Intergovernmental Authority on Development (IGAD). ESRI Society for Conservation GIS conferences provides exposure to technology trends and networking with different types of industries such as military and urban planners. NASA has an in-depth introduction to using remote sensing on their web site (ARSET). Due to the disparate nature of these courses, it is difficult to assess how wide an outreach they have. Indeed, it is mainly through our connections with those running these programs as well as individuals who have attended them that we are aware of them so it is likely that there are many more that we are not aware of. We are unaware of a central place where all of these courses and training options are documented.

Discussion

That remote sensing can make a contribution to conservation monitoring is well established, and is underlined

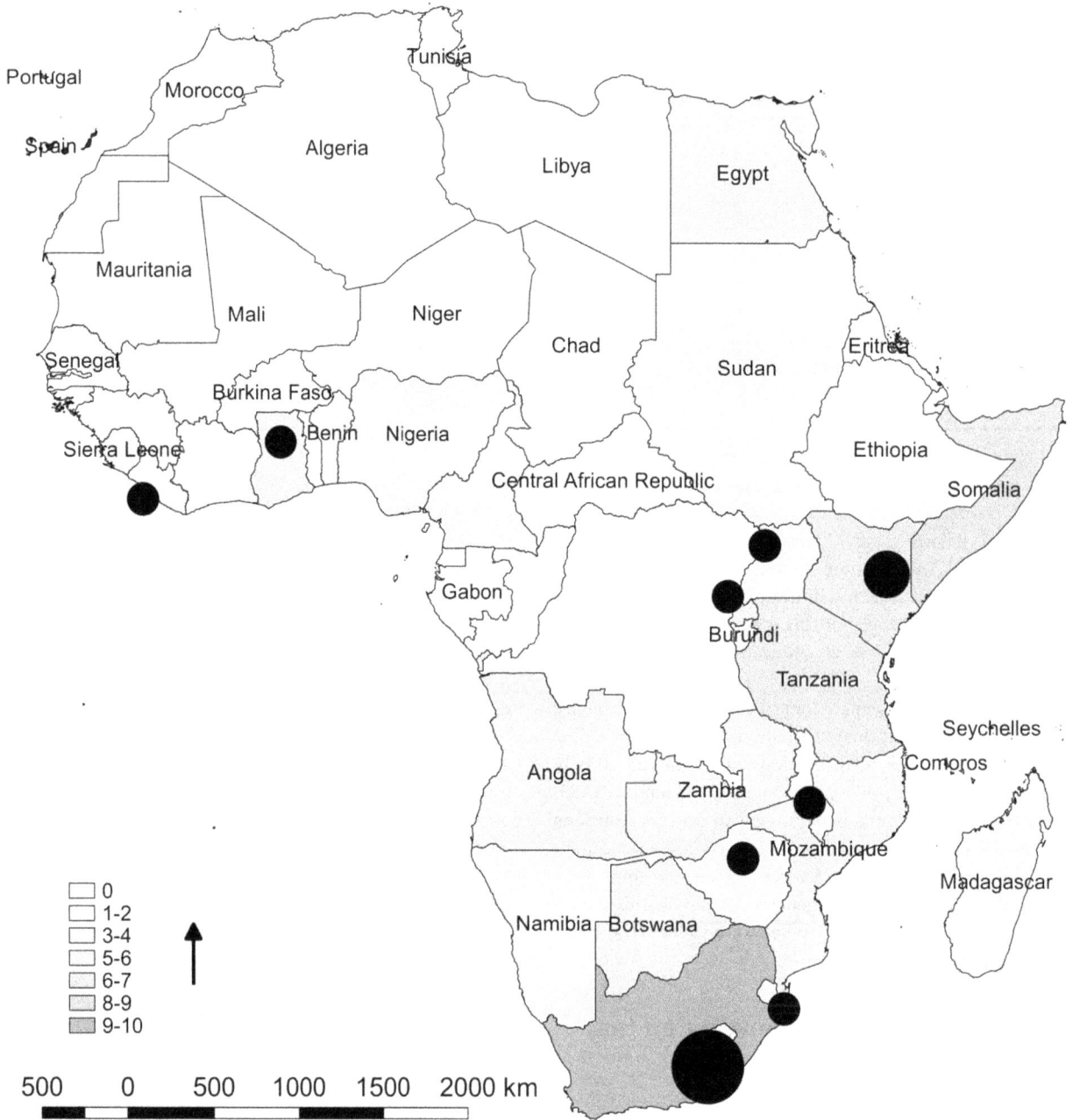

Figure 6. The number of academic institutions registered with the Association of African Universities surveyed per country (grey shades) and the number that have RS teaching included in a conservation or environmental program (size of circles).

here by the increasing number of studies that have applied these data in Africa. Increased capacity within the conservation community in African countries would be most beneficial as it would develop the skills closest to point of use. Strong links between those doing the analysis and those managing the resources, whether a park or a species, are likely to result in the best outcomes for conservation. However, the majority of studies published in the peer review literature have been led by authors or institutions based outside the continent. We cannot assess the degree to which these studies developed capacity within Africa, but given the relatively low level of interaction generated from the subsequent interviews with the conservation community, it is possible that in many cases the majority of analysis was undertaken outside Africa. Undertaking image analysis within Africa, or at least training people to do so, could increase the capacity to do analysis locally, resulting in both better skills and a shorter distance between analysis and use. Given that there is a feeling in the conservation community that not all analysis using remote sensing are answering the questions that the conservation community need answered, we suggest that having the skills closer to the point of use should result in the analysis being targeted toward the most pressing needs.

There is evidence that the capacity to undertake such studies within Africa is increasing. South Africa has increased the number of published articles that have used remote sensing for conservation between 2000 and 2015. The increase in just one country, however, might indicate that there remains a gap in skills and barriers to growth in post-graduate teaching programs and facilities. International funding could be used to develop these programs and facilities, enabling African countries to undertake a larger proportion of their own research. The target audience for research might also be responsible for the spatial disparity in output – conservation and research organizations based outside of Africa might have agendas which differ from national organizations within Africa, resulting in the difference in the location from which studies were undertaken. The analysis described was based on just the academic output. As such, it will not capture the use of remote sensing data for ongoing monitoring or in cases where studies have not been written up or published. Consequently, it may present a slightly skewed picture.

We considered the responses from conservation staff in Africa to identify what the opportunities and limitations on uptake of remote sensing were. The free data policy for the Landsat program, Landsat 8 in particular, and MODIS have impacted positively on use of remote sensing in conservation organizations in Africa. And most probably on academic research too. The importance of free observation data to all users has been recognized (Turner et al. 2003), and this applies as much (if not more) in Africa as in other parts of the world. Encouragingly, other providers are increasing access to free data. The Copernicus program of European Space Agency (ESA), through which Sentinel and other data are already, and will continue to be, available, and the continuing and NASA programs, will feed a significant portion of this need. However, high-resolution data are not always available through these initiatives, and some respondents indicated that the inclusion of, for example, SPOT 6 data would be very useful as the costs of these images are still prohibitive and are only accessed through partnership projects or very localized funded studies. But as Landsat resolution data are perhaps the most appropriate for mapping land cover change (Mayaux et al. 2005;

Wegmann et al. 2016), the availability of these data at around 30 m resolution continues to be an important step forward. However, it is important to communicate with those in Africa that these data are available. Recent experience has indicated that not all conservationists who use remote sensing data are aware that Landsat images are now free, and have not been taking advantage of these data. It is essential that these programs undertake outreach to ensure that as many potential users as possible are aware that these data are available and are free.

Initiatives do exist through which African countries are trying to improve and maintain access to data. Notable is AfriGEOSS. This describes itself as an "initiative of the intergovernmental Group on Earth Observations (GEO), [which] was formed in 2014 to coordinate access to and use of Earth observations – from satellites, airborne and ground- and marine-based systems – across the African continent. The 27 AfriGEOSS members are focusing their efforts on data access and dissemination, forest management, food security, urban planning and water resources management, as well as contributing to achieving the 2030 Sustainable Development Agenda in Africa." The Regional Visualization and Monitoring System, SERVIR, is another, aimed at improving Geo-Information access to improve sustainable development with 20 contracting member states (https://servirglobal.net/Regions/ESAfrica). The South African National Space Agency (SANSA, http://www.sansa.org.za/) serves as a repository of imagery that is also available from the original suppliers, such as Landsat 8 and historical SPOT 5 imagery. It provides a download station for MODIS data, but these can also be accessed from USGS.

In addition to these initiatives, many images are now also available rapidly in Google Earth Engine (GEE), an online environment for the analysis of satellite images and other spatial data. This platform, which is currently free, allows users to access and process images in a cloud environment. Consequently, the need for expensive desk top or cluster processers is obviated. GEE users can write code in multiple languages including Python and JavaScript, which while allowing for a high degree of flexibility in processing and automation of processes, might be a block for casual users. But development of skills in these languages could pay big dividends. As this platform allows users to analyze images in the cloud, the need for good internet connections to download images is also obviated. This is an online, cloud-based platform. Consequently, internet connectivity is required to run processes and extract results. The current internet stability and speed in many African countries might mean that this potentially useful tool is, unfortunately, out of reach. The improvement of connectivity would therefore be of great value in increasing the accessibility of this tool.

It would appear that, theoretically at least, access to data will be less of an obstacle in the future than it has been. The open data policies of NASA and ESA should be welcomed. However, the issue of downloading and processing these data remains for many. Internet speeds in many parts of Africa, including within many institutions surveyed, can be slow. This might become an issue when multiple images are required, something that might become an important issue as images are captured with a greater frequency. In addition, historically, the use and manipulation of images required software which can be expensive. Open Source software such as QGIS, and programming languages such as Python, and R have helped with the visualization and manipulation of imagery, although data still need to be downloaded. Advances in R, together with better access to appropriate code (Wegmann et al. 2016) should increase the uptake of R and Python for image analysis as, theoretically, there are no obstacles to undertaking advanced analysis using these programming languages. As already noted, Google Earth Engine requires users to have an internet connection to use the interface. In addition to user processing, GEE also supports data produced by others, such as the global forest loss map (Hansen et al. 2013). Access to these data, and future data (e.g. water bodies, Pekel et al. 2014), will reduce the need for individuals to undertake their own classifications, making detection and quantification of land cover change much easier. Some of these data, with the addition of alerts for forest loss, are also available online through Global Forest Watch. Users of GEE can use code developed by others, and placed online (e.g. through GitHub). These data have already been used for global conservation purposes (e.g. Tracewski et al. 2016).

Countries are recognizing that access to data is key to empowering society to participate in decision making in various arena's, including conservation and town and regional planning; promotes transparency and accountability, public management and policy of government agencies (Davies 2014). South Africa has legislated the public's right to access data through the Promotion of Access to Information Act, Act 2 of 2000 (Republic of South Africa 2000). This is mostly facilitated through data portals (e.g. https://web1.capetown.gov.za/web1/OpenData Portal/; accessed 24 June 2016). Where internet access is not available, or insufficient for downloading large files, or only intermittent, plans are being made to distribute key datasets via DVD at public facilities (or nodes) such as libraries. This will sound archaic to some, but it is a pragmatic solution in many situations and previous iterations of the distribution of town planning products as CD or DVD for public use were successful as the public engaged with these products (e.g. Maree and Vromans's

2010 biodiversity sector plan for the Witzenberg, Breede Valley and Langeberg Municipalities compiled DVDs that provided land-use planners with data layers on land use, conservation priorities, infrastructure and potential conservation corridors). Older initiatives include the Aster Spectral Library (version .2) distributed on CD in 1998 by the California Institute of Technology; The Water Resources eAtlas: watersheds of the world CD distributed by IUCN, WRI, Ramsar and IWMI; and the Earth science reference handbook and data products handbooks CD distributed by NASA in 2006. The supply of off-line versions of the web-based IMPACT (Szantoi et al. 2016; The satellite IMagery ProCessing Toolbox) and eStation+, which uses a parabolic antenna (Clerici et al. 2013), continues the duel online/off-line to cater for different internet capabilities.

Many African academic institutions have recognized the need for young conservation professionals to be equipped with remote sensing and GIS tools and either offer tailor-made remote sensing and GIS courses into their conservation programs (both at undergraduate and post-graduate levels), or make use of 'service modules' presented by other departments, typically Geography or Engineering departments, to provide RS and GIS teaching into the conservation program. Many of the RS teaching offerings are substantive and cover a wide range of important topics from image registration and atmospheric correction to classification and modeling. Retention of individuals with these skills within conservation might be an issue, but the greater availability of this type of training to more people will potentially reduce the flow of skilled individuals from conservation jobs. In addition to the formal university education we also documented a range of courses and workshops that are run by a diverse range of parties. These span those involved in conservation through multinational governmental organizations to software companies. The number of these courses and workshops that have been run remains unknown, but it is likely that there are many more than those which we know about as we have based our description on personal knowledge. The establishment of a central register of these courses could prove useful for those interested in these topics. It is understandable that in many cases they are targeted to specific end users, but a wider knowledge of the type of workshops being run could not only improve access to courses through demand but also potentially reduce overlap and make for a more efficient use of resources.

New and emerging technology will also play a part in the increased uptake of the use of remote sensing data. For example, PeaceParks's Sustainable and Safe Environmental Travels (SENSA) is being piloted in Kgalagadi Transfrontier Park. The app can facilitate tourist trip planning, safety and tourist management. The tourist user can plan their trip and upload the places they will stay and routes they will use onto their cellphone or android device. There are communications at the TweeRivieren check-in site where they hire a Rock Star unit that connects to the Iridium satellite constellation and to the android device via Bluetooth. This enables communication between the tourist and park staff about safety issues, such as a flat tyre; if the tourist are going to run late for the camp closing time, and compliance issues, such as if the tourist drives off road, which are picked up with a geofence on the unit. Visitors are also provided with information of what vegetation type they are traveling through and heat maps of species sightings over last 60 days. PeaceParks provides awareness of the tool and assists with deployment in campsites/hotspot hubs to support satellites, base mapping and training on use and deployment of units. Development of similar systems that allow local conservation groups to report incidents of degradation to sites they visit to a central system could be developed in a similar way, especially where mobile phone network coverage is extensive (e.g. Kenya). In return, data from remote sensing alert systems (e.g. fire alerts) could be sent to local stakeholders for their near real-time investigation. Thus, a two-way flow of information could be developed.

In summary, the uptake of remote sensing in conservation is increasing, and while African countries are leading on some of this expansion, the majority are either not participating or not expanding output. But there are a number of universities that are delivering some component of remote sensing and GIS in environmental conservation courses. Ideally the impact of these courses will, over time, be increased use of remote sensing in conservation, and possibly published scientific papers. With the increase in freely available remote sensing data, it is now perhaps access to suitable internet to download images, and the cost of software which are the major obstacles. Free software or online cloud processing platforms might ultimately be what liberates the use of remote sensing in African conservation.

Acknowledgements

Marie Theron of Stellenbosch University Library and Information Service is thanked for help with the bibliometrics analysis. Andrew Turner and Therese Forsyth of CapeNature, Craig Beech of PeaceParks Foundation, Faustin Gashakamba of the Albertine Rift Conservation Society, Yilma Abebe of the Ethiopian Wildlife and Natural History Society helped with user surveys. Dr. George Eshiamwata helped with the survey and provided background information. Two anonymous Reviewers and the Editor are thanked for comments that improved this manuscript. This project received ethical clearance from Stellenbosch University's Departmental Ethical Screening Committee (SU-HSD- 003362).

References

de Araujo Barbosa, C. C., P. M. Atkinson, and J. A. Dearing. 2015. Remote sensing of ecosystem services: a systematic review. *Ecol. Ind.* **52**, 430–443. Available at: http://linkinghub.elsevier.com/retrieve/pii/S1470160X15000084 (Accessed: 13 December 2016).

Buchanan, G. M., A. Nelson, P. Mayaux, A. Hartley, and P. F. Donald. 2009. Delivering a global, terrestrial, biodiversity observation system through remote sensing. *Conserv. Biol.* **23**, 499–502. doi:10.1111/j.1523-1739.2008.01083.x.

Buchanan, G. M., G. W. Eshiamwata, and P. F. Donald. 2011. Using satellite imagery for African bird conservation. *Bulletin of the African Bird Club* **18**, 68–73.

Buchanan, G. M., A. B. Brink, A. K. Leidner, R. Rose, and M. Wegmann. 2015. Advancing terrestrial conservation through remote sensing. *Ecol. Inform.* **30**, 318–321. doi:10.1016/j.ecoinf.2015.05.005.

Clerici, M. B., J. F. Combal, G. Pekel, J. Dubois, J. O. van't Klooster, E. Skøien, et al. 2013. The eStation, an Earth Observation processing service in support to ecological monitoring. *Ecol. Inform.* **18**, 162–170. doi:10.1016/j.ecoinf. 2013.08.004.

Davies, T. G. 2014. Open data policies and practice: an international comparison. social science research network [Online Help]. Available at: http://papers.ssrn.com/sol3/papers.cfm?abstract_id=2492520 (Accessed 06 June 2016).

Green, R. E., G. M. Buchanan, and R. Almond. 2011. Cambridge Conservation Initiative Report, Cambridge, UK. Avalaible at: http,//www.conservation.cam.ac.uk/resource/working-papers-and-reports/ccireport-what-do-conservation-practitioners-want-remote.(Accessed: 13 December 2016).

Hansen, M. C., P. V. Potapov, R. Moore, M. Hancher, S. A. Turubanova, and A. Tyukavina. 2013. High-resolution global maps of forest cover change. *Science* **342**, 850–853. doi:10.1126/science.1244693.

de Klerk, H. 2008. A pragmatic assessment of the usefulness of the MODIS (Terra and Aqua) 1-km active fire (MOD14A2 and MYD14A2) products for mapping fires in the fynbos biome. *Int. J. Wildl. Fire* **17**, 166–178.

de Klerk, H. M., A. Wilson, and K. Steenkamp. 2011. Evaluation of satellite-derived burned area products for the fynbos, a Mediterranean shrubland. *Int. J. Wildl. Fire* **21**, 36–47. doi:10.1071/WF11002.

Lück-Vogel, M., P. J. O'Farrell, and W. Roberts. 2013. Remote sensing based ecosystem state assessment in the Sandveld Region, South Africa. *Ecol. Ind.* **33**, 60–70. doi:10.1016/j.ecolind.2012.11.007.

Maree, K., and D. Vromans. 2010. The Biodiversity Sector Plan for the Witzenberg, Breede Valley and Langeberg Municipalities: Supporting land-use planning and decision-making in Critical Biodiversity Areas and Ecological Support Areas. Produced by CapeNature as part of the C.A.P.E. Fine-scale Biodiversity Planning Project. Kirstenbosch.

Mayaux, P., P. Holmgren, F. Achard, H. Eva, H.-J. Stibig, and A. Branthomme. 2005. Tropical forest cover change in the 1990s and options for future monitoring. *Philos. Trans. Royal Society B* **360**, 373–384.

O'Connor, B., C. Secades, J. Penner, R. Sonnenschein, A. Skidmore, N. D. Burgess, et al. 2015. Earth observation as a tool for tracking progress towards the Aichi Biodiversity Targets. *Remote Sens. Ecol. Conserv.* **1**, 19–28.Avaliable at: http://doi.wiley.com/10.1002/rse2.4. (Accessed: 13 December 2016).

Pekel, J. F., C. Vancutsem, L. Bastin, and P. Defourny. 2014. A near real-time water surface detection method based on HSV transformation of MODIS multi-Spectral time series data. *Remote Sens. Environ.* **140**, 704–716. doi:10.1016/j.rse. 2013.10.008.

Pettorelli, N., K. Safi, and W. Turner. 2014. Satellite remote sensing, biodiversity research and conservation of the future. *Philos. Trans. Royal Soc. B: Biol. Sci.*, **369**: 20130190. Available at : http://rstb.royalsocietypublishing.org/cgi/doi/ 10.1098/rstb.2013.0190. (Accessed: 13 December 2016)

Piel, A. K., N. Cohen, S. Kamenya, S. A. Ndimuligo, L. Pintea, and F. Stewart. 2015. Population status of chimpanzees in the Masito-Ugalla Ecosystem, Tanzania. *Am. J. Primatol.* **77**, 1027–1035. doi:10.1002/ajp.22438.

Republic of South Africa. 2000. Promotion of Access to Information Act, No. 2 of 2000.

Rose, R. A., D. Byler, J. R. Eastman, E. Fleishman, G. Geller, S. Goetz, et al. 2015. Ten ways remote sensing can contribute to conservation. *Conserv. Biol.* **29**, 350–359. doi:10.1111/ cobi.12397.

Stuckenberg, T., Z. Münch, and A. V. Niekerk. 2012. Multi-temporal remote sensing land-cover change detection as tool for biodiversity conservation in the Berg River catchment. *GISSA Ukubuzana* **2**, 189–205.

Szantoi, Z., A. Brink, G. Buchanan, L. Bastin, A. Lupi, D. Simonetti, et al. 2016. A simple remote sensing based information system for monitoring sites of conservation importance. *Remote Sens. Ecol. Conserv.* **2**, 16–24. doi:10. 1002/rse2.14.

Tracewski, L., S. H. M. Butchart, P. F. Donald, M. Evans, L. D. C. Fishpool, and G. M. Buchanan. 2016. Patterns of twenty-first century forest loss across a global network of important sites for biodiversity. *Remote Sens. Ecol. Conserv.* **2**, 37–44. Avaiable at: http://doi.wiley.com/10.1002/rse2. 13.(Accessed: 13 December 2016)

Turner, W. 2014. Sensing biodiversity. *Science* **346**, 301–302. doi:10.1126/science.1256014.

Turner, W., S. Spector, N. Gardiner, M. Fladeland, E. Sterling, and M. Steininger. 2003. Remote sensing for biodiversity science and conservation. *Trends Ecology Evol.* **18**, 306–314.

Verhulp, J., and A. Van Niekerk. 2016. Effect of inter-image spectral variation on land cover separability in heterogeneous areas. *Int. J. Remote Sens.* **37**, 1639. Avaliable at: http://dx.doi.org/10.1080/01431161.2016.1160300 (Accessed: 13 December 2016).

Wallin, D. O., C. C. H. Elliott, H. H. Shugart, C. J. Tucker, and F. Wilhelmi. 1992. Satellite remote sensing of breeding habitat for an African Weaver-bird. *Landscape Ecol.* **7**, 87–99.

Wegmann, M., B. Leutner, and S. Dech. 2016. *Remote Sensing and GIS for Ecologists: using Open Source Software.* Pelagic Publishing, Exeter.

Wicht, C. L. 1945. *Preservation of vegetation of the SouthWestern Cape.* Special Publication of the Royal Society of South Africa, Cape Town.

8

Building capacity in remote sensing for conservation: present and future challenges

Ilaria Palumbo[1], Robert A. Rose[2], Rachel M. K. Headley[3], Janet Nackoney[4], Anthony Vodacek[5] & Martin Wegmann[6]

[1]European Commission – Joint Research Centre, Institute for Environment and Sustainability, 21027 Ispra (VA), Italy
[2]Center for Geospatial Analysis, College of William & Mary, Williamsburg, Virginia 23185
[3]Cobblestone Science, Spearfish, South Dakota 57783
[4]Department of Geographical Sciences, University of Maryland, College Park, Maryland 20742
[5]Rochester Institute of Technology, Chester F. Carlson Center for Imaging Science, Rochester, New York 14623
[6]Department of Remote Sensing, Remote Sensing and Biodiversity Research, Institute of Geography, University of Wuerzburg, Oswald Külpe Weg 86, 97074 Wuerzburg, Germany

Keywords
Capacity building, conservation, open-source software, remote sensing, survey, training

Correspondence
Ilaria Palumbo, European Commission – Joint Research Centre, Via E. Fermi, Ispra (VA) 21027, TP 261, Italy. E-mails: ilaria.palumbo@ext.jrc.ec.europa.eu; ilaria.palumbo@remote-sensing-conservation.org

Funding Information
This work was funded by the European Commission – Joint Research Centre.

Editor: Harini Nagendra
Associate Editor: Duccio Rocchini

Abstract

Remote sensing (RS) has made significant contributions to conservation and ecology; however, direct use of RS-based information for conservation decision making is currently very limited. In this paper, we discuss the reasons and challenges associated with using RS technology by conservationists and suggest how training in RS for conservationists can be improved. We present the results from a survey organized by the Conservation Remote Sensing Network to understand the RS expertise and training needs of various categories of professionals involved in conservation research and implementation. The results of the survey highlight the main gaps and priorities in the current RS data and technology among conservation practitioners from academia, institutions, NGOs and industry. We suggest training to be focused around conservation questions that can be addressed using RS-derived information rather than training pure RS methods which are beyond the interest of conservation practitioners. We highlight the importance of developing essential biodiversity variables (EBVs) and how this can be achieved by increasing the RS capacity of the conservation community. Moreover, we suggest that open-source software is adopted more widely in the training modules to facilitate access to RS data and products in developing countries, and that online platforms providing mapping tools should also be more widely distributed. We believe that improved RS capacity among conservation scientists will be essential to improve conservation efforts on the ground and will make the conservation community a key player in the definition of future RS-based products that serve conservation and ecological needs.

Introduction

Conservation managers, practitioners and policymakers increasingly rely on geospatial information and analysis to assess habitat status and pressures, understand species distribution and vulnerability, monitor external threats and more effectively plan conservation action and response. As a consequence, geographic information systems (GIS) have been applied widely for conservation prioritization and planning as it is well documented in the biological sciences literature (Myers et al. 2000; Brooks et al. 2006). Spatially explicit, systematic conservation planning (Margules and Pressey 2000; Groves et al. 2002; Pressey and Bottrill 2008) has been extensively used to set conservation priorities, assessing measures of anthropological threat and biological significance, and identify and map locations where conservation actions are needed (Wilson et al. 2007; Trombulak and Baldwin 2010).

Increasingly, a large amount of geospatial information is derived from satellite and aerial image processing and analysis, also known as remote sensing (RS), and these data hold tremendous potential for conservation application (Turner et al. 2003; Buchanan et al. 2008; Rose et al. 2015). Today, access to remotely sensed data has been vastly improved, and many aerial and satellite data are freely available [e.g. moderate resolution imaging spectro-radiometer (MODIS), Landsat and Sentinel]. Data providers are increasingly able to serve imagery that is already pre-processed, thereby eliminating much of the work required to prepare the data for analysis. Moreover, increased technological capability of some personal computer systems allows users to better manage and analyse RS data. However, direct use of RS-based information for conservation decision making remains limited. Reasons for this might include (1) remotely sensed data are large in volume, more complex than standard GIS data and, therefore, require more advanced systems with high storage and processing capacity; (2) the acquisition of satellite data has historically been too costly for most organizations and institutions to afford and (3) as a consequence, conservation organizations and institutions have not invested in building capacity for RS in terms of personnel and technological requirements, and this has limited further the use of remotely sensed datasets.

As a result, many conservation-based organizations and institutions have developed capacity catered more to building GIS rather than RS expertise. Furthermore, RS training opportunities for conservation professionals, that are aligned to conservation applications, are not widely offered; yet, an example of this is given by the lack of RS modules dedicated to ecology and conservation in the wide range of training available at ConservationTraining (https://www.conservationtraining.org/course/index.php, accessed on 24 August 2016).

A lot of the current RS training is provided by degree programs at institutes of higher education, but we believe that, beside what offered by academia, professional trainings in conservation applications of RS are greatly needed to increase the capacity of those working in the conservation field.

In this paper, we identify current needs and challenges for building RS capacity in conservation and present results from a survey of RS conservation professionals designed to better understand the vision, needs and priorities for conservation RS capacity development. We argue that the best and most cost-effective approach for building such capacity should focus on increased understanding of RS basic principles, increased data access and production of relevant conservation datasets, and the development of best practices and data analysis tools for their use in conservation. Increasing

remote sensing capacity among conservation practitioners and decision makers will also allow these users to actively contribute to the definition and production of new remotely sensed datasets and indicators that address conservation needs.

Role of Remote Sensing in Conservation

Remote sensing has made significant contributions to conservation globally as satellite observations first highlighted an increase in forest loss in the Amazon during the 1970s. Since then, studies have demonstrated meaningful use of RS data analysis for ecology and conservation (Kennedy et al. 2010; Pettorelli et al. 2011; Rose et al. 2015). They provide examples that can be grouped thematically into topics that include (1) identifying and mapping undisturbed terrestrial habitat (Potapov et al. 2008; Tyukavina et al. 2016) and species-specific suitable habitat (Goetz et al. 2007; Bergl et al. 2012), (2) analysing species-specific resource use (Stoner et al. 2016), (3) investigating multi-temporal changes in habitat quality (Buchanan et al. 2008; Nackoney et al. 2014) and (4) developing dynamic RS-based decision support systems to support wildlife conservation and management (Jantz et al. 2016). Recent advances in high-end data computing have for the first time allowed huge archives of imagery to be leveraged, resulting in multi-temporal data on global tree cover loss (Hansen et al. 2013), mangroves (Giri et al. 2011) and global fire activity (Chuvieco et al. 2008).

Rose et al. (2015) highlighted 10 *critical questions* in conservation that could be best solved through RS. These questions, which covered 10 broad conservation themes, defined a conceptual framework for using RS to improve conservation outcomes. Very recently, the Group on Earth Observations Biodiversity Observation Network (GEO BON) similarly identified a set of variables necessary for long-term biodiversity monitoring (Skidmore et al. 2015). Out of the 22 *essential biodiversity variables*, or EBVs, that were proposed, 14 can be monitored using RS data. Both efforts, the 10 critical questions and the EBVs, provide clear arguments for increasing the capacity for integrating RS into conservation practice.

Training Needs and Challenges

Access to RS methods relevant to conservation applications and the applied use of conservation-related RS data in training courses is critical for conservation NGO employees, national parks managers and other conservation professionals. During a workshop on Remote Sensing for Conservation, organized at the Joint Research Centre of the European Commission in 2013, several working groups discussed the use of RS by conservation scientists and practitioners and the main needs and challenges in using RS data to address conservation questions (Leidner et al. 2013). One of the key points highlighted was a lack of understanding about how RS could contribute to solving certain conservation-related problems in addition to the particular capabilities and limitations of RS observation in monitoring and mapping specific environmental phenomena. Taking into account the differences in objectives and use of RS methods and data in conservation policy and research, RS trainings for conservation professionals need to be structured to overcome these knowledge gaps. Training a RS scientist can involve years worth of technical, engineering and applied training, including courses on the physics of light and light–object interactions, atmospheric science, sensors and advanced image processing and analysis (e.g. signal interpretation, classification techniques and algorithm development). Applied training in conservation RS, however, should instead target the goals and needs of a conservation practitioner and be shaped around answering pertinent ecological and conservation questions that utilize methods and data derived from RS observation. In this context, the standard RS modules that teach image (pre-)processing and the methods applied to derive spectral indices [e.g. leaf area index (LAI), normalized difference vegetation index (NDVI), fraction of absorbed photosynthetically active radiation (FAPAR)] become less relevant. Conservationists instead need to learn the ecological meaning of vegetation spectral indices and become aware of the basic RS principles behind image acquisition that affect the quality of the derived products. Training should therefore focus on how to use information derived from RS technology in order to conduct scientific analysis and make informed decisions.

Through targeted training, conservation professionals can benefit from the rapid growth of RS-based information and derived datasets and learn how these products can be used to drive conservation analysis and decision making. Trainings should provide (1) understanding of basic RS principles and RS-derived information, (2) knowledge about how to access and use these data and products for environmental analysis, (3) information about how the accuracy and resolution (i.e. spectral, spatial and temporal) of the original raw data may affect a particular study and (4) basic principles of RS data analysis (e.g. time-series) and data formats (raster and vector datasets). The use of open-source mapping software (e.g. QGIS, R, GRASS) is also essential to open RS to all categories of users. With this focus, capacity development tools for conservation practitioners could be more cost- and time-effective and better engage the conservation community in the same way that GIS training did over the past 10 years.

Conservation Remote Sensing Network Survey

Considering the growing attention dedicated to RS data and products, and their improved quality and availability, there is strong need to build a network of users in the conservation community with appropriate knowledge and understanding of the opportunities and limitations of RS technology and products in order to improve conservation monitoring, implementation and planning. The Conservation Remote Sensing Network (CRSNet, http://remote-sensing-conservation.org/about/, accessed on 16 June 2016) was established in 2013 as a direct response to this need. This community aims to (1) improve the dialogue between the conservation-based users of RS technology and products and the scientists that develop those products, (2) increase RS capacity in conservation programs and (3) increase collaboration between RS experts and ecologists. CRSNet includes over 500 members, encompassing individuals from various disciplines and professional affiliations that range from academia and NGOs to industry and space agencies.

Recently, CRSNet conducted a survey (see Appendix S1) to better understand members' main needs and expectations concerning RS technology and data. The survey asked about members' level of RS expertise and training needs, and asked them to identify main gaps and priorities in the current RS data and technology as applied to conservation. In total, 140 people participated. The survey participants were summarized according to the general location(s) of their study regions (Fig. 1) and their professional affiliations (Fig. 2). In both cases, it was possible that participants were associated with multiple geographic areas and professional affiliation categories. Both terrestrial and marine regions of the world were represented as study regions, and several participants had a global-level focus. Participants from academic and NGO institutions were the most represented in the group.

Answers to questions that asked about participants' specific RS expertise were based on the participants' self-assessment. The average values were summarized by professional affiliation class (Fig. 3), with values ranging between 1 and 5, value 1 indicating a beginner level and 5 a proficient user. Categories of RS expertise included data use, image pre-processing and image processing. Although there was a general balance in terms of overall expertise across affiliation type, the private sector self-assessed to have slightly higher expertise in most of the image pre-processing and processing techniques.

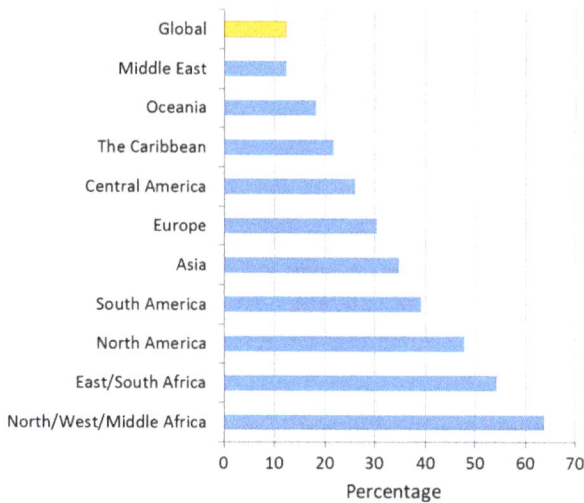

Figure 1. The distribution of the survey participants according to their study region(s). Note that one person can be working on more than one region.

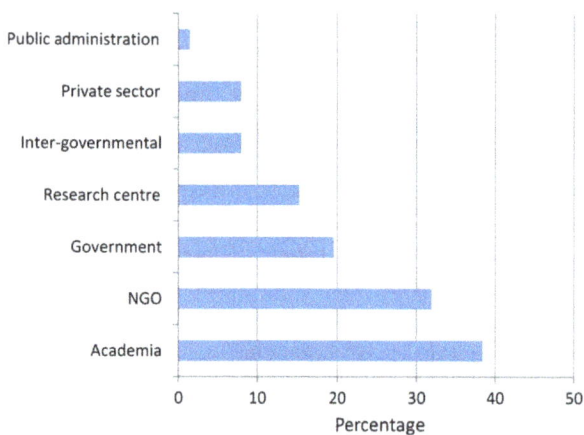

Figure 2. The distribution of the survey participants according to their professional affiliation. Note that one person can have multiple affiliations.

The survey also asked questions related to RS training access, ranging from online to in-person courses by affiliation type. Results showed much higher access to training opportunities for participants working in academia, especially trainings conducted in-person by a live instructor (Fig. 4).

Besides access to training, the survey asked about participants' use of commercial and open-source software for image processing and spatial data analysis. Results showed a general tendency in using software for spatial data analysis rather than image processing, commercial software are more spread than open-source regardless the affiliation type (Fig. 5), with ArcGIS being the most widely used package. The open-source software, R, was reported

to be used by up to 40% of the survey participants. Only in the academic and research affiliations did other open-source software (i.e. GRASS GIS, QGIS) reach a similar level.

The CRSNet survey also asked participants to identify priorities in specific topics related to conservation RS. Results included (in the order of high to low priority) the need to acquire software and processing tools, production of publications and reports, development of guidelines about satellite data products and where to download them, funding opportunities, understanding policy impacts and highlighting success stories, organizing and attending community meetings (in-person society meetings, webinars, etc.) and learning about member news and projects. Additionally, high importance was also given to the preparation of guidelines, reviews and recommendations about the (1) use of satellite products in conservation, (2) existing software for the analysis of satellite-based products and (3) data collection to validate RS products.

The participants' priorities as identified through the survey provide especially valuable information for designing trainings in RS for conservationists. We believe that greater access to training could improve the overall capacity of the conservation community in making use of RS data and tools and promote appropriate methods and datasets for analysis and decision making. The survey results suggested that CRSNet could play a substantial role in leading increased collaboration between scientists and practitioners in the conservation community. This can potentially foster a new generation of RS datasets specifically tailored to ecology and conservation needs, thereby improving current conservation decision making and effectiveness.

Current Remote Sensing Training and Opportunities

Current training opportunities in RS are often focused on teaching the use of software packages for image processing and data analysis, with no specific objectives related to ecology and conservation. Perhaps one of the most prominent centres for formal postgraduate academic training (>20,000), based primarily on the number of graduated students in RS application and spatial science, is the Faculty of Geo-Information Science and Earth Observation (ITC) at the University of Twente. ITC provides degree, diploma and certificates in geo-information science and earth observation using RS and GIS with a focus on natural resource management and training for environmental managers from developing countries. In addition, a number of short-course training opportunities (often 1 or 2 weeks in duration) have more recently

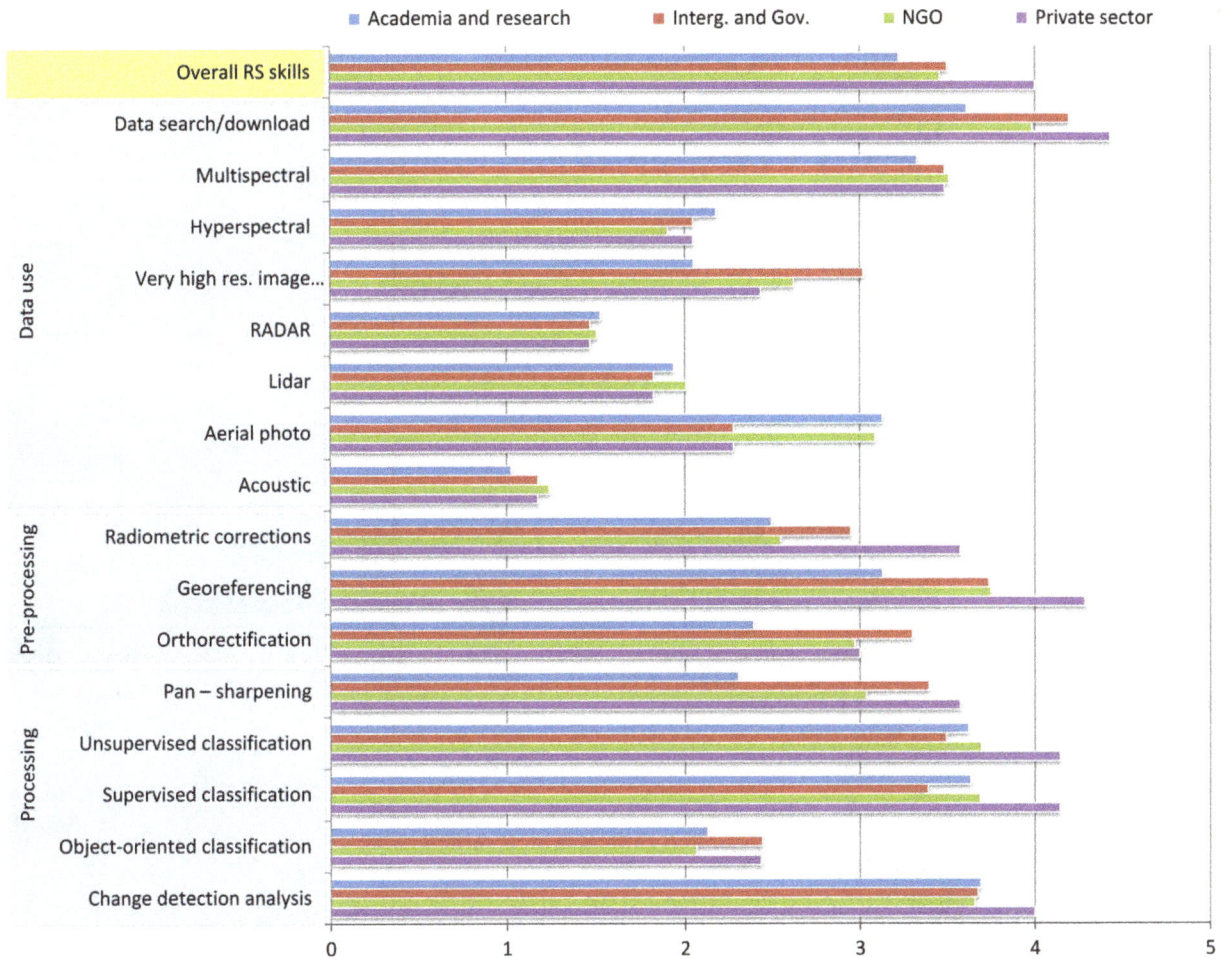

Figure 3. Remote sensing skills by affiliation type based on participants' self-assessment (level 1: beginner, level 5: proficient).

emerged that are specifically dedicated to conservation RS. The Smithsonian-Mason School of Conservation (http://smconservation.gmu.edu, accessed on 16 June 2016) offers courses in species distribution modelling and ecological geospatial statistics, among others. NASA's Applied Remote Sensing Training (ARSET) program (http://arset.gsfc.nasa.gov/eco, accessed on 16 June 2016) has offered a series of webinars devoted to the fundamentals of RS for conservation application as well as land management and wildfire monitoring. These webinars covered topics such as satellite sensors and their applicability to different environment-related problems and featured real-world application of data and tools for monitoring animal habitats and animal movement, land cover change, fire detection, among others. Another RS training opportunity, that although is not specifically focused on conservation has components very relevant to it, is the joint NASA and USAID program, called SERVIR (http://www.servirglobal.net, accessed on 16 June 2016).

SERVIR provides 2-week trainings on the use of NASA data for sustainable development-related activities and policies; examples include global navigation systems, introduction to GIS and RS, land cover mapping from satellite image data using eCognition and the use of the software program R for REDD+ applications. To complement these tools, SERVIR also provides training in web-based services (e.g. Google Earth Engine, GPOD (http://gpod.eo.esa.int, accessed on 16 June 2016).

The Remote Sensing for Biodiversity & Conservation website (http://remote-sensing-biodiversity.org/, accessed on 16 June 2016) is a source of news and information about upcoming conservation RS training opportunities and conferences, it gathers the networks focused on conservation RS and lists a wealth of both terrestrial and marine RS data resources. Similarly, the Spatial Ecology Wiki (http://www.spatial-ecology.net, accessed on 16 June 2016) provides a platform for posting trainings and tutorials geared towards open-source software for ecological data analysis.

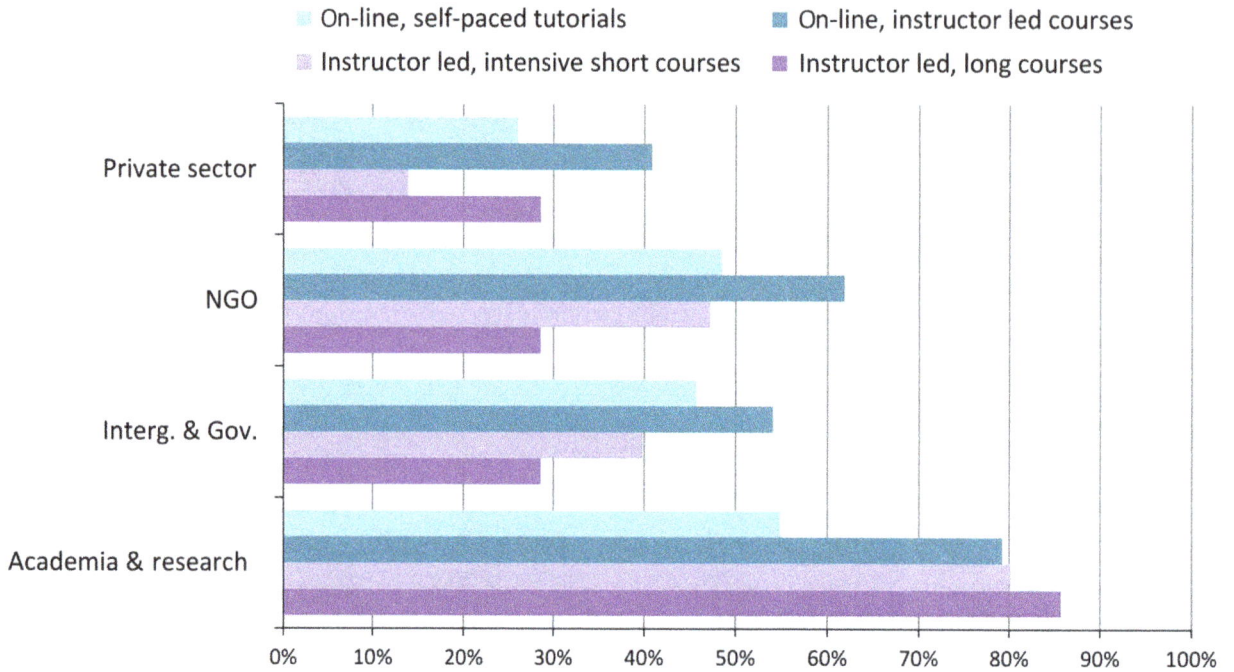

Figure 4. Access to different types of training by affiliation type.

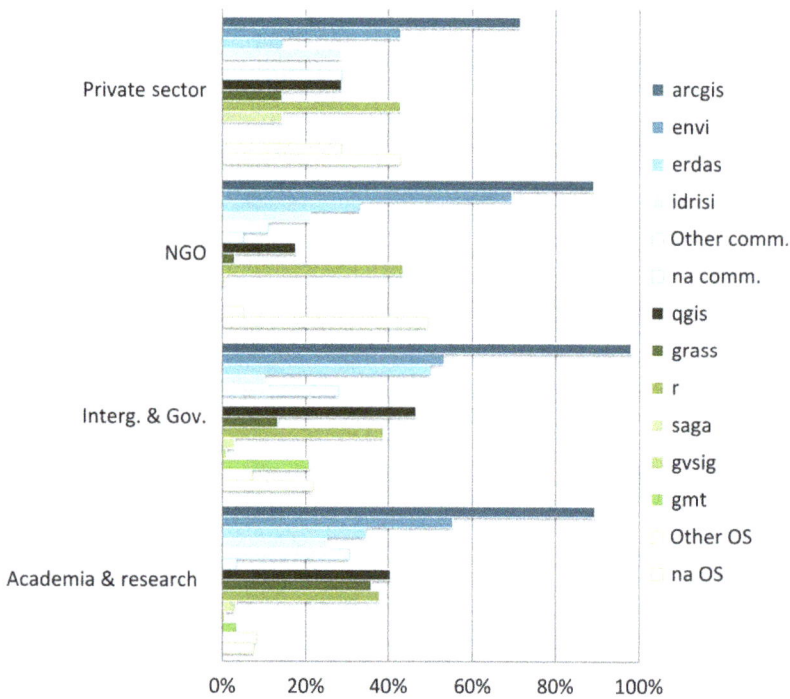

Figure 5. Use of commercial (comm.) versus open-source (OS) software by professional affiliation.

Next Steps for Training

In order to increase the current use of remotely sensed information by conservation scientists and decision makers, targeted approaches are needed. Trainings should focus on the use of the growing pool of derived products rather than raw satellite data and image processing; moreover, training modules should be shaped around answering specific ecological questions and teach how to derive indicators and relevant environmental variables (e.g. EBVs) from current RS products. More emphasis on the application of RS data analysis methods for conservation purposes is needed, such as time-series analysis with Landsat and Sentinel datasets (e.g. detection of land cover and land use change and fire occurrence). Similarly, webinars can serve a wide user community, highlighting new RS products such as deforestation data and demonstrating how to use these data to answer pertinent conservation questions.

The use and continued development of open-source software packages are crucial for teaching applied data analysis methods. Open-source software packages, which are available free of charge, can increase the use of RS data among non-RS experts at no cost; in addition, open-source software is much more accessible especially to users in developing countries who lack the financial resources to purchase commercial software packages (Rocchini and Neteler 2012; Wegmann et al. 2016). Another advantage of using open-source software is that the programming code used to develop the software packages is shared openly, so that the software can be modified to suit individual needs. Many open-source software packages have online user forums that are actively maintained and promote collaboration across the globe. It is likely that RS experts will continue to rely on proprietary software such as Esri products, ERDAS, ENVI/IDL and eCognition; however, open-source software such as the RStoolbox package in R (see Wegmann et al. 2016), GRASS, QGIS, OTB, SAGA and GDAL provide much of the same functionality, at no cost.

An opportunity to enrich current trainings is also offered by the new map-based platforms that host online public data and allow dynamic mapping. Such platforms have increased in recent years, including those primarily directed toward ecologists and conservation practitioners. These online platforms contribute to a wide dissemination of RS data and product visualization, and generally facilitate broader access to RS-based information. They are easier to use by non-RS experts as they do not require image processing, mapping capability or high-end computer capacity. Some examples of such platforms are the IMPACT toolbox developed by JRC–European Commission (Szantoi et al. 2016), the Global Forest Watch (http://www.globalforestwatch.org/, accessed on 16 June 2016)

developed by the World Resources Institute, and the FIRMS Web Fire Mapping (https://firms.modaps.eosdis.nasa.gov/firemap/, accessed on 16 June 2016) developed by NASA. Finally, by taking a collaborative approach, conservation NGOs, government scientists and RS specialists working in ecology could pool their resources to develop materials for online tutorials or instructor-led courses. Instructor-led courses are often preferred over online courses because they facilitate direct interaction between the students and the teacher and often provide networking opportunities as well. However, they generally require more resources than online trainings and, unless they are recorded, have a lower chance of being repeated in the future.

Standardizing conservation-based RS training curriculum and methods would be useful to provide comparable training content across courses and also deliver course content on a wide range of topics, like those described by Rose et al. (2015). This might lead to a 'Sourcebook of Remote Sensing Approaches for Conservation' that would allow a transfer of methods and approaches across different regions and organizations, and also allow conservation practitioners to compare how various methods are implemented and integrated into their unique decision-making processes. Harmonization of the topics relevant to the most common RS tasks in environmental analysis could also be beneficial, such as (1) basic RS principles (image acquisition, spatial, temporal and spectral resolution, data accuracy), (2) the use and development of vegetation indices, land cover classifications and other datasets important for environmental analysis and (3) fundamentals of conducting analysis with RS-based datasets, including their application to solving environmental problems and their main challenges and limitations. From a technical perspective, harmonization of methods could be achieved by developing user-friendly add-ons that could be launched from an open-source software platform to provide a suite of conservation-related tools such as land cover change analysis or retrieval of data from online sources. Trainings should be built around commonly used and freely available datasets such as tree canopy and tree cover loss data provided by Hansen et al. (2013), MODIS-based vegetation data, fire products and other data derived from Landsat and Sentinel satellite programmes. Different applications of data, ranging from high to low spatial resolution, should also be featured. It is important to acknowledge that the RS needs of conservation practitioners, who are engaged in applied problems on the ground, might differ significantly from the RS needs of ecological researchers. Rather than analysing the distribution of a single species across decades, conservation practitioners might focus instead on identifying locations of deforestation hotspots within a species range that might require more rapid response. Training curricula should be sensitive to these differences and introduce tools and data accordingly so that multiple needs and situations are addressed.

Similarly, to prepare a new generation of conservationists, academia should offer to both undergraduate and graduate level students more interdisciplinary curricula and courses to integrate conservation and RS topics. This will also facilitate the creation of future interdisciplinary research centres that focus on conservation topics.

Outlook

Although the enormous potential for RS in ecology and conservation is well known, knowledge of pertinent datasets, methodological approaches and examples of direct application of RS to conservation problems is still limited. Trainings focused on ecological and conservation applications are greatly needed to improve and increase the application of RS in these disciplines. In light of the recently launched European Space Agency's (ESA) Sentinel satellites, the new upcoming Sentinels (planned for launch between 2017 and 2020) and other satellite missions such as Landsat 9, hyperspectral missions and the GEDI Lidar mission, we expect a rapid increase in datasets valuable for conservation studies. The coordination of the ESA Sentinel programme with NASA will increase satellite data availability and, more importantly, access to a number of improved products useful for monitoring both terrestrial and marine ecosystems. Both programmes are making data available as subsets with smaller file sizes, which will make on-demand RS application more viable for users in regions with limited bandwidth. Because conservation practitioners will need to increasingly understand how to use new products with improved spatial, temporal and spectral resolutions, training curricula will need to keep pace with these technological improvements. In addition, as RS technology develops in scope, it will be increasingly important to foster a RS-informed conservation community that can provide important feedback to satellite programme managers and RS experts about what products are most needed and how existing data can be improved in order to increase conservation effectiveness. The Conservation Remote Sensing Network (CRSNet) provides a valuable platform to help achieve this goal; in addition, it provides a diverse information network to share updates on existing and new datasets, information on training opportunities, methodological approaches and valuable lessons learned on the ground.

Continued development and advances in making RS software open access and more user-friendly will facilitate the use of RS products among conservation practitioners

around the globe. This should be paired with efforts to make RS-derived data freely available and easily accessible and usable. Online mapping platforms that allow users to visualize and map data relevant to environmental monitoring are key; in addition, the data should be freely downloadable for more experienced users who wish to conduct further analysis.

Our outlook for conservation RS and capacity building is very positive. Events centred around the theme of conservation and environment with RS are increasing, such as the Zoological Society of London's (ZSL) Remote Sensing for Conservation symposium held in 2014 and the ESA Living Planet Symposia, recently held this past May 2016. Satellite and aerial sensors and technology are evolving rapidly, and data providers now have greater opportunity and experience higher urgency to share their data openly. The use of drones and unmanned aerial vehicles (UAVs) is also growing among conservation managers, providing a wealth of high-resolution data for targeted geographic areas that can help monitor environmental pressures that might be more difficult to assess otherwise (e.g. wildlife poaching). As a consequence, awareness of RS capabilities and interest in applying new RS technology and data are growing enormously within the conservation community and are generating momentum for a coordinated action to develop dedicated RS trainings for conservation. True success will depend on addressing the variety of users' needs and expectations to ensure that capacity development efforts generate increased RS expertise in the conservation community so that on-the-ground conservation actions and decision making become most effective.

References

Bergl, R. A., Y. Warren, A. Nicholas, A. Dunn, I. Imong, J. L. Sunderland-Groves, et al. 2012. Remote sensing analysis reveals habitat, dispersal corridors and expanded distribution for the Critically Endangered Cross River gorilla *Gorilla gorilla diehli*. *Oryx* **46**, 278–289.

Brooks, T. M., R. A. Mittermeier, G. A. B. da Fonseca, J. Gerlach, M. Hoffmann, J. F. Lamoreux, et al. 2006. Global biodiversity conservation priorities. *Science* **313**, 58–61.

Buchanan, G. M., S. H. Butchart, G. Dutson, J. D. Pilgrim, M. K. Steininger, K. D. Bishop, et al. 2008. Using remote sensing to inform conservation status assessment: estimates of recent deforestation rates on New Britain and the impacts upon endemic birds. *Biol. Conserv.* **141**, 56–66.

Chuvieco, E., L. Giglio, and C. Justice. 2008. Global characterization of fire activity: toward defining fire regimes from Earth observation data. *Glob. Change Biol.* **14**, 1488–1502.

Giri, C., E. Ochieng, L. L. Tieszen, Z. Zhu, A. Singh, T. Loveland, et al. 2011. Status and distribution of mangrove forests of the world using earth observation satellite data. *Glob. Ecol. Biogeogr.* **20**, 154–159.

Goetz, S., D. Steinberg, R. Dubayah, and B. Blair. 2007. Laser remote sensing of canopy habitat heterogeneity as a predictor of bird species richness in an eastern temperate forest, USA. *Remote Sens. Environ.* **108** 254–263.

Groves, C. R., D. B. Jensen, L. L. Valutis, K. H. Redford, M. L. Shaffer, J. M. Scott, et al. 2002. Planning for biodiversity conservation: putting conservation science into practice. *Bioscience* **52**, 499–512.

Hansen, M. C., P. V. Potapov, R. Moore, M. Hancher, S. A. Turubanova, A. Tyukavina, et al. 2013. High-resolution global maps of 21st-century forest cover change. *Science* **342**, 850–853.

Jantz, S., L. Pintea, J. Nackoney, and M. C. Hansen. 2016. Employing remotely sensed datasets to map and monitor range-wide chimpanzee (Pan troglodytes) habitat suitability to aid conservation efforts. *Remote Sens.***8**, 427; doi:10.3390/rs8050427.

Kennedy, R. E., Z. Yang, and W. B. Cohen. 2010. Detecting trends in forest disturbance and recovery using yearly Landsat time series: 1. LandTrendr – Temporal segmentation algorithms. *Remote Sens. Environ.* **114**, 2897–2910.

Leidner, A., A. Brink, and Z. Szantoi. 2013. Leveraging Remote Sensing for Conservation Decision Making. *EOS* **94**, 508.

Margules, C. R., and R. L. Pressey. 2000. Systematic conservation planning. *Nature* **405**, 243–253.

Myers, N., R. A. Mittermeier, C. G. Mittermeier, G. A. B. da Fonseca, and J. Kent. 2000. Biodiversity hotspots for conservation priorities. *Nature* **403**, 853–858.

Nackoney, J., G. Molinario, P. Potapov, S. Turubanova, M. C. Hansen, and T. Furuichi. 2014. Impacts of civil conflict on primary forest habitat in northern Democratic Republic of the Congo, 1990-2010. *Biol. Conserv.* **170**, 321–328.

Pettorelli, N., S. Ryan, T. Mueller, N. Bunnefeld, B. Jedrzejewska, M. Lima, et al. 2011. The Normalized Difference Vegetation Index (NDVI): unforeseen successes in animal ecology. *Clim. Res.* **46**, 15–27.

Potapov, P., A. Yaroshenko, S. Turubanova, M. Dubinin, L. Laestadius, C. Thies, et al. 2008. Mapping the world's intact forest landscapes by remote sensing. *Ecol. Soc.* **13**, 51.

Pressey, R. L., and M. C. Bottrill. 2008. Opportunism, threats, and the evolution of systematic conservation planning. *Conserv. Biol.* **22**, 1340–1345.

Rocchini, D., and M. Neteler. 2012. Let the four freedoms paradigm apply to ecology. *Trends Ecol. Evol.* **27**, 310–311.

Rose, R. A., D. Byler, J. R. Eastman, E. Fleishman, G. Geller, S. Goetz, et al. 2015. Ten ways remote sensing can contribute to conservation. *Conserv. Biol.* **29**, 350–359.

Skidmore, A. K., N. Pettorelli, N. C. Coops, G. N. Geller, M. Hansen, R. Lucas, et al. 2015. Environmental science: Agree on biodiversity metrics to track from space. *Nature* **523**, 403–405.

Stoner, D. C., J. O. Sexton, J. Nagol, H. H. Bernales, and T. C. Jr Edwards. 2016. Ungulate Reproductive Parameters Track Satellite Observations of Plant Phenology across Latitude and Climatological Regimes. *PLoS ONE* **11**.

Szantoi, Z., A. Brink, G. Buchanan, L. Bastin, A. Lupi, D. Simonetti, et al. 2016. A simple remote sensing based information system for monitoring sites of conservation importance. *Remote Sens. Ecol. Conserv.* **2**, 16–24.

Trombulak, S. C., and R. F. Baldwin, eds. 2010. *Landscape-scale Conservation Planning.* Springer, New York.

Turner, W., S. Spector, N. Gardiner, M. Fladeland, E. Sterling, and M. Steininger. 2003. Remote sensing for biodiversity science and conservation. *Trends Ecol. Evol.* **18**, 306–314.

Tyukavina, A., M. C. Hansen, P. V. Potapov, A. M. Krylov, and S. J. Goetz. 2016. Pan-tropical hinterland forests: mapping minimally disturbed forests. *Glob. Ecol. Biogeogr.* **25**, 151–163.

Wegmann, M., B. Leutner, and S. Dech (Eds.). 2016. *Remote sensing and GIS for ecologists: using open source software.* Pelagic Publishing, Exeter.

Wilson, K. A., E. C. Underwood, S. A. Morrison, K. R. Klausmeyer, W. W. Murdoch, B. Reyers, et al. 2007. Conserving biodiversity efficiently: what to do, where, and when. *PLoS Biol.* **5**, e223. doi:10.1371/journal.pbio.0050223.

9

Remote sensing of species dominance and the value for quantifying ecosystem services

Stephanie Pau[1] & Laura E. Dee[2]

[1]Department of Geography, Florida State University, Tallahassee, Florida 32306
[2]Institute on the Environment, University of Minnesota, St. Paul, Minnesota 55108

Keywords

Biodiversity monitoring, conservation, ecosystem functions, essential biodiversity variables, species abundance, species composition

Correspondence

Stephanie Pau, Department of Geography, Florida State University, Tallahassee, FL 32306. E-mail: spau@fsu.edu

Funding Information

L.D. was supported by a grant from the Institute on the Environment at University of Minnesota to P.B. Reich and S. Polasky.

Editor: Duccio Rocchini
Associate Editor: Martin Wegmann

Abstract

Remote sensing (RS) is a powerful tool to measure and monitor Essential Biodiversity Variables (EBVs) and their environmental drivers. Despite this potential, stronger integration between remote sensing experts and the ecological community could better support biodiversity initiatives. Here we highlight opportunities to harness remote sensing technology to better understand biodiversity patterns, ecological processes and the consequences for ecosystem services (ESs). We argue that tracking many EBVs using remote sensing should prioritize the monitoring of dominant species, a scalable property across multiple EBV classes, for several reasons. First, a few dominant species in an ecological community disproportionately contribute to the satellite spectral signature. Second, a focus on dominance would enable a stronger links to ecological research, as dominance reflects the ecological community context (i.e. relative abundance of coexisting species). For example dominant species should be especially important contributors to many ecosystem functions and services that rely on abundance or biomass, such as carbon storage or nutrient cycling, because of their greater representation in a community. Furthermore, global change impacts on communities may be reflected in changing dominance structure before the losses of species, thus tracking dominance provides an early-warning sign of community change for EBVs. Finally, focusing on dominant species should improve understanding of spatial and temporal dynamics of dominance-driven ESs through RS mapping. Given the importance of dominant species to ecological communities and ESs, monitoring dominance under changing environmental conditions and human impacts should be a global priority.

Introduction

Global environmental change is rapidly transforming the Earth's ecosystems. The rate of biodiversity loss may have already crossed a threshold that will degrade ecosystem functions and have unexpected consequences on interrelated subsystems of the planet (Rockstrom et al. 2009). To capture the major dimensions of biodiversity change, monitoring efforts have recently identified a set of global Essential Biodiversity Variables (EBV). EBVs are measurements required for studying, reporting and managing biodiversity change (Pereira et al. 2013). The six broad classes of EBVs – genetic composition, species populations, species traits, community composition, ecosystem/habitat structure and ecosystem functions – are intended to complement each other and observations of environmental change (Pereira et al. 2013). Remote sensing (RS) is a powerful tool to measure many EBVs and environmental change across time and space (O'Connor et al. 2015). Hundreds of satellites orbiting in space together provide global observations of the Earth's surface at repeated intervals. Increasing use of airborne sensors, unmanned aerial vehicles (UAVs) and in situ sensors also enhance the capability to detect changes in EBVs (Turner 2014). RS data should be complementary to time- and site-specific field observations by emphasizing scalable properties. By measuring scalable properties, RS can be used to extrapolate local data using models or proxies and provide information on how EBVs respond to environmental drivers (Pereira et al. 2013; Pettorelli et al. 2014a).

Despite the potential for RS to effectively monitor EBVs, recent calls for RS experts to better support biodiversity initiatives highlighted the need to form stronger links to the ecological community (Pereira et al. 2013; Pettorelli et al. 2014b; Skidmore et al. 2015). Discrepancies in scale between ecological data and pixel resolution, as well as incongruity between ecological metrics and satellite-derived proxies have been identified as barriers to integration (Turner et al. 2003; Skidmore et al. 2015). Rather than compromising to find overlap, efforts should aim for co-production of new knowledge beginning with the initial research questions. Toward that goal, here we highlight opportunities to integrate RS and ecology to address shared questions about biodiversity patterns, ecological processes and the consequences for ecosystem services (ESs). In particular, we argue that tracking many EBVs using RS should more explicitly focus on the causes and consequences of dominance, a scalable property across multiple EBV classes. Dominance reflects both species abundances and community composition (i.e. species that represent a disproportionate number of individuals and/or biomass in the community). Such a focus also enables spatially and temporally dynamic mapping of dominance-driven ecosystem functions and services (e.g. provisioning and regulating services).

In most optical RS applications, a few dominant species in an ecological community disproportionately contribute to the satellite spectral signature (Fig. 1). Dominance (or its counterpart, evenness) considers the abundance of species relative to the other species in a community (Hillebrand et al. 2008). Much of the biodiversity literature has focused on species richness – strictly the number of species within a community (Purvis and Hector 2000; Hillebrand et al. 2008). Yet, even in species-rich tropical forests, only a subset of species are large, emergent trees that comprise the canopy and thus the spectral signal of a pixel. Indeed, many advances in optical RS have focused on spectral mixture analysis, where the fraction of different surfaces contributing to the remotely sensed spectra is decomposed into linear combinations of pure spectra for each type of surface ('endmembers') (Roberts et al. 1998; Asner and Martin 2009). Spectrally mixed pixels can be a technical limitation of remote sensing that is problematic for studying local species richness and tracking losses of rare species; however, it provides an opportunity to link species dominance to EBVs. Given the highly skewed abundance of species in most natural systems (Whittaker 1975; Ricklefs 1990; Schwartz et al. 2000), focusing on dominance should enable a scalable, ecologically meaningful measure (rather than partitioning spectral endmembers) of EBVs using remotely sensed data.

Despite the importance of dominant species to ecological communities and ecosystem functioning, there is much to learn about the causes and consequences of dominance, especially under changing environmental conditions and human impacts (Hillebrand et al. 2008). RS is particularly well-suited to track changes in dominance over time and space and to quantify associated changes in ecosystem functions and services. In the following sections, we outline how a focus of RS on dominant species provides several insights for monitoring multiple classes of EBVs and better connects RS to ecological research priorities. Identifying which species are dominant in different locations or at different time periods should help inform our understanding of the environmental conditions that confer ecological success, and whether patterns and processes vary consistently across environmental gradients. Finally, individuals of dominant species are disproportionately represented in communities, often leading to a larger representation of their trait values and larger contributions to ecosystem structure and function (Hillebrand et al. 2008). Therefore, tracking dominant species in an ecosystem can provide a proxy for ES supply and greater insight into spatiotemporal dynamics of ESs.

Remote Sensing of Biodiversity and Ecological Dominance

Several review papers identify various approaches for using remote sensing to track biodiversity and ecosystem functions (Nagendra 2001; Kerr and Ostrovsky 2003; Turner et al. 2003; Gillespie et al. 2008; Kuenzer et al. 2014; Skidmore et al. 2015). Many remote sensing studies have been devoted to detecting local species richness (i.e. alpha diversity). One approach is using land cover classifications and thematic mapping of habitats to monitor habitat extent and conversion. Habitat extent can then be used to estimate species loss based on species-area relationships (Turner et al. 2003). In addition to land-cover or habitat mapping, remote sensing provides various measures of habitat conditions, including proxies for habitat complexity (e.g. LAI; % tree cover) and primary productivity (e.g. NDVI). These proxies have been commonly used to indicate plant or bird species richness because of diverse resources provided by more productive and complex habitats (Waring et al. 2002; Gillespie 2005; Rompré et al. 2007; Phillips et al. 2010; Pau et al. 2012) or to test the species-energy hypothesis, which proposes that energy is a limiting resource to the number of species in an area (Currie and Paquin 1987; Hawkins et al. 2003).

Remote sensing has been widely used to understand species distributions (reviewed in He et al. 2015). Bioclimatic variables derived from satellites are increasingly included in species distribution models as environmental predictors of species occurrence (Bradley and Fleishman 2008; Buermann et al. 2008; Pau et al. 2013; Deblauwe et al. 2016). While species occurrence is the presence of a species in

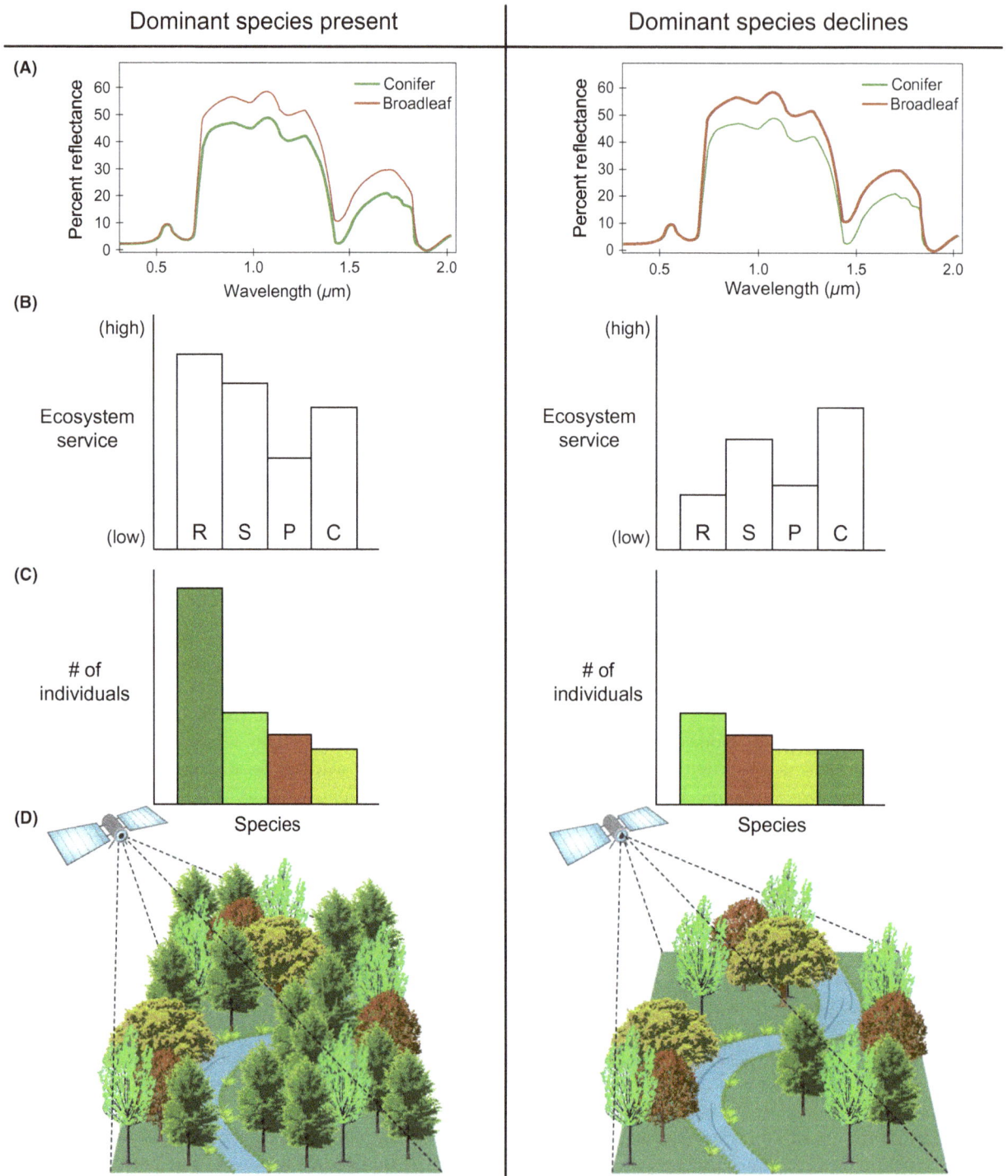

Figure 1. Dominant species disproportionately contribute to the satellite spectral signature, whereas rare species, included in species richness counts, contribute relatively little (A). These dominant species are also primary contributors to abundance-driven ecosystem services (ESs) (B), often provisioning, regulating and supporting services ('P' = provisioning, 'R' = regulating, 'S' = supporting, and 'C' = cultural; Millennium Ecosystem Assessment 2005), because dominant species comprise more individuals and biomass in a community (C). Each species can potentially contribute to more than one type of service (e.g. carbon storage that regulates the climate as well as supporting biodiversity habitat). Rank abundance curves (C) showing an example of a dominant conifer species (left) and the loss of this dominant under changing conditions (right). Following the loss of a dominant species, the total biomass of a community (C) and total amount of some ESs (B) will be reduced, corresponding to different spectral signatures (A).

particular sites or locations, species turnover quantifies the replacement of one species by another species at different sites. Remote sensing has been particularly useful for understanding global patterns of species turnover by providing continuously observed data at a global extent (Gaston et al. 2007; Buckley and Jetz 2008; Qian and Ricklefs 2012). By examining environmental variation across large spatial gradients, these studies have disentangled the importance of environmental drivers (abiotic tolerances) from dispersal limitations and evolutionary histories in determining a species distribution. Another metric of biodiversity, beta diversity, is closely related to species turnover. Beta diversity can refer to a mathematical partitioning (additive) or scaling (multiplicative) of diversity between local (alpha) and landscape (gamma) scale diversity; or beta diversity can be a measure of community dissimilarity among paired sites (see Socolar et al. 2016 for a full discussion). It provides a useful measure for identifying how geographic or environmental gradients structure changes in species composition across sites, rather than turnover in individual species (e.g. Rocchini 2007). Dissimilarly metrics include abundance-weighted community composition or simply species presence–absence (Socolar et al. 2016). Species composition and abundance are thought to be more robust to satellite discrimination than species richness (Tuomisto et al. 2003; He et al. 2009; Oldeland et al. 2010; Rocchini et al. 2010).

Remote sensing also provides data on species themselves (reviewed in He et al. 2015). Images with high spatial resolution can directly detect species by delineating individual tree crowns or through traits associated with particular species (e.g. Weisberg et al. 2007; Palace et al. 2008; He et al. 2011). This is generally most successful for species or functional types with distinct growth forms, phenologies or biochemical traits. For example woody encroachment along a forest-savanna boundary (Mitchard et al. 2009) or into tundra biomes (Stow et al. 2004) has been detected by distinguishing woody versus non-woody species. Distinct phenologies have been used to quantify an upward shift of a hardwood (deciduous)-boreal (evergreen) forest ecotone (Beckage et al. 2008) and to detect the relative abundance of functional groups, such as C_3 and C_4 grass species (Goodin and Henebry 1997; Davidson and Csillag 2003; Foody and Dash 2010). Phenological differences are also commonly used to identify invasive species (Bradley 2014). For example the distinct annual phenology of cheatgrass was used to quantify the extent it is replacing native perennial grasses and shrubs (Bradley and Mustard 2006). Similarly, species-specific leaf biochemical properties, detected using hyperspectral reflectance data (>100 spectral bands), has helped identify species invasions (e.g. Asner et al. 2008; Asner and Martin 2009; He et al. 2011).

While large advances have been made using remote sensing to address different measures of biodiversity, a body of RS studies centered on the causes and consequences of dominance is lacking. An explicit focus on dominance, as we advocate here, would facilitate methodological and conceptual advances in the study of dominance. This focus would enable stronger links to the ecological community because most communities are dominated by only a few species (Whittaker 1975; Ricklefs 1990; Schwartz et al. 2000). Dominance is also ecologically relevant because species dominance reflects biotic and abiotic ecological relationships (e.g. Lohbeck et al. 2014) and many ecosystem functions (EF) and associated ESs depend heavily on dominant species (e.g. Walker et al. 1999; Schwartz et al. 2000; Smith and Knapp 2003; Dangles and Malmqvist 2004; Hillebrand et al. 2008; Kleijn et al. 2015). For example aboveground net primary productivity was reduced in a tallgrass prairie when dominant species abundances were experimentally altered, whereas declines of less common species did not affect productivity (Smith and Knapp 2003). Importantly, species richness and dominance (or evenness) do not always co-vary and thus dominance provides independent information on ecological functioning (Stirling and Wilsey 2001).

Some of the RS studies reviewed above illustrate remote sensing's ability to detect shifting species dominance, such as those addressing biomes shifts (shifting dominant vegetation), species invasions (increasing dominance of a new species) and beta diversity (when using abundance-weighted composition). Other work may implicitly address species dominance by identifying range distributions where species are assumed to have the highest density in the core of their range. However, these studies do not address which species in a community is more abundant or has greater biomass than coexisting species. Furthermore, relationships between species occurrence and dominance can be inconsistent depending on species and spatial scale (Schroeder et al. 2010). The most robust methods for using RS to detect dominant species will require ground-based field surveys of relative abundance. These data can be used as training data for classification approaches that allow extrapolation across space and time when RS provides data on species themselves (in contrast to RS providing data on environmental or ecological predictors). Moderate to coarse resolution multi-spectral RS will be more effective discriminating structurally and functionally distinct species (i.e. conifer vs. broadleaf; Fig. 1A); although pixel dissimilarity is also linked to abundance-weighted species turnover (e.g. Tuomisto et al. 2003; He et al. 2009; Oldeland et al. 2010).

In addition to detecting and characterizing communities based on dominant species, RS can help elucidate drivers of dominance by identifying how communities change along environmental gradients. RS provides valuable observations of environmental gradients and drivers that can influence dominance, including primary productivity, land-use/land cover and disturbances, in addition to climatic variables (see Table 1). More deliberately tracking dominance across environmental gradients should help inform our understanding of the relative importance of environmental drivers of ecological success (e.g. climate, land-use, disturbances) in comparison to biotic factors (competition or consumption), life-history strategies (dispersal and colonization) or historical processes (e.g. relict populations). The role of each of these drivers may change with spatial and temporal scale (i.e. grain size and extent), highlighting further how RS could facilitate new research directions in understanding species dominance. In the next section, we outline the value of remotely sensing dominant species for quantifying ESs.

Remote Sensing of Ecosystem Services

There is increasing evidence that dominant species are important contributors to many ESs that rely on abundance or biomass. For instance in many natural forests, dominant species store most above-ground carbon under current conditions (Balvanera et al. 2005; Ngo et al. 2013; Fauset et al. 2015). In the Amazon, a 'hyperdominant' 1% of all estimated 16 000 tree species provide around 50% of total tree carbon storage (ter Steege et al. 2013; Fauset et al. 2015). Recent syntheses show a similar pattern for pollination services: the most abundant pollinators provide the bulk of crop pollination services across several crop types and continents (Kleijn et al. 2015; Winfree et al. 2015). Similarly, in coastal ecosystems, the biomass of salt marsh vegetation and its density/complexity contribute to coastal protection by stabilizing the shoreline and attenuating wave energy (Sullivan and Zedler 1999; Zedler et al. 2001; Callaway et al. 2003; Sullivan et al. 2007; Shepard et al. 2011). As for these services, when the amount of an ES depends on abundance or biomass, tracking dominant species with RS offers a proxy for ES supply. ES supply refers to the amount of that ES that an ecological community can provide (also known as its capacity; Villamagna et al. 2013). Supply differs from how much people use or demand a service (known as ES 'flow'; Martínez-Harms and Balvanera 2012).

RS can track changes in populations of dominant species across changing conditions (Fig. 1), providing a scalable measure to monitor ESs. Some provisioning services (e.g. timber and food production) are derived directly from these populations, and regulating services depend on the EFs that populations support (e.g. carbon storage by tree populations). RS can estimate the size or extent of dominant populations or habitats/land-uses defined by dominant species, providing quantitative measures of ES supply (Table 1; Fig. 1). In the case of wetlands, which are characterized by their dominant vegetation (Zedler and Kercher 2005; Frieswyk et al. 2007), water regulation or flood control is associated with the size and distribution of the wetland (Troy and Wilson 2006). One study showed that freshwater wetlands (defined by the % cover of indicator species) yielded the highest ecosystem values compared to other habitats when their extents were mapped (Troy and Wilson 2006). Remote sensing of species' traits is another approach for detecting dominance-driven EFs and ESs. The effect of dominant species on EFs and ESs can be mediated through their respective traits, many of which are observable from space (reviewed in Jetz et al. 2016). Trait values of dominant species are disproportionately represented in communities leading to larger contributions to EFs and ESs like productivity (Smith and Knapp 2003 add Mason et al. 2016) and carbon storage (Fauset et al. 2015). However, some ESs, for example supporting or cultural services, are often provided by ecosystem integrity rather than functional traits of particular species.

Recent research shows the importance of considering spatiotemporal dynamics of ESs, rather than considering ES supply as static (Renard et al. 2015). While many ES studies consider static maps (Tallis et al. 2008), which generate an average ES value based on different land-cover types, considering spatiotemporal dynamics can lead to the different conclusions about the amount of an ES or trade-offs among multiple ESs (Bennett et al. 2009; de Groot et al. 2010; Renard et al. 2015; Tomscha and Gergel 2016). For instance Renard et al. (2015) analyze the spatiotemporal dynamics of livestock production, recreation, hunting, flood control and carbon sequestration, revealing not only that the magnitude of these ESs differs over time but also that trade-offs or synergies among ESs can fade or increase. Inaccurate estimates of ES value or of trade-offs among multiple ESs can result in suboptimal management decisions.

Despite the importance of considering spatiotemporal dynamics to account for and manage ESs, these data are costly and difficult to collect in the field. RS can aid these efforts by providing greater sampling and coverage. Remote sensing is especially useful for detecting changes in ESs (or ES proxies) relative to previous measurements. For example RS has reliably estimated annual food production from crop yields of the dominant crop species. Consecutive years of crop extent, yield, and planting dates were modeled for wheat fields in an intensive agricultural region in northwest Mexico (Lobell et al. 2003). Such changes in quantity can then be associated with different drivers, like variation in abiotic conditions (e.g. climate),

Table 1. Commonly used remotely sensed proxies for, or abiotic/biotic predictors of, ecosystem functions (modified from He et al. 2015). Some proxies measure biomass or abundance-driven ecosystem functions (e.g. plant productivity or standing tree biomass; forest structural complexity) that support ecosystem services (ESs; e.g. carbon storage and ultimately climate regulation; biodiversity habitat associated with dominant vegetation), while other proxies more directly estimate ESs (e.g. crop provisioning). Some variables could be considered both abiotic and biotic (land-use/land cover and land surface temperature) depending on the surface of interest (we note that the concept of a niche acknowledges that species alter their own environment; Holt 2009). Examples of ESs are listed according to the Millennium Ecosystem Assessment (2005) classifications (R = regulating, S = supporting, P = provisioning, C = cultural).

Predictor or proxy for ecosystem functions and services	Abiotic variables	Sources
• Agriculture (P,C) • Standing tree biomass (R,S) • Biodiversity habitat (S,C) • Recreation, sense of place & esthetics (C)	Land-use/Land cover (biotic)	MODIS, Landsat, Landsat ETM+
• Flood control (R)	Topography, elevation, slope	SRTM, LiDAR, WorldView-2, ASTER, GTOPO30, GMTED2010, UAVs
• Climate regulation (R) • Biodiversity habitat (S,C)	Land Surface Temperature (LST)	Landsat-8, MODIS
• Plant productivity (S) • Climate regulation (R)	Cloud cover and frequency, quality	MODIS, GOES, ISCCP
• Freshwater input as groundwater recharge or surface water (R,S,P,C)	Rainfall	TRMM, GPM, MODIS, NASA SMAP
• Plant productivity (S) • Biodiversity habitat (S,C)	Soil moisture	NASA SMAP
• Biodiversity habitat (S,C)	Fire (presence/burned area, frequency)	MODIS
	Biotic variables	
• Plant productivity (S) • Crop yield (P)	Normalized Difference Vegetation Index (NDVI)	AVHRR, Landsat, MODIS, Quickbird
• Plant productivity (S) • Crop yield (P)	Vegetation phenology	MODIS, Landsat
• Plant productivity (S) • Biodiversity habitat (S,C)	Leaf Area Index (LAI)	MODIS
• Plant productivity (S)	fPAR (fraction of absorbed photosynthetically active radiation)	MODIS, Landsat
• Biodiversity habitat (S,C) • Climate regulation (R) • Carbon storage (R) • Recreation, sense of place & esthetics (C)	Percent tree cover	MODIS VCF
• Biodiversity habitat (S,C) • Climate regulation (R) • Carbon storage (R)	Canopy/tree height	LiDAR, RADAR, MODIS VCF
• Biodiversity habitat (S,C) • Carbon storage (R)	Stem density	LiDAR, RADAR
• Microclimate (S)	Canopy moisture	Hyperspectral sensors, QSCAT
• Biodiversity habitat (S,C)	Canopy roughness/surface complexity	QSCAT
• Biodiversity habitat (S,C) • Carbon storage (R)	3-D habitat structural profile	LiDAR, RADAR
• Productivity (S)	Leaf water content	Hyperspectral sensors
• Productivity (S) • Nutrient cycling (R)	Leaf nitrogen content	Hyperspectral sensors

human impacts or management changes (e.g. land-use change and deforestation rates) (De Araujo Barbosa et al. 2015). For instance RS enabled global mapping of cropland distribution and their sensitivity to climate change (Ramankutty and Foley 1998; Ramankutty et al. 2002; Leff et al. 2004). Carbon stocks in tropical forests were estimated using RS measures of tree height and forest structure and then extrapolated across three continents based on their relationships with environmental predictors (i.e. NDVI, LAI, canopy roughness and topography; Saatchi et al. 2011). Assessing the effects of management and policy changes, Griffiths et al. (2012) used a Landsat time-series to examine how changing forest ownership and natural factors affected forest disturbance and recovery rates in Romania, with important implications for several ESs (e.g. carbon, water quality, timber production).

Ecosystem responses to external drivers are often nonlinear and exhibit threshold dynamics (Scheffer and Carpenter 2003; DeFries et al. 2004; Rockstrom et al. 2009). Species abundances and population dynamics are commonly characterized by population crashes or periodic irruptions that in turn alter their environment and population trajectories (Holt 2009). These unexpected and rapid changes will have important consequences for quantifying ESs (Gordon et al. 2008). RS can contribute to our ability to track nonlinear or threshold responses by providing independent information on the driver and/or response, which importantly can operate on different spatial and temporal scales. Forest insects and pathogens have resulted in massive disturbances affecting forest structure, function and ESs (Kurz et al. 2008). In the case of mountain pine beetles, the spatial and temporal characteristics of outbreaks are the result of complex feedbacks between beetle population sizes (endemic, epidemic, post-epidemic) and tree mortality or damage, which affect the spread of the beetle. Remote sensing can map and quantify the spatial, temporal and spectral damage from these outbreaks, such as lagged effects of different types of foliage damage (Wulder et al. 2006). The increasing frequency of extreme climatic events will also result in nonlinear and threshold ecosystem response (Jentsch et al. 2007). As just one example, RS allowed the tracking of mangroves across the east coast of Florida over several decades (Cavanaugh et al. 2014). This work showed that mangroves (dominated by one to three species) were insensitive to mean annual temperature and precipitation; rather, their distribution exhibited a threshold response to fewer cold events. This implies that associated ESs provided by mangroves – wood production, coastal protection and habitat for commercially harvested fish and invertebrates – might also exhibit nonlinear responses to environmental change.

Remotely sensing dominant species can detect early signs of future changes in ecological communities and the ESs they supply. Reorganization of ecological communities following global change may more immediately manifest as altered patterns of dominance and evenness rather than changes in species richness (Hillebrand et al. 2008). Climate warming, altered biogeochemical cycles, the addition or removal of species and land-use change all alter dominance structure (Walker et al. 2006; Hillebrand et al. 2008; Naeem 2009; Kardol et al. 2010; Forrestel et al. 2015). Therefore, detecting changes in dominant species may provide insight into future changes in both an ecological community and the ES it provides – before species losses, species turnover or changes in beta diversity occur. Furthermore, RS can measure flows and rates of ecosystem processes (e.g. change over space or time), offering early signs of changes in a population or future trends in ES supply.

The proposed focus on dominant species complements a growing body of work applying remote sensing to map and quantify ESs (reviewed in Ayanu et al. 2012; De Araujo Barbosa et al. 2015). One approach uses RS to derive biophysical parameters associated with an EF or ES in radiative transfer models, which establish relationships between an EF or ES proxy and how satellite measured radiation is transferred through different atmospheric and canopy conditions. Alternatively, an ES or EF is empirically linked to RS data using regression models. RS proxies can be used to map an ES, or provide information on predictors of an ES (e.g. abiotic variables and land-use change; Table 1; Ayanu et al. 2012). In particular, studies have linked RS data on ecosystem functions or species' traits, such as productivity, leaf biochemistry or growth forms, to field measurements (e.g. biomass and carbon storage) and ultimately to ESs (Table 1; Feng et al. 2010; Ayanu et al. 2012; De Araujo Barbosa et al. 2015). Empirical relationships between an ES and RS data can be modeled continuously: for each pixel, a unit change in RS data (e.g. % reflectance, land surface temperature (Kelvin); fire frequency (counts)) is associated with a unit change in an ES. More commonly, pixels with similar spectral information are grouped into classes to create thematic maps of land-use/land cover classifications (e.g. habitat type). Often studies map ESs by assigning an average amount of an ES to land cover type, or using land cover information as inputs into models that estimate ES supply for a location (Chan et al. 2006; Troy and Wilson 2006; Naidoo et al. 2008; Nelson et al. 2009a; Feng et al. 2010; Martínez-Harms and Balvanera 2012; Costanza et al. 2014; see Tallis and Polasky 2009 for a detailed explanation). For instance Nelson et al. (2009) generated maps of soil conservation as a function of land-use/land-cover and other inputs that determine soil retention (i.e. soil type, topography, rainfall). Despite the value of these existing studies, future work should develop and test meaningful RS proxies and RS drivers of dominance, and their link to ESs, to effectively contribute to ES assessments.

Conclusions

Rapid global change creates an urgent need for biodiversity monitoring to inform conservation and the management of ESs. Here we argue that RS can better support biodiversity monitoring by matching RS metrics with dominant species. We argue that dominance is an ecologically relevant and scalable measure (of population abundance, community composition, traits and ecosystem function), can provide an early-warning sign of ecological change, and overwhelmingly contributes to some biomass-driven ESs. RS metrics can be used as a response or predictors of EFs and ESs, or provide environmental correlates or predictors of dominance. Considering changes in dominant species is a new research agenda that extends current work on RS mapping of ESs.

RS mapping of dominance-driven ESs is a promising new direction to quantify how ESs change across space and time in an increasingly dynamic world. Focusing on dominant species contributes to ES assessments by harnessing the inherent bias of dominant species in optical RS, and providing cost-effective and near-continuous data to extrapolate point measurements. Indeed, RS can detect changes in dominance due to global change, which furthers our understanding of what structures ecological communities (e.g. biotic, abiotic, dispersal limitation) as well as the consequences for the supply of EFs and ESs. It is important to note that not all EFs and ESs are biomass or abundance-driven. A diversity of species, including rare species, are highly valued for various EFs and ESs, thus our proposition here should complement other EBV monitoring efforts.

Acknowledgments

We thank three anonymous reviewers for their helpful comments on a previous version of this paper. L.D. was supported by a grant from the Institute on the Environment at University of Minnesota to P.B. Reich and S. Polasky.

Conflict of Interest

None declared.

References

Asner, G. P., and R. E. Martin. 2009. Airborne spectranomics: mapping canopy chemical and taxonomic diversity in tropical forests. *Front. Ecol. Environ.* **7**, 269–276.

Asner, G. P., R. F. Hughes, P. M. Vitousek, D. E. Knapp, T. Kennedy-Bowdoin, J. Boardman, et al. 2008. Invasive plants transform the three-dimensional structure of rain forests. *Proc. Natl Acad. Sci. USA* **105**, 4519–4523.

Ayanu, Y., C. Conrad, T. Nauss, M. Wegmann, and T. Koellner. 2012. Quantifying and mapping ecosystem services supplies and demands: a review of remote sensing applications. *Environ. Sci. Technol.* **46**, 8529–8541.

Balvanera, P., C. Kremen, and M. Martínez-Ramos. 2005. Applying community structure analysis to ecosystem function: examples from pollination and carbon storage. *Ecol. Appl.* **15**, 360–375.

Beckage, B., B. Osborne, D. G. Gavin, C. Pucko, T. Siccama, and T. Perkins. 2008. A rapid upward shift of a forest ecotone during 40 years of warming in the Green Mountains of Vermont. *Proc. Natl Acad. Sci. USA* **105**, 4197–4202.

Bennett, E. M., G. D. Peterson, and L. J. Gordon. 2009. Understanding relationships among multiple ecosystem services. *Ecol. Lett.* **12**, 1394–1404.

Bradley, B. A. 2014. Remote detection of invasive plants: a review of spectral, textural and phenological approaches. *Biol. Invasions* **16**, 1411–1425.

Bradley, B. A., and E. Fleishman. 2008. Can remote sensing of land cover improve species distribution modelling? *J. Biogeogr.* **35**, 1158–1159.

Bradley, B. A., and J. F. Mustard. 2006. Characterizing the landscape dynamics of an invasive plant and risk of invasion using remote sensing. *Ecol. Appl.* **16**, 1132–1147.

Buckley, L. B., and W. Jetz. 2008. Linking global turnover of species and environments. *Proc. Natl Acad. Sci. USA* **105**, 17836–17841.

Buermann, W., S. Saatchi, T. B. Smith, B. R. Zutta, J. A. Chaves, B. Milá, et al. 2008. Predicting species distributions across the Amazonian and Andean regions using remote sensing data. *J. Biogeogr.* **35**, 1160–1176.

Callaway, J. C., G. Sullivan, and J. B. Zedler. 2003. Species-rich plantings increase biomass and nitrogen accumulation in a wetland restoration experiment. *Ecol. Appl.* **13**, 1626–1639.

Cavanaugh, K. C., J. R. Kellner, A. J. Forde, D. S. Gruner, J. D. Parker, W. Rodriguez, et al. 2014. Poleward expansion of mangroves is a threshold response to decreased frequency of extreme cold events. *Proc. Natl Acad. Sci. USA* **111**, 723–727.

Chan, K. M. A., M. R. Shaw, D. R. Cameron, E. C. Underwood, and G. C. Daily. 2006. Conservation planning for ecosystem services. *PLoS Biol.* **4**, 2138–2152.

Costanza, R., R. de Groot, P. Sutton, S. van der Ploeg, S. J. Anderson, I. Kubiszewski, et al. 2014. Changes in the global value of ecosystem services. *Glob. Environ. Change* **26**, 152–158.

Currie, D., and V. Paquin. 1987. Large-scale biogeographical patterns of species richness of trees. *Nature* **329**, 326–327.

Dangles, O., and B. Malmqvist. 2004. Species richness-decomposition relationships depend on species dominance. *Ecol. Lett.* **7**, 395–402.

Davidson, A., and F. Csillag. 2003. A comparison of three approaches for predicting C4 species cover of northern mixed grass prairie. *Remote Sens. Environ.* **86**, 70–82.

De Araujo Barbosa, C. C., P. M. Atkinson, and J. A. Dearing. 2015. Remote sensing of ecosystem services: a systematic review. *Ecol. Ind.* **52**, 430–443.

Deblauwe, V., V. Droissart, B. Sonke, J.-C. Svenning, J. Wieringa, B. R. Ramesh, et al. 2016. Remotely sensed temperature and precipitation data improve species distribution modelling in the tropics. *Glob. Change Biol.* **25**, 443–453.

DeFries, R. S., J. A. Foley, and G. P. Asner. 2004. Land-use choices: balancing human needs and ecosystem function. *Front. Ecol. Environ.* **2**, 249–257.

Fauset, S., M. O. Johnson, M. Gloor, T. R. Baker, M. A. Monteagudo, R. J. Brienen, et al. 2015. Hyperdominance in Amazonian forest carbon cycling. *Nat. Commun.* **6**, 6857.

Feng, X., B. Fu, X. Yang, and Y. Lü. 2010. Remote sensing of ecosystem services: an opportunity for spatially explicit assessment. *Chin. Geogr. Sci.* **20**, 522–535.

Foody, G. M., and J. Dash. 2010. Estimating the relative abundance of C3 and C4 grasses in the Great Plains from multi-temporal MTCI data: issues of compositing period and spatial generalizability. *Int. J. Remote Sens.* **31**, 351–362.

Forrestel, E. J., M. J. Donoghue, and M. D. Smith. 2015. Functional differences between dominant grasses drive divergent responses to large herbivore loss in mesic savanna grasslands of North America and South Africa. *J. Ecol.* **103**, 714–724.

Frieswyk, C. B., C. A. Johnston, and J. B. Zedler. 2007. Identifying and characterizing dominant plants as an indicator of community condition. *J. Great Lakes Res.* **33**, 125–135.

Gaston, K. J., R. G. Davies, C. D. L. Orme, V. A. Olson, G. H. Thomas, T. S. Ding, et al. 2007. Spatial turnover in the global avifauna. *Proc. Biol. Sci.* **274**, 1567–1574.

Gillespie, T. W. 2005. Predicting woody-plant species richness in tropical dry forests: a case study from south Florida, USA. *Ecol. Appl.* **15**, 27–37.

Gillespie, T. W., G. M. Foody, D. Rocchini, A. P. Giorgi, and S. Saatchi. 2008. Measuring and modelling biodiversity from space. *Prog. Phys. Geogr.* **32**, 203–221.

Goodin, D. G., and G. M. Henebry. 1997. Monitoring ecological disturbance in tallgrass prairie using seasonal NDVI trajectories and a discriminant function mixture model. *Remote Sens. Environ.* **61**, 270–278.

Gordon, L. J., G. D. Peterson, and E. M. Bennett. 2008. Agricultural modifications of hydrological flows create ecological surprises. *Trends Ecol. Evol.* **23**, 211–219.

Griffiths, P., T. Kuemmerle, R. E. Kennedy, I. V. Abrudan, J. Knorn, and P. Hostert. 2012. Using annual time-series of Landsat images to assess the effects of forest restitution in post-socialist Romania. *Remote Sens. Environ.* **118**, 199–214.

de Groot, R. S., R. Alkemade, L. Braat, L. Hein, and L. Willemen. 2010. Challenges in integrating the concept of ecosystem services and values in landscape planning, management and decision making. *Ecol. Complex.* **7**, 260–272.

Hawkins, B., R. Field, H. Cornell, D. Currie, J. Guegan, D. Kaufman, et al. 2003. Energy, water, and broad-scale geographic patterns of species richness. *Ecology* **84**, 3105–3117.

He, K. S., J. Zhang, and Q. Zhang. 2009. Linking variability in species composition and MODIS NDVI based on beta diversity measurements. *Acta Oecol.* **35**, 14–21.

He, K. S., D. Rocchini, M. Neteler, and H. Nagendra. 2011. Benefits of hyperspectral remote sensing for tracking plant invasions. *Divers. Distrib.* **17**, 381–392.

He, K. S., B. A. Bradley, A. F. Cord, D. Rocchini, M. N. Tuanmu, S. Schmidtlein, et al. 2015. Will remote sensing shape the next generation of species distribution models? *Remote Sens. Ecol. Conserv.* **1**, 4–18.

Hillebrand, H., D. M. Bennett, and M. Cadotte. 2008. Consequences of dominance: a review of evenness effects on local and regional ecosystem processes. *Ecology* **89**, 1510–1520.

Holt, R. D. 2009. Bringing the Hutchinsonian niche into the 21st century: ecological and evolutionary perspectives. *Proc. Natl Acad. Sci.* **106**, 19659–19665.

Jentsch, A., J. Kreyling, and C. Beierkuhnlein. 2007. A new generation of climate change experiments: events, not trends. *Front. Ecol. Environ.* **5**, 365–374.

Jetz, W., J. Cavender-Bares, R. Pavlick, D. Schimel, F. W. Davis, G. P. Asner, et al. 2016. Monitoring plant functional diversity from space. *Nat. Plants* **2**, 16024.

Kardol, P., C. E. Campany, L. Souza, R. J. Norby, J. F. Weltzin, and A. T. Classen. 2010. Climate change effects on plant biomass alter dominance patterns and community evenness in an experimental old-field ecosystem. *Glob. Change Biol.* **16**, 2676–2687.

Kerr, J. T., and M. Ostrovsky. 2003. From space to species: ecological applications for remote sensing. *Trends Ecol. Evol.* **18**, 299–305.

Kleijn, D., R. Winfree, I. Bartomeus, L. G. Carvalheiro, M. Henry, R. Isaacs, et al. 2015. Delivery of crop pollination services is an insufficient argument for wild pollinator conservation. *Nat. Commun.* **6**, 7414.

Kuenzer, C., M. Ottinger, M. Wegmann, and H. Guo. 2014. Earth observation satellite sensors for biodiversity monitoring: potentials and bottlenecks. *Int. J. Remote Sens.* **35**, 6599–6647.

Kurz, W. A., C. C. Dymond, G. Stinson, G. J. Rampley, E. T. Neilson, A. L. Carroll, et al. 2008. Mountain pine beetle and forest carbon feedback to climate change. *Nature* **452**, 987–990.

Leff, B., N. Ramankutty, and J. A. Foley. 2004. Geographic distribution of major crops across the world. *Global Biogeochem. Cycles* **18**, 1009.

Lobell, D. B., G. P. Asner, J. I. Ortiz-Monasterio, and T. L. Benning. 2003. Remote sensing of regional crop production in the Yaqui Valley, Mexico: estimates and uncertainties. *Agric. Ecosyst. Environ.* **94**, 205–220.

Lohbeck, M., L. Poorter, M. Martínez-Ramos, J. Rodriguez-Velázquez, M. van Breugel, and F. Bongers. 2014. Changing drivers of species dominance during tropical forest succession. *Funct. Ecol.* **28**, 1052–1058.

Martínez-Harms, M. J., and P. Balvanera. 2012. Methods for mapping ecosystem service supply: a review. *Int. J. Biodiv. Sci. Ecosys. Serv. Manage.* **8**, 17–25.

Mason, N. W., K. Orwin, S. Lambie, S. L. Woodward, T. McCready, and P. Mudge. 2016. Leaf economics spectrum-productivity relationships in intensively grazed pastures depend on dominant species identity. *Ecol. Evol.* **6**, 3079–3091.

Millennium Ecosystem Assessment. 2005. *Ecosystems and human well-being: synthesis.* Island Press, Washington, DC.

Mitchard, E. T. A., S. S. Saatchi, F. F. Gerard, S. L. Lewis, and P. Meir. 2009. Measuring woody encroachment along a forest-savanna boundary in Central Africa. *Earth Interact.* **13**, 1–29.

Naeem, S. 2009. Gini in the bottle. *Nature* **458**, 2–3.

Nagendra, H. 2001. Using remote sensing to assess biodiversity. *Int. J. Remote Sens.* **22**, 2377–2400.

Naidoo, R., A. Balmford, R. Costanza, B. Fisher, R. E. Green, B. Lehner, et al. 2008. Global mapping of ecosystem services and conservation priorities. *Proc. Natl Acad. Sci.* **105**, 9495–9500.

Nelson, E., G. Mendoza, J. Regetz, S. Polasky, H. Tallis, D. R. Cameron, et al. 2009. Modeling multiple ecosystem services, biodiversity conservation, commodity production, and tradeoffs at landscape scales. *Front. Ecol. Environ.* **7**, 4–11.

Ngo, K. M., B. L. Turner, H. C. Muller-Landau, S. J. Davies, M. Larjavaara, N. F. Bin Nik Hassan, et al. 2013. Carbon stocks in primary and secondary tropical forests in Singapore. *For. Ecol. Manage.* **296**, 81–89.

O'Connor, B., C. Secades, J. Penner, R. Sonnenschein, A. Skidmore, N. D. Burgess, et al. 2015. Earth observation as a tool for tracking progress towards the Aichi Biodiversity Targets. *Remote Sens. Ecol. Conserv.* **1**, 19–28.

Oldeland, J., D. Wesuls, D. Rocchini, M. Schmidt, and N. Jürgens. 2010. Does using species abundance data improve estimates of species diversity from remotely sensed spectral heterogeneity? *Ecol. Ind.* **10**, 390–396.

Palace, M., M. Keller, G. P. Asner, S. Hagen, and B. Braswell. 2008. Amazon forest structure from IKONOS satellite data and the automated characterization of forest canopy properties. *Biotropica* **40**, 141–150.

Pau, S., T. W. Gillespie, and E. M. Wolkovich. 2012. Dissecting NDVI-species richness relationships in Hawaiian dry forests. *J. Biogeogr.* **39**, 1678–1686.

Pau, S., E. J. Edwards, and C. J. Still. 2013. Improving our understanding of environmental controls on the distribution of C3 and C4 grasses. *Glob. Change Biol.* **19**, 184–196.

Pereira, H. M., S. Ferrier, M. Walters, G. N. Geller, R. H. G. Jongman, R. J. Scholes, et al. 2013. Essential biodiversity variables. *Science* **339**, 277–278.

Pettorelli, N., W. F. Laurance, T. G. O'Brien, M. Wegmann, H. Nagendra, and W. Turner. 2014a. Satellite remote sensing for applied ecologists: opportunities and challenges. *J. Appl. Ecol.* **51**, 839–848.

Pettorelli, N., K. Safi, and W. Turner. 2014b. Satellite remote sensing, biodiversity research and conservation of the future. *Phil. Trans. R. Soc. Biol. Sci.* **369**, 1–5.

Phillips, L. B., A. J. Hansen, C. H. Flather, and J. Robison-Cox. 2010. Applying species-energy theory to conservation: a case study for North American birds. *Ecol. Appl.* **20**, 2007–2023.

Purvis, A., and A. Hector. 2000. Getting the measure of biodiversity. *Nature* **405**, 212–219.

Qian, H., and R. E. Ricklefs. 2012. Disentangling the effects of geographic distance and environmental dissimilarity on global patterns of species turnover. *Glob. Ecol. Biogeogr.* **21**, 341–351.

Ramankutty, N., and J. A. Foley. 1998. Characterizing patterns of global land use: an analysis of global croplands data. *Global Biogeochem. Cycles* **12**, 667–685.

Ramankutty, N., J. A. Foley, J. Norman, and K. McSweeney. 2002. The global distribution of cultivable lands: current patterns and sensitivity to possible climate change. *Glob. Ecol. Biogeogr.* **11**, 377–392.

Renard, D., J. M. Rhemtulla, and E. M. Bennett. 2015. Historical dynamics in ecosystem service bundles. *Proc. Natl Acad. Sci. USA* **112**, 13411–13416.

Ricklefs, R. E. 1990. *Ecology*, 3rd ed. Freeman, New York.

Roberts, D. A., M. Gardner, R. Church, S. Ustin, G. Scheer, and R. O. Green. 1998. Mapping chaparral in the Santa Monica mountains using multiple spectral mixture models. *Remote Sens. Environ.* **65**, 267–279.

Rocchini, D. 2007. Distance decay in spectral space in analysis ecosystem β-diversity. *Int. J. Remote Sens.* **28**, 2635–2644.

Rocchini, D., N. Balkenhol, G. A. Carter, G. M. Foody, T. W. Gillespie, K. S. He, et al. 2010. Remotely sensed spectral heterogeneity as a proxy of species diversity: Recent advances and open challenges. *Ecol. Inform.* **5**, 318–329.

Rockstrom, J., W. Steffan, K. Noone, A. Persson, F. S. Chapin, E. F. Lambin, et al. 2009. A safe operating space for humanity. *Nature* **461**, 472–475.

Rompré, G., W. D. Robinson, A. Desrochers, and G. Angehr. 2007. Environmental correlates of avian diversity in lowland Panama rain forests. *J. Biogeogr.* **34**, 802–815.

Saatchi, S. S., N. L. Harris, S. Brown, M. Lefsky, E. T. Mitchard, W. Salas, et al. 2011. Benchmark map of forest carbon stocks in tropical regions across three continents. *Proc. Natl. Acad. Sci. USA* **108**, 9899–9904.

Scheffer, M., and S. R. Carpenter. 2003. Catastrophic regime shifts in ecosystems: linking theory to observation. *Trends Ecol. Evol.* **18**, 648–656.

Schroeder, T. A., A. Hamann, T. Wang, and N. C. Coops. 2010. Occurrence and dominance of six Pacific Northwest conifer species. *J. Veg. Sci.* **21**, 586–596.

Schwartz, M. W., C. A. Brigham, and J. D. Hoeksema. 2000. Linking biodiversity to ecosystem function: implications for conservation ecology. *Oecologia* **122**, 297–305.

Shepard, C. C., C. M. Crain, and M. W. Beck. 2011. The protective role of coastal marshes: a systematic review and meta-analysis. *PLoS One* **6**, e27374.

Skidmore, A. K., N. Pettorelli, N. C. Coops, G. N. Geller, M. Hansen, R. Lucas, et al. 2015. Agree on biodiversity metrics to track from space. *Nature* **523**, 403–405.

Smith, M. D., and A. K. Knapp. 2003. Dominant species maintain ecosystem function with non-random species loss. *Ecol. Lett.* **6**, 509–517.

Socolar, J. B., J. J. Gilroy, W. E. Kunin, and D. P. Edwards. 2016. How should beta-diversity inform biodiversity conservation? *Trends Ecol. Evol.* **31**, 67–80.

ter Steege, H., N. C. A. Pitman, D. Sabatier, C. Baraloto, R. P. Salomão, J. E. Guevara, et al. 2013. Hyperdominance in the Amazonian tree flora. *Science* **342**, 1243092.

Stirling, G., and B. Wilsey. 2001. Empirical relationships between species richness, evenness, and proportional diversity. *Am. Nat.* **158**, 286–299.

Stow, D. A., A. Hope, D. McGuire, D. Verbyla, J. Gamon, F. Huemmrich, et al. 2004. Remote sensing of vegetation and land-cover change in Arctic tundra ecosystems. *Remote Sens. Environ.* **89**, 281–308.

Sullivan, G., and J. B. Zedler. 1999. Functional redundancy among tidal marsh halophytes: a test. *Oikos* **84**, 246–260.

Sullivan, G., J. C. Callaway, and J. B. Zedler. 2007. Plant assemblage composition explains and predicts how biodiversity affects salt marsh functioning. *Ecol. Monogr.* **77**, 569–590.

Tallis, H., and S. Polasky. 2009. Mapping and valuing ecosystem services as an approach for conservation and natural-resource management. *Ann. N. Y. Acad. Sci.* **1162**, 265–283.

Tallis, H., P. Kareiva, M. Marvier, A. Chang, and A. H. Mwinyi. 2008. An ecosystem service framework to support both practical conservation and economic development. *Proc. Natl Acad. Sci. USA* **105**, 9457–9465.

Tomscha, S. A., and S. E. Gergel. 2016. Ecosystem service trade-offs and synergies misunderstood without landscape history. *Ecol. Soc.* **21**, 43.

Troy, A., and M. A. Wilson. 2006. Mapping ecosystem services: practical challenges and opportunities in linking GIS and value transfer. *Ecol. Econ.* **60**, 435–449.

Tuomisto, H. A., A. D. Poulsen, K. Ruokolainen, and R. Moran. 2003. Linking floristic patterns with soil heterogeneity and satellite imagery in Ecuadorian Amazonia. *Ecol. Appl.* **13**, 352–371.

Turner, W. 2014. Sensing biodiversity. *Science* **346**, 301–302.

Turner, W., S. Spector, N. Gardiner, M. Fladeland, E. Sterling, and M. Steininger. 2003. Remote sensing for biodiversity science and conservation. *Trends Ecol. Evol.* **18**, 306–314.

Villamagna, A. M., P. L. Angermeier, and E. M. Bennett. 2013. Capacity, pressure, demand, and flow: a conceptual framework for analyzing ecosystem service provision and delivery. *Ecol. Complex.* **15**, 114–121.

Walker, B., A. Kinzig, and J. Langridge. 1999. Plant attribute diversity, resilience, and ecosystem function: the nature and significance of dominant and minor species. *Ecosystems* **2**, 95–113.

Walker, M. D., C. H. Wahren, R. D. Hollister, G. H. R. Henry, L. E. Ahlquist, J. M. Alatalo, et al. 2006. Plant community responses to experimental warming across the tundra biome. *Proc. Natl Acad. Sci. USA* **103**, 1342–1346.

Waring, R. H., N. C. Coops, J. L. Ohmann, and D. A. Sarr. 2002. Interpreting woody plant richness from seasonal ratios of photosynthesis. *Ecology* **83**, 2964–2970.

Weisberg, P. J., E. Lingua, and R. B. Pillai. 2007. Spatial patterns of Pinyon–Juniper woodland expansion in central Nevada. *Rangeland Ecol. Manag.* **60**, 115–124.

Whittaker, R. H. 1975. *Communities and ecosystems*, 2nd ed. MacMillan, New York.

Winfree, R., J. W. Fox, N. M. Williams, J. R. Reilly, and D. P. Cariveau. 2015. Abundance of common species, not species richness, drives delivery of a real-world ecosystem service. *Ecol. Lett.* **18**, 626–635.

Wulder, M. A., C. C. Dymond, J. C. White, D. G. Leckie, and A. L. Carroll. 2006. Surveying mountain pine beetle damage of forests: a review of remote sensing opportunities. *For. Ecol. Manage.* **221**, 27–41.

Zedler, J. B., and S. Kercher. 2005. Wetland resources: status, trends, ecosystem services, and restorability. *Ann. Rev. Environ. Res.* **30**, 39–74.

Zedler, J. B., J. C. Callaway, and G. Sullivan. 2001. Declining biodiversity: why species matter and how their functions might be restored in Californian tidal marshes. *Bioscience* **51**, 1005.

10

Observing ecosystems with lightweight, rapid-scanning terrestrial lidar scanners

Ian Paynter[1], Edward Saenz[1], Daniel Genest[1], Francesco Peri[1], Angela Erb[1], Zhan Li[1], Kara Wiggin[1], Jasmine Muir[2], Pasi Raumonen[3], Erica Skye Schaaf[4], Alan Strahler[5] & Crystal Schaaf[1]

[1]School for the Environment, University of Massachusetts Boston, Boston, Massachusetts
[2]Department of Science, Information Technology and Innovation, Queensland, Brisbane, Australia
[3]Department of Mathematics, Tampere University of Technology, Tampere, Finland
[4]Department of Geography, McGill University, Quebec, Montreal, Canada
[5]Department of Earth and Environment, Boston University, Boston, Massachusetts

Keywords
Compact biomass lidar (CBL), ecosystem properties, lidar, quantitative structure models, terrestrial lidar scanners, validation

Correspondence
Ian Paynter, School for the Environment, University of Massachusetts Boston, 100 Morrissey T. Blvd, Boston, MA 02125.
E-mail: ian.paynter@umb.edu

Funding Information
We also acknowledge financial support from the Oracle Graduate Fellowship, the NASA Harriet Jenkins Graduate Fellowship, the UMass Boston International Research Initiative Seed Grants Program, and NASA grants NNX14AK12G and NNX14AI73G.

Editor: Harini Nagendra
Associate Editor: Doreen Boyd

Abstract

A new wave of terrestrial lidar scanners, optimized for rapid scanning and portability, such as the Compact Biomass Lidar (CBL), enable and improve observations of structure across a range of important ecosystems. We performed studies with the CBL in temperate and tropical forests, caves, salt marshes and coastal areas subject to erosion. By facilitating additional scanning points, and therefore view angles, this new class of terrestrial lidar alters observation coverage within samples, potentially reducing uncertainty in estimates of ecosystem properties. The CBL has proved competent at reconstructing trees and mangrove roots using the same cylinder-based Quantitative Structure Models commonly utilized for data from more capable instruments (Raumonen et al. 2013). For tropical trees with morphologies that challenge standard reconstruction techniques, such as the buttressed roots of *Ceiba* trees and the multiple stems of strangler figs, the CBL was able to provide the versatility and the speed of deployment needed to fully characterize their unique features. For geomorphological features, the deployment flexibility of the CBL enabled sampling from optimal view-angles, including from a novel suspension system for sampling salt marsh creeks. Overall, the practical aspects of these instruments, which improve deployment logistics, and therefore data acquisition rate, are shown to be emerging capabilities, greatly increasing the potential for observation, particularly in highly temporally dynamic, inaccessible and geometrically complex ecosystems. In order to better analyze information quality across these diverse and challenging ecosystems, we also provide a novel and much-needed conceptual framework, the microstate model, to characterize and mitigate uncertainties in terrestrial lidar observations.

Introduction

Light detection and ranging (lidar) instruments are rapidly developing to provide robust estimates of important ecosystem properties at new spatial and temporal scales. Uncertainty in these estimates arises from characteristics of the instrument, complexities of the ecosystem and weather conditions during sampling. We provide a new conceptual framework, the *microstate model*, to

establish these uncertainties. Through the application of the microstate model to a series of data-driven case studies of diverse ecosystems, we explore the unique advantages that these newly emerging rapid-scanning and lightweight lidar instruments may offer for reliably observing ecosystems.

Lidar instruments emit pulses of light energy. These pulses are reflected from objects, and lidar instruments record the time-of-flight and intensity of the returning energy. This process provides information about the three-dimensional structure of an object or environment. Since many aspects of an ecosystem's condition are manifest in properties of its structure, lidar can monitor the condition of ecosystems through observations of their structure. As a result, lidar data are increasingly being used to augment and refine traditional passive remote sensing in the accurate assessment and reliable monitoring of ecosystems (Tang et al. 2014).

Lidar observations can be made at a range of spatial and temporal scales (Dubayah and Drake 2000; Lefsky et al. 2002; Hurtt et al. 2004; Zhao et al. 2012; Wulder et al. 2012), and from a variety of platforms (Lovell et al. 2003; Pfeifer and Briese 2007; Zolkos et al. 2013; Lucas et al. 2015; Woodhouse et al. 2011). This study focuses on terrestrial lidars, which typically scan their surroundings from a fixed point, using rotating mirrors and rotary stages to emit pulses in the surrounding directions. Terrestrial lidars have proven particularly useful for observing ecosystem structural properties in fine detail. This characterization of near-surface variation has established terrestrial lidars as important validation tools for airborne and satellite observations (Tang et al. 2014), as well as tools capable of producing independent observations of ecosystem structural properties.

At the present time, terrestrial lidar scanners are developing iteratively along two parallel pathways, which can be loosely termed *capability* and *practicality*. Capability is characterized by improvements in the attributes of individual lidar pulses, such as range, resolution and emitted wavelengths, increasing the maximum capabilities of an instrument. Contemporary archetypes include the research instruments, the Dual-Wavelength Echidna Lidar (DWEL) (Strahler et al. 2008; Douglas et al. 2012; Howe et al. 2015; Li et al. 2016) and the SALford Advanced Canopy Analyzer (SALCA) (Danson et al. 2014), and commercially available instruments such as the Riegl VZ-400 (Calders et al. 2015). While some of these highly capable instruments allow for variable scanning settings to decrease resolution and increase scanning speed, small, practical terrestrial lidar instruments offer the means to augment and extend the extensive, high-caliber information acquired with more capable scanners. Therefore, in this context, practicality is characterized by refinement, generally in the form of miniaturization and efficiency, to produce instruments that are lighter, faster and more resilient. Prominent examples include the research Compact Biomass Lidar (CBL) used in this study, and the commercially available Zebedee (Bosse et al. 2012) In the context of terrestrial lidar used for ecological investigation, the practical attributes of acquisition speed, portability and resilience can in themselves be considered important capabilities, necessary to properly characterize aspects of complex ecosystems.

There are many uncertainties and challenges involved in making terrestrial lidar observations of ecosystems. Prominently, lidar observations of structure are restricted to line-of-sight, creating a challenge when assessing the structural properties of an ecosystem. Since constrained areas of ecosystems are typically used as samples, it is expected that these areas will be fully observed with detailed lidar acquisitions. Yet, the line-of-sight limitations of lidar mean that certain regions or objects within ecosystem samples will be occluded, and therefore will not be observed, resulting in an incomplete characterization of the sample.

Lidar observations are further complicated as terrestrial lidar information density is range-dependent (Côté et al. 2009, 2011; Dassot et al. 2011), due to the angular separation between emitted pulses, resulting in larger gaps between observations over distance. Additionally, the quality of captured information can also be range-dependent, since individual lidar beams diverge with distance, resulting in lower specificity and accuracy of object detection. This means that even observed regions of samples are unlikely to have equal density and quality of information.

In response to these problems, it is common to perform multiple terrestrial lidar scans within an ecosystem sample, thus increasing the density of observations and variety of view angles. However, even the most intensive scanning cannot guarantee complete observation of an ecosystem sample. Additionally, capturing more scans considerably increases observation time, challenging the ability of terrestrial lidar to capture a complete sample, especially in ecosystems with high levels of temporal variation. Therefore, a comprehensive systematic understanding of the challenges to terrestrial lidar observations of ecosystems is urgently needed, accompanied by practical approaches to quantifying and mitigating these challenges.

The microstate model: a framework for understanding uncertainty in terrestrial lidar observations

Terrestrial lidar observations of ecosystem samples are typically made to assess one or more properties of interest, such as the number or volume of trees, the volume of a creek, or the surface structure of an eroding bluff. To understand and help mitigate the uncertainties in such observations of properties of interest within ecosystem samples, we have developed a conceptual framework called the *microstate model* (Fig. 1). This framework contextualizes aspects of documented sampling uncertainties in remote sensing (Phinn 1998) particularly for lidar sampling of ecosystems, and provides the systematic relationships between these aspects.

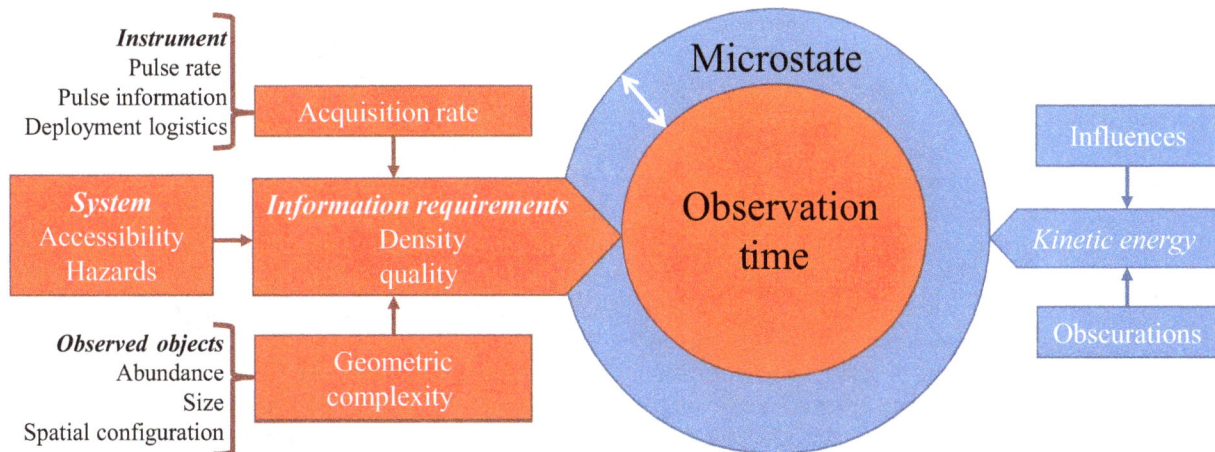

Figure 1. Conceptual diagram for the microstate model, designed to describe and provide a pathway to mitigate uncertainties and errors that arise from sampling ecosystem properties with lidar.

The microstate model comprises two major components. The first component describes how ecosystem temporal dynamism bounds the time available for sample observations. The second component concerns the propagation of error from individual lidar pulses, through interactions with ecosystem geometry, and ultimately to the uncertainty in an entire sample. Overall, the microstate model describes the requirements for terrestrial lidar observations of ecosystem samples to be appropriately constrained, complete and comparable.

Microstate model component 1: ecosystem temporal dynamism

Ecosystems are subjected to many temporally varying factors. These result from the import, export, movement or state change of matter within the ecosystem (Côté et al. 2009, 2011; Hancock et al. 2014). Intuitive examples arise from tidal cycles, seasonal influences, weather conditions, and other natural and anthropogenic disturbances. These temporal factors should be understood relative to the ecosystem properties of interest, such as tree height or creek volume. We can then categorize temporal factors as either obscurations of a property of interest, or influences upon it.

Influences are events that change the true value of a property of interest. They can take the form of import or export of relevant material from the sample space, or a state change of material within the sample space. *Obscurations,* on the other hand, are events that prevent, or increase uncertainty in, the observation of a property of interest, without changing its true value. One form of obscuration is the introduction of material to the sample space, such as a rising tide occluding geomorphological features. Obscurations can also be events that change the

position of a terrestrial lidar relative to objects, or vice versa. Wind, for example, can alter the position of terrestrial lidars or the position of flexible objects such as tree branches during a scan, increasing uncertainty or preventing observation.

Appropriate observations of ecosystems can only be made during windows of time in which conditions are stable. Therefore, in the terminology just established, observations can only be made when obscurations and influences are absent. To define these stable spans of time, we can borrow the concept of 'microstate' from thermodynamics, where it refers to an instantaneous energetic state of a molecular-scale system (Hilser et al. 2006). A microstate in the context of ecosystems is a temporal window in which the ecosystem permits a discrete and accurate observation of a property of interest. By extension, observations made in equivalent microstates can be compared without concern. In practical terms, a microstate is the time *available* to complete an observation of one or more properties of interest of an ecosystem sample.

Microstate model component 2: sampling requirements

Lidar uncertainties originate at the pulse level, due to multiple factors which are dependent on the instrument's attributes and specifications. Primarily, uncertainties occur in the trajectory of an emitted pulse, and the range to an object, resulting in error between the observed and true position of a scattering object. Other sources of uncertainty include the shape and power of the outgoing pulse (Hosoi and Omasa 2007; Côté et al. 2009, 2011; Danson et al. 2014; Jupp et al. 2009); beam divergence; detector noise (Hancock et al. 2014; Douglas et al. 2015); and the precision of change detectable by the instrument.

Several of the factors that introduce uncertainty into each lidar pulse return can be range-dependent. This makes the uncertainty in ecosystem sampling partly dependent on the ecosystem's spatial properties and the geometry of objects in an ecosystem sample. An illustrative example is beam divergence, which is the increase in the width of a lidar beam with distance (Béland et al. 2014). Any interaction between an outgoing lidar pulse and an object is placed on the central trajectory of a pulse, which is inaccurate when an object only intersects a non-central portion of the beam. When lidar data are used to reconstruct objects, these misplaced returns can change the overall dimensions and other geometric properties of the objects, and this potential for positional error increases as the beam divergence increases with range.

The geometry and spatial configuration of objects in an ecosystem contribute greatly to the challenge of observing that ecosystem. The integration of the size, abundance, relative position and orientation of objects within the sample space of an ecosystem can be described as its *geometric complexity*. The geometric complexity of an ecosystem interacts with the line-of-sight nature of lidar to create inconsistent information gain and quality, causing the time required to observe an ecosystem sample to vary non-linearly with the overall size of the sample.

Additionally, to decide whether an ecosystem sample has been fully observed, we require a measurable threshold of the adequate terrestrial lidar information quantity and quality that must be met for each region and object in the sample for the sample to be considered valid. These thresholds can be collectively termed the *information requirements* for the sample. Appropriate values for information requirements would depend on the desired level of uncertainty in the final estimation of an ecosystem property. Preliminary experimentation, simulation or reference to similar studies can be used to suggest appropriate information requirements.

Different terrestrial lidars vary in the speed at which they can make observations of an ecosystem, according to their pulse rate and the space observed by each pulse. Also of importance are an instrument's deployment logistics, which are the time it takes to move and deploy the instrument. The combination of these factors gives each instrument an *acquisition rate*. The acquisition rate of an instrument, integrated with the information requirements of a study, and the geometric complexity of an ecosystem, determine an overall *observation time*. To summarize, the observation time is the time required to complete a valid observation of an ecosystem sample, while the microstate is the time available to complete an observation. If the observation time is shorter than the microstate, then the ecosystem sample can be validly observed (Fig. 1).

Materials and Methods

Hereafter, we describe several studies involving structural modeling of diverse ecosystems, conducted with terrestrial lidars optimized for acquisition speed, portability and resilience. These studies explore many challenges arising from the interactions described in the microstate model.

The terrestrial lidar data utilized for this study were provided by the first and second iterations of the CBL1 and CBL2 produced the by University of Massachusetts Boston (UMB) (Fig. 2). The instrument design is based on an initial concept by Katholieke Universiteit Leuven (Van der Zande et al. 2006), realized by the Rochester Institute of Technology (Kelbe et al. 2013, 2015) and extensively refined by UMB. The detailed specifications of the instruments are included in Figure 2. Note that the vertical angular resolution is the only data-relevant difference between the CBL1 and CBL2.

Unless otherwise stated, the instruments were deployed on a lightweight tripod such that the optical center was approximately 1.3 m above the ground. However, the light weight, compactness and resilience of the CBL provides immense flexibility for mounting the instrument on nontraditional platforms, such as tripods taller than 10 m, walk-up towers and overhead suspension systems. Where relevant, these are described below.

When multiple scans are acquired within a sample, these scans must be co-located (adjusted for relative position and rotation). For the CBL data in this study, an initial unsupervised co-location was performed when the relative location of scan positions was known. The final co-location of scans was supervised, involving visual assessment of common objects and features. Any returns not corresponding to the object of interest were manually removed.

Case Studies and Results

Norway maple

Trees are objects with extremely high geometric complexity, creating complex patterns of occlusion, and considerable uncertainty in their structural modeling with lidar. CBL2 scans were obtained at a distance of approximately 5 m in the four cardinal directions around a Norway maple (leaves on) in managed parkland in Boston, Massachusetts. Quantitative Structure Models (QSMs) for the tree were formed from the extracted tree points, using a reconstruction algorithm specifically designed for terrestrial lidar observations of trees (Raumonen et al. 2013; Kaasalainen et al. 2014; Krooks et al. 2014; Calders et al. 2015). The algorithm uses small point clusters to first segment the data into individual branches, and then fits a network of cylinders to the segmented data. Properties of

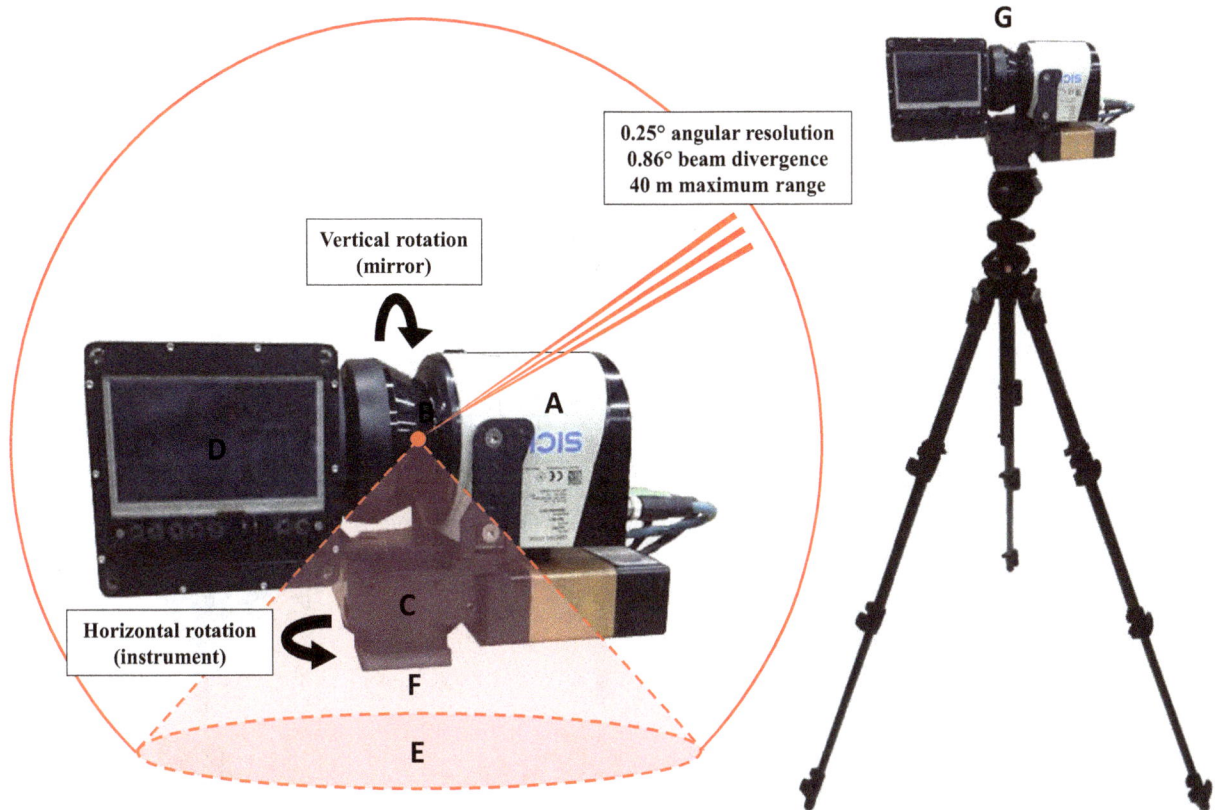

Figure 2. Compact Biomass Lidar 2 (CBL2) overall specifications: low weight (3.4 kg without tripod); small dimensions (40 cm × 15 cm × 25 cm); waterproof (IPv68); rapid scan time (33 sec); long battery life (1.5 h active use). (A) SICK LMS 151 Lidar unit: 905 nm wavelength; 40 m maximum range; 0.86° beam divergence; discrete returns (first and second); uncalibrated intensity. (B) Rotating mirror and detector: 0.25° vertical resolution (CBL1 0.5° vertical resolution). (C) Motor-driven rotary stage: 0.25°horizontal resolution; 360° horizontal view. (D) Beagle Bone Computer: Onboard data ingest; Onboard storage; Wireless control; Touchscreen control. (E) 270° vertical view (F) Flexible attachment for mounting on diverse platforms. (G) Mounted on tripod: 1.3 m optical center height.

the tree, including woody volume by branching order, can then be estimated using the geometric properties of the cylinders. The successful application of QSMs has been well-documented, particularly with highly capable terrestrial lidars (Calders et al. 2015; Raumonen et al. 2015). Since the clustering component of the algorithm involves random search patterns, 100 repeats of the reconstruction process were conducted with the CBL data (Fig. 3). The CBL produced reasonable and adequately stable volume estimates, which is encouraging given its considerable advantage in observation time, resulting from the instrument's high acquisition rate of 33 sec per scan (Fig. 2).

However, uncertainty in volume estimates between the QSM repeats, shown by the coefficient of variation, did notably increase with branching order (Table 1). This increased uncertainty could be attributed to several factors. Beam divergence creates range-dependent

uncertainty in object position and geometry, which could translate into proportionally higher volume uncertainty in smaller objects such as the higher order branches, which are also generally farther from the terrestrial lidar when the instrument is deployed near ground-level. The information quantity and quality could also be decreased by occlusion from the lower order branches. Finally, although weather conditions were stable in this study, any obscuration by wind would have the greatest effect on the smaller, higher order branches, changing their position within and between scans, and increasing uncertainty.

Mangrove tree

Mangroves trees have exposed root systems, which increases their geometric complexity. Mangrove trees also reside in ecosystems with extremely high temporal dynamism, as tides can result in microstates of as little as 2 h

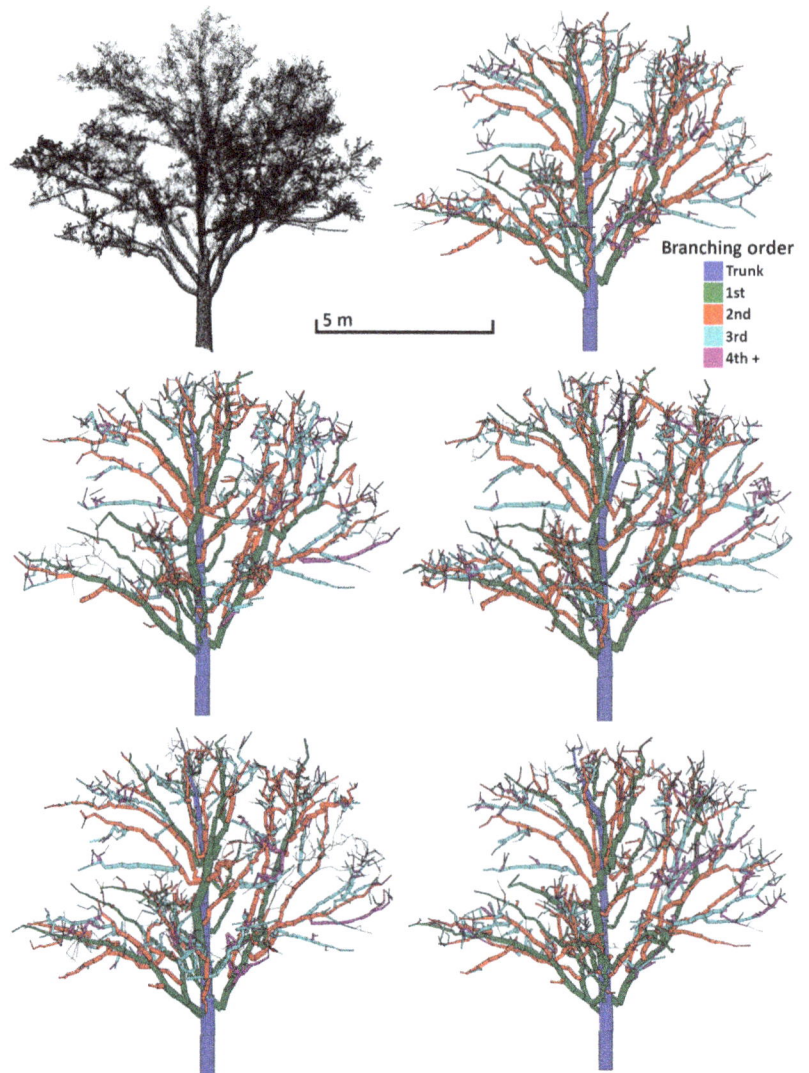

Figure 3. Point cloud (top left) of Norway Maple (Boston, MA), assembled from four CBL scans. The other panels are five Quantitative Structure Models (QSM). QSM reconstruction utilizes random selection of points during its clustering analysis, which creates variation between individual models, even while parameters remain the same. While some small differences between these cylinder models are visible, the overall structural consistency, and the quality of the representation of the point cloud are encouraging. Based on preliminary experimentation, QSM was utilized without filtering and parameterized as follows: $d1 = 0.2$; $r1 = 0.22$; $n1 = 3$; $d2 = 0.05$; $r2 = 0.23$; $n2 = 3$; $l = 6$; $f = 0.5$ (Raumonen et al. 2013, 2015). CBL, compact biomass lidar; QSM, quantitative structure models.

during which the full tree structure can be observed. In addition, mangroves are slow and hazardous to traverse, increasing observation time.

CBL1 scans were obtained of a 400 m² plot of red mangroves in Drake's Bay, Sarapiqui, Costa Rica, as part of a NASA EcoSAR (Rincon et al. 2011) field campaign. The plot was divided according to a 5 m × 5 m grid and scans were taken on the vertices of the grid. The entire 25 scan observation was conducted during a single low tide, leveraging the CBL's high acquisition rate (33 sec, 40 m range). For the reconstruction of the mangrove stem and root structure presented, the four closest scans were utilized. A QSM for the roots was reconstructed according to the principles used to reconstruct branches.

A hundred QSMs were successfully generated from the mangrove roots point cloud (Fig. 4), but higher coefficients of variation (Table 2) than in the Norway Maple tree (Table 1) suggested a higher overall uncertainty. This could

be attributed to the need to adjust the parameterization of the QSM reconstruction algorithm when modeling roots, or that the CBL1's lower vertical (zenithal) resolution resulted in less certainty in object structure.

Deployment of terrestrial lidar in mangrove ecosystems has only previously been possible in areas with boardwalks (Knight et al. 2009; Kamal et al. 2015) or with deployment techniques that greatly increase observation time, such as scanning from a boat (Feliciano et al. 2014). The water-proofing of the CBL, coupled with its low weight, and favorable deployment logistics, enabled the observation of the particularly inaccessible mangroves in this study (Fig. 4).

Ceiba tree trunk

Tropical forest ecosystems have high geometric complexity at multiple scales. All trees are complex objects, but many tropical species have unique morphological features, such

Table 1. Volume estimates for the Norway Maple (Boston, MA) from QSMs.

	Mean (m³)	S.D (m³)	S.D/Mean (coeff. var.)
Total	3.766	0.153	0.041
Trunk	0.654	0.026	0.040
1st order branches	1.166	0.090	0.077
2nd order branches	1.153	0.093	0.081
3rd order branches	0.604	0.061	0.100
4th+ order branches	0.189	0.045	0.235

The mean, standard deviation and coefficient of variation (standard deviation divided by the mean) are presented for the 100 QSMs. Volumes are divided by branching order, with 1st order branches being those that are directly attached to the trunk, 2nd order directly branching off from 1st order, and so on. Note that the coefficient of variation increases through the branching orders, suggesting more variation between QSMs, and therefore higher uncertainty.

as buttressed roots, which further increase their geometric complexity. In addition, vertical stratification of vegetation, typically including dense understory, greatly increases occlusion and therefore observation time. CBL2 scans were obtained of the lower trunk of a large (approximately 40 m height) tree of the genus *Ceiba* within a Carbono study site (Clark and Clark 2000) at La Selva Biological Research Station (Sarapiqui, Costa Rica, 2014). Six scans were taken consecutively, at an approximate distance of 4 m from the tree center and at intervals of approximately 60°. An additional scan was taken inside a large hollow at the base of the trunk at an optical center height of 0.4 m. Observing atypical morphological features such as this hollow could improve non-destructive volume estimates for trees.

The outside of the trunk was captured with (Fig. 5) little apparent occlusion. The high acquisition rate of the CBL facilitated the observation of the many view angles necessary to fully capture the complex structure of the base of the *Ceiba* tree including its characteristic buttressed roots. The internal hollow occupied a substantial proportion of the tree base, and was a convoluted structure (Fig. 5). The compactness of the CBL, as well as the flexibility of its deployment platform, enabled the observation of this geometrically complex region of the object, which could impact volume, and therefore biomass, estimates from airborne and forestry measures.

There were still localized missing portions of the external and internal structure due to occlusion (Fig. 5). These could have easily been observed during the deployment with additional scans, given the high acquisition rate of the CBL. However, recognizing the need for additional scans would have required that information gain be characterized between scans, allowing iterative assessment of the occlusion.

Strangler fig tree

In addition to the previously described geometric complexity inherent to tropical forest ecosystems, strangler fig trees consist of many closely grouped stems, which are impossible to fully observe with lidar from the perimeter of the tree. CBL1 scans were obtained of a large, free-standing strangler fig in Corcovado National Park (Costa Rica, 2014) (Fig. 6) as part of a NASA EcoSAR (Rincon et al. 2011) field campaign. Scans were taken at intervals of approximately 12° around the strangler fig center (totaling 30 scans). Four additional scans were taken in the most accessible locations between the stems of the tree. Scanning between the stems of the strangler fig was only possible due to the compactness and portability of the CBL, and many structural components of the strangler fig were only captured by these internal scans (Fig. 7).

This study shows how the practical attributes of a terrestrial lidar can interact with the geometric complexity of an ecosystem sample. The scans taken around the strangler fig achieved higher information density in the lower regions of the tree (Fig. 7), even though the upper regions had a much greater extent. This suggests that internal regions of the canopy were occluded, which could be addressed in similar situations by deploying the CBL on a platform offering higher scanning positions. The lower vertical (zenithal) resolution of the CBL1 (0.5 degrees) compared to the CBL2 (0.25 degrees), and the range-dependent uncertainty resulting from beam divergence, may also contribute to the coarser definition of upper-canopy features.

Cave

Caves possess geometric complexity both in the form of large and frequently inaccessible occluded regions, and smaller features such as stalactites which contribute inter-object occlusion patterns similar to tree branches. CBL1 scans were obtained throughout a system of showcaves (Treak Cliff Caverns, Derbyshire, UK, 2014), for mapping. The scanning positions were determined ad hoc to maximize coverage, with consecutive scans in mutual line of sight (totaling 36 scans).

The favorable acquisition rate of the CBL permitted rapid capture of data throughout the entire cave system, providing excellent representation of the superstructure (Fig. 8). However, the resolution of the CBL1 limited its ability to characterize fine-scale structures such as stalactites. Higher resolution terrestrial lidar have proved capable of observing cave geomorphological structure in great detail (Hoffmeister et al. 2015; Mohammed Oludare and Pradhan 2016), and the Treak Cliff Caverns, which have constructed pathways, would certainly have been

Remote Sensing Techniques for Environmental Studies

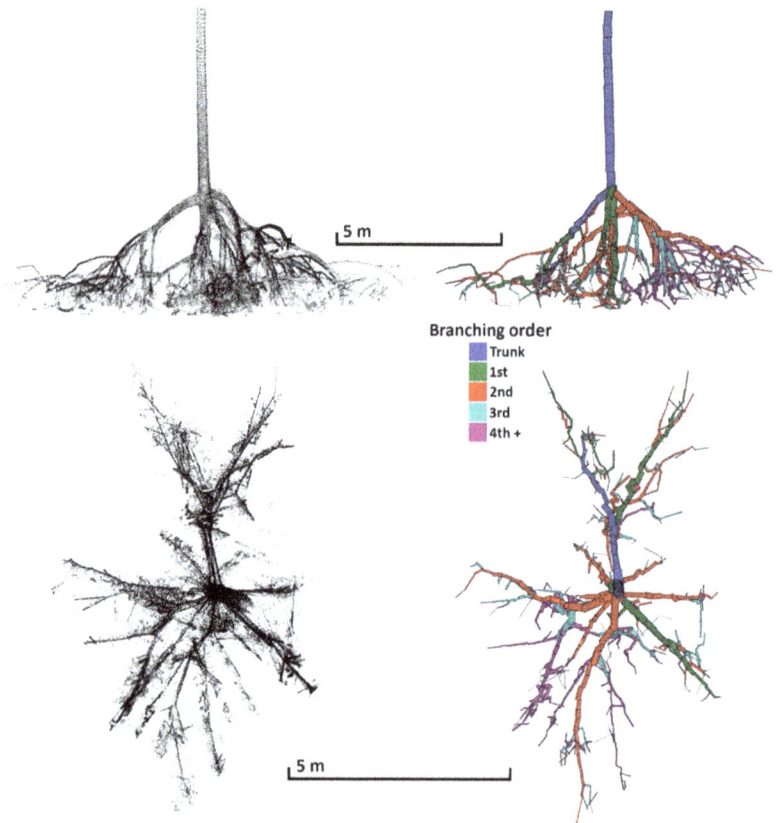

Figure 4. Point cloud (left) and QSM (right) of mangrove roots (Drake's Bay, Costa Rica). The cylinder model seems to represent the point cloud very well, even down to the fine roots. The overhead view (bottom) suggests that the four scans captured information to a reasonably high density with little residual occlusion. QSMs were reconstructed without filtering and parameterized as follows: d1 = 0.2; r1 = 0.22; n1 = 3; d2 = 0.05; r2 = 0.23; n2 = 3; l = 6; f = 0.5 (Raumonen et al. 2013, 2015). QSM, quantitative structure models.

Table 2. Volume estimates for mangrove tree roots from QSM.

	Mean (m^3)	S.D. (m^3)	S.D./Mean (coeff. var.)
Total	1.450	0.225	0.155
Trunk	0.465	0.054	0.115
1st order roots	0.363	0.144	0.396
2nd order roots	0.350	0.134	0.382
3rd order roots	0.183	0.059	0.328
4th+ order roots	0.088	0.056	0.632

The mean, standard deviation and coefficient of variation (standard deviation divided by the mean) are presented for 100 runs of QSM. Volumes are divided by root branching order, with 1st order branches being those roots that are directly attached to the trunk, 2nd order directly branching off from 1st order, and so on. Note that the coefficient of variation increases through the root branching orders, suggesting more variation between QSM runs, and therefore higher uncertainty. QSM, quantitative structure models.

accessible to larger, heavier instruments. However, for characterization of more inaccessible caves, or exploratory surveying where deployment flexibility is important, the CBL and similar practical instruments would offer considerable benefit. Utilizing platforms that provided a variety of scanning heights, as facilitated by the CBL's low weight, could also help mitigate occlusion resulting from cave geometric complexity.

Eroding bluff

Geomorphological features which are subject to erosion can have extremely high temporal dynamism during events such as storms. The low observation time and resilience of the CBL1 instrument enabled scanning within the short microstates and challenging conditions during a storm. Two CBL1 scans of an eroding bluff at Lovell's Island, Massachusetts, were obtained minutes apart during a storm event in April 2014 (Fig. 9). A third scan of the bluff was obtained by the CBL2 in January 2015. This dataset exemplifies influences on properties of interest, namely, the influence of wind and water on the position and shape of the bluff between time points (Fig. 10).

There was low magnitude, localized erosion observed in the minutes between scans during the storm in 2014, while erosion was widespread and of high magnitude across seasons, as revealed by the 2015 scan (Fig. 10). The bluff in 2015 was comparatively featureless, having fewer protrusions and indentations on its surface than in 2014 (Fig. 10).

These data also illuminate some considerations of information requirements. Gaps from a lack of information can be seen in the bluff data (Fig. 11), which could have been

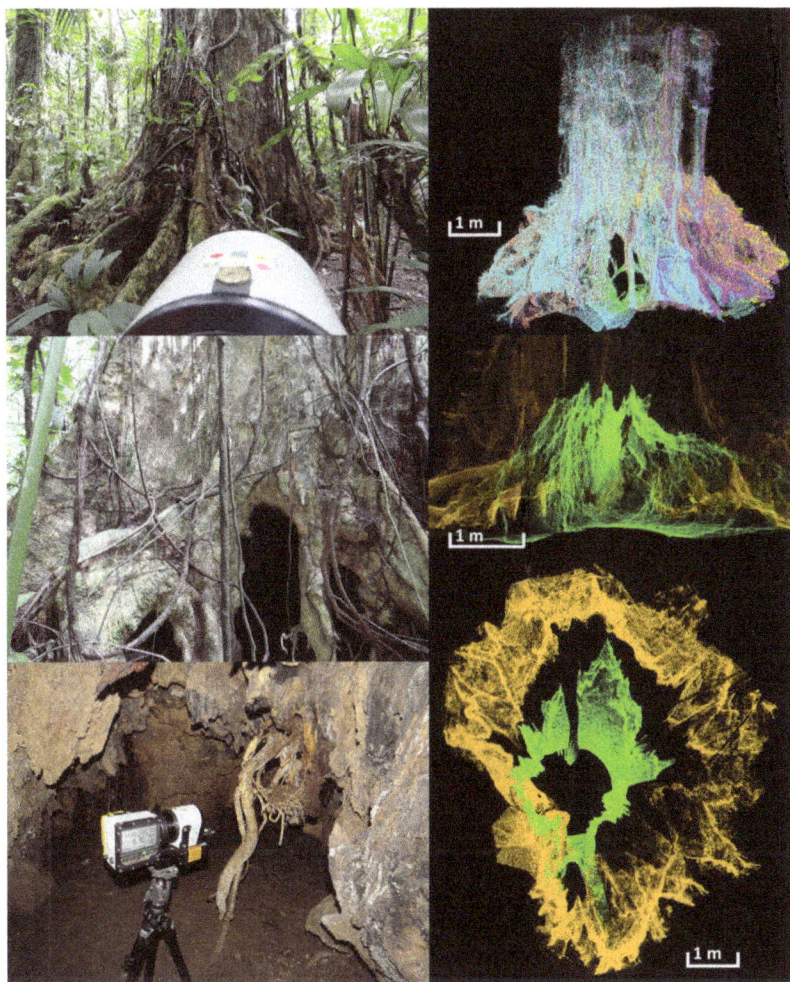

Figure 5. Photographs (left, top to bottom) of CBL2 scanning the exterior of the *Ceiba* tree (Costa Rica); the exterior of the hollow inside the trunk; and the CBL2 scanning the interior of the hollow. Point clouds (right, top to bottom) of the exterior of the *Ceiba* tree (colors denote source scans); structure from interior (green) and exterior (orange) CBL2 scans; and an overhead view of the trunk exterior (orange) and hollow interior (green). CBL, compact biomass lidar.

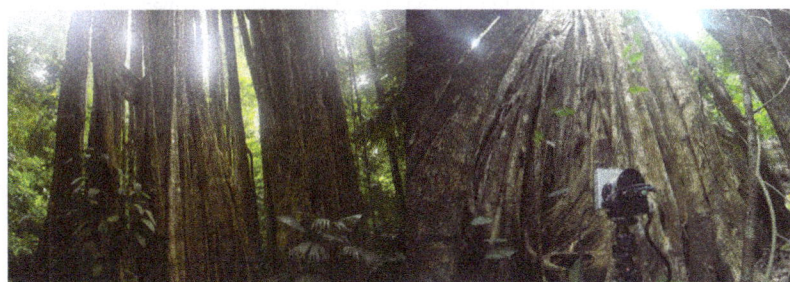

Figure 6. Photographs of the exterior (left) and interior with CBL1 (right) of the strangler fig (Costa Rica). The geometric complexity of the many separate stems created challenging patterns of occlusion that could only be mitigated by leveraging the compactness and rapid scanning of the CBL. CBL, compact biomass lidar.

precluded by establishing a required minimum information density. The importance of being able to monitor information gain during observation deployment is highlighted by the information density in the CBL2 observation at time point 3. More gaps would be expected in the lower density CBL1 data at time points 1 and 2. However, occlusion from a fallen tree in the foreground in the CBL2 data at time point 3 (Fig. 10) creates a locale of low information density (Fig. 11). This is also a clear example of the dependence of observations on the geometric complexity of the ecosystem sample.

Salt marsh creek

Salt marshes, like mangroves, are subject to tidal cycles that obscure observation of their geomorphological features and vegetation, and limit their accessibility. Microstates for salt marsh observation are typically very short, and vegetation growth or snow can also obscure geomorphology for large parts of the year. CBL2 scans were obtained of a tidal creek in the Plum Island Estuary, Massachusetts, Long Term Ecological Research (LTER) site, in 2015. The instrument was

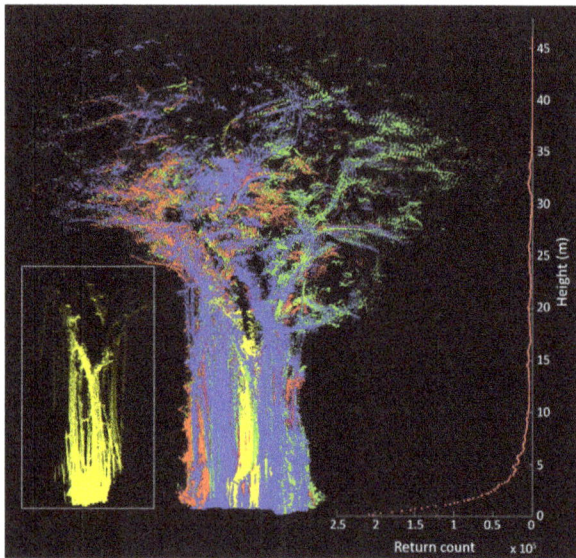

Figure 7. The point cloud (center) is colored according to contributions by external (red, green and blue) and internal (yellow) CBL1 scans. The unique structural features observed by the internal scans are also shown in isolation (left, yellow). The overall information density, in terms of the number of lidar returns, declined rapidly with height (right) due to the substantial occlusion of internal canopy structure, and angular separation between beams over the distance to the canopy of the tall tree. The strangler fig tree in this study completely collapsed during the following rainy season, with the CBL1 scans the only record of its structure. CBL, compact biomass lidar.

mounted upside down on a small tram, suspended above the creek by a cable tensioned by a pair of portable towers (Fig. 12). Five scans were taken at intervals of approximately 2 m along the creek.

The structure of the creek was captured with little occlusion, and reasonably even information density (Fig. 13). However, a raster analysis of the geomorphology (Fig. 14) shows some regions are still lacking information. Even though these regions are few in number and each less than 20 cm^2 in extent, these regions would still ideally have been observed. These data also demonstrate an additional obscuration which perturbed the scanner position. The effect of wind on the suspended scanner caused an accumulation of low magnitude ranging errors (Fig. 13). However, the optimal view angle for the ecosystem geometric complexity, achieved by mounting the lightweight and remotely operated CBL2 on a non-traditional suspended platform, greatly reduced occlusion and therefore observation time for the ecosystem sample.

It should be noted that the full absorption of lidar pulses by water in the bottom of the creek completely obscured observation of the underlying region. This is an illustrative application for the microstate model, since the extreme temporal dynamism of the salt marsh means that the microstates for observing creek geomorphology completely

Figure 8. Point clouds of the complete structure of Treak Cliff Caverns (UK) (top); and particular regions (middle and bottom). The superstructure of the entire caverns were represented in the 36 CBL1 scans. However, there were some regions of occlusion, mostly associated with upward-facing surfaces, and variable information density, particularly in the taller caverns. CBL, compact biomass lidar.

may occur only briefly and at the lowest of low tides. The short observation time of the CBL and similar instruments, resulting from their acquisition rate and favorable deployment logistics, can meet the temporal constraints of such short microstates. Since vegetation growth and snow are additional obscuring factors for salt marsh creek geomorphology, observations are best made during late fall or early spring, facilitated by the wide operating temperature range, and IP68 waterproofing of the CBL.

Discussion

The trade-offs of practical terrestrial lidar scanners

These studies show how the practical attributes of terrestrial lidars such as speed, portability and resilience, are

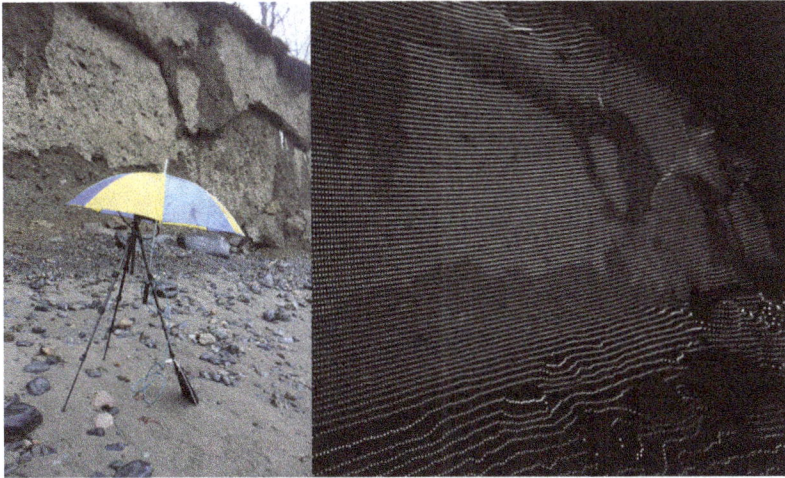

Figure 9. Photograph of eroding bluff on Lovell's Island (Boston, MA) (left) during April 2014 storm. CBL1 was covered with an umbrella between scans to protect from debris. The point cloud (right) shows the structural features of the bluff are clearly represented. The intensity (grayscale) of the CBL1 returns are sensitive to the varied water saturation of the bluff. CBL, compact biomass lidar.

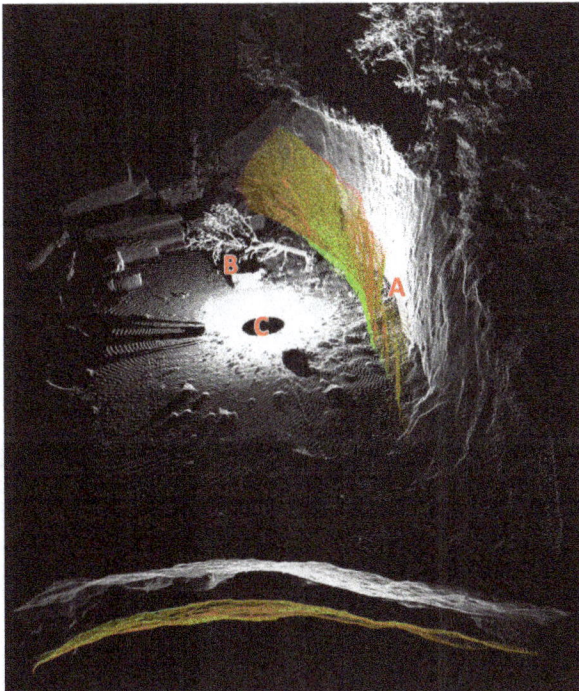

Figure 10. (Top) Extracted bluff sections of two scans taken minutes apart during April 2014 storm (red and green). These are overlaid on the full point cloud from January 2015 (white). (Bottom) Overhead view of extracted bluff sections for April 2014 (red and green) and January 2015 (white). Note the localized, minor erosion between the April 2014 scans, then the major erosion and retreat by January 2015 (A) including the deposition of a tree onto the beach (B). The footprint of the CBL scanning location can be seen (C). CBL, compact biomass lidar.

essential when observing properties of inaccessible, highly dynamic and geometrically complex ecosystems. Summarized in the terminology of the *microstate model*, low scan times and favorable deployment logistics improve the *acquisition rate* of practical terrestrial lidars, reducing

observation time. Low weight, compactness and resilience help meet the accessibility challenges and hazards of ecosystems. Combined, these attributes enable fulfillment of *information requirements* even when sampling inaccessible ecosystems with high *geometric complexity* and short microstates.

There are trade-offs for these practicality optimizations, typically in the form of limitations to the quality and variety of information these instruments provide. Prominently, the angular resolution of the CBL is substantially lower than contemporary instruments such as the DWEL, SALCA and particularly the Riegl VZ-400, resulting in lower information density per scan. Additionally, the Riegl VZ-400 has a measurement range of up to 600 m, while the maximum range for the CBL is approximately only 40 m, meaning the CBL observes a much smaller space with each pulse.

These drawbacks are somewhat mitigated by the favorable deployment logistics of the CBL, enabling many scans to be taken from different locations within a sample, in the same time frame as a single scan from a higher resolution, longer range instrument. In fact, in *geometrically complex* environments such as tropical forests, the advantage of longer maximum range is reduced by the high degree of occlusion.

There are, however, still clear information benefits offered by highly capable commercial and research terrestrial lidars. For example, DWEL and SALCA use lasers of different wavelengths (1064 nm and 1548 nm) to separately identify returns from leaves and branches (Danson et al. 2014; Howe et al. 2015). More capable instruments such as DWEL, SALCA and Riegl may also record all or parts of the return waveform, at the very least providing more returns per pulse than the CBL, which only records the first and last. In addition, Riegl instruments utilize automatic co-alignment of scans based on deployed calibration targets, which add to deployment time, but

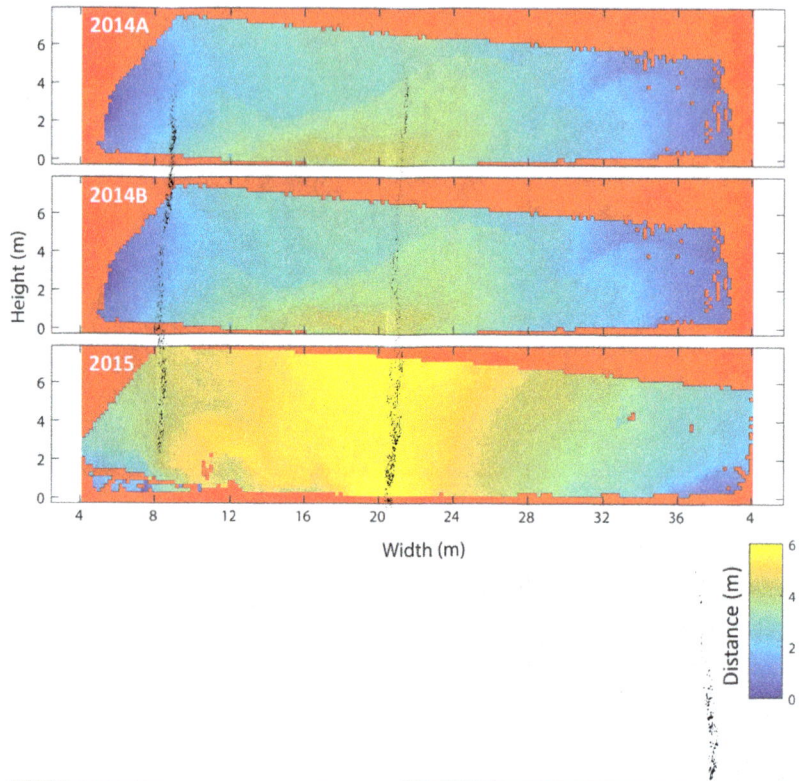

Figure 11. 20 cm rasters of extracted point clouds of bluff from CBL1 scans in April 2014 (top and middle) and January 2015 (bottom). Coloration denotes distance from plane intersecting scans position and parallel to bluff. Minor gaps (red) appear in areas further from the scanner, due to the lower information density with angular separation between beams over distance. These gaps are reduced slightly by the higher vertical resolution of the CBL2, utilized in January 2015 (bottom), but persist due to the increased distance to the retreating bluff, and the occlusion resulting from the fallen tree (Fig. 10, C). CBL, compact biomass lidar.

Figure 12. Photographs of salt marsh creeks at Plum Island (MA) (top) during winter months at low tide when vegetation and tide are not obscuring geomorphological features. The lightweight tower system for deploying CBL2 (bottom) is demonstrated during summer, when vegetation obscures creek banks. CBL, compact biomass lidar.

decreases post-processing time compared to the manual co-alignment of the CBL.

Furthermore, the CBL's beam divergence of 15 mrad (0.86 degrees), is substantially higher than that of most terrestrial lidars currently employed for ecosystem sampling, and even exceeds the angular resolution, creating overlap in the sampled volume. As previously mentioned, beam divergence can add considerable uncertainty to the bounds of objects, especially in small and finely clustered objects, such as higher order tree branches and leaves. Given the range-dependence of uncertainty from beam divergence, this uncertainty might be mitigated by filtering returns from beyond a certain range, and in tandem, setting information requirements to demand shorter range observations for all regions of the sample.

A final, essential context to this discussion is that the practical improvements in terrestrial lidars have been

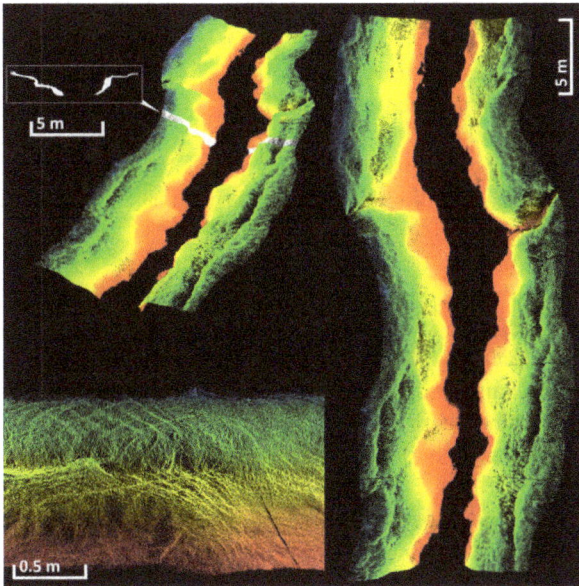

Figure 13. Point cloud of salt marsh creek from CBL2 scans (top left, colored by height), including extracted 20 cm profile (top, white and box). Creek point cloud viewed overhead (bottom left) and perpendicularly (bottom right). Note minor disagreements in bank structure (bottom right) resulting from wind influencing deployment platform during scans. Information density and representation of structure is excellent overall, with little evidence of occlusion. CBL, compact biomass lidar.

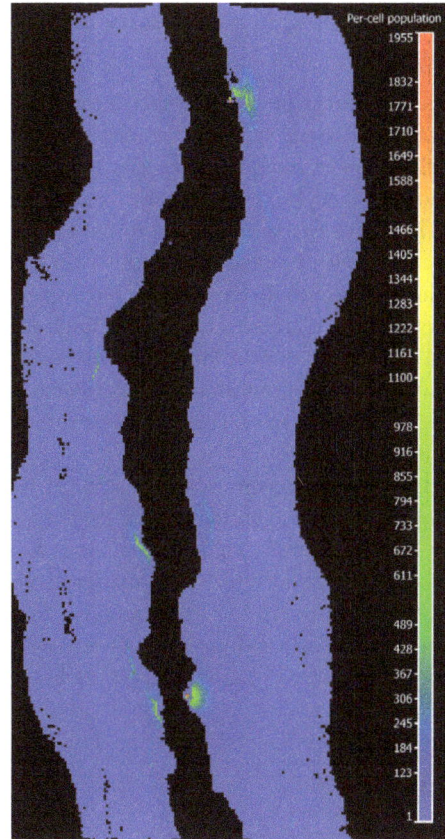

Figure 14. 10 cm raster image of return density for overhead view of salt marsh creek. Note the minor regions of missing information (black) within the bank structure, typically associated with upward facing surfaces and changes in elevation (Fig. 13).

accompanied by a notable reduction in cost. In addition, optimization of instruments such as the CBL for durability and resilience reduces the risk and therefore the functional cost of their deployment. Although financial cost is not a factor in the absolute scientific value of a method, it does influence information quality, since the cost of an observation method determines the spatial coverage achieved.

Evaluation of the microstate model

In these studies, the microstate model described the necessary conditions for observing diverse ecosystems. Using the microstate model, we were able to interpret the observations from each case study to suggest deficiencies and propose improvements. The microstate model responds dynamically to both the ecosystem-level factors, and changes to technologies and methodologies. For example, improvements in terrestrial lidar information quality or acquisition rate could enable observation of lower magnitude changes of an ecosystem property. This would, in turn, change the bounds of the microstate, as they are partly dependent on the possible resolution of observation.

The concept of information requirements provides an essential, measurable standard for terrestrial lidar observations. However, appropriate information requirements are specific both to an ecosystem property of interest and the method used to extract it from lidar data. Generally, information requirements for sampling will be derived from a desired maximum level of uncertainty in the downstream estimation of an ecosystem property. This is most easily achieved for problems with pre-existing data or the potential for simulation, but otherwise begets preliminary experimentation.

When determining observation time in the microstate model, it is useful to quantify a lidar's information quality. However, for spatially complex ecosystems, manufacturers' specifications may not accurately describe information quality. Therefore, independent calibration of instruments is recommended, utilizing objects with equivalent geometry and reflective properties to those found in the ecosystem to be sampled. Fortunately, pulse-level uncertainties can be robustly established with ever-improving methods (Kaasalainen et al. 2009; Li et al. 2016).

Dynamic sampling is a necessity and a challenge

Information requirements must be met for an observation to be considered complete. However, the unique geometry of each ecosystem sample means that the number and position of scans to meet information requirements will change. Therefore, a sampling design to meet information requirements cannot be predetermined. Instead, the necessary scan positions will have to be discovered during the observation of a sample, based on information gain reassessed after each scan. Additional scan positions can then be suggested based on accumulated information and projected information gain. This iterative process can be termed *dynamic sampling*.

Given that observations must be constrained to microstates, dynamic sampling methods must be very efficient, especially in ecosystems with high temporal dynamism. This presents a considerable computational challenge. However, ongoing work by the authors integrating improvements in lidar hardware and software with spatial algorithms places operational dynamic sampling close at hand.

Conclusions

Through the diverse studies presented herein, the practical attributes of terrestrial lidars are suggested to be true capabilities, establishing new potential possibilities in ecosystem sampling. The low weight and compactness of the Compact Biomass Lidar (CBL), and similar instruments, enable compatibility with diverse deployment platforms and greatly facilitate observations of geometrically complex ecosystems. This is particularly true in partitioned and stratified ecosystems such as tropical forests where the ability to mount terrestrial lidars on mobile towers allows scanning within occluded upper canopy regions. Utilizing such deployment platforms is only possible with remote instrument operation, which the CBL achieves through wireless control from mobile devices.

Improvements in lidar deployment logistics also enable the investigation of shorter microstates, as well as sampling in direct response to influential, episodic events such as storms. Combining these capabilities can even facilitate the partitioning of influential events into observable microstates. For example, we demonstrated the ability to observe erosion of a bluff within a storm (Fig. 12–14). This could lead to the coupling of specific erosion effects to specific and measurable components of the storm event.

Terrestrial lidars have shown great promise as validation tools for airborne and satellite estimations of ecosystem properties (Cook et al. 2009; Perroy et al. 2010; De Sy et al. 2012; Avitabile et al. 2016), and shorter observation times provide the ability to expand ecosystem sample volumes and increase information gain. These improvements in the validation capability of lidar instruments for diverse ecosystems is particularly timely with the Global Ecosystem Dynamics Investigation (GEDI) lidar scheduled to begin acquisitions from the International Space Station in 2018.

Ultimately, terrestrial lidar scanning is a rapidly developing technology, with a lot to offer to ecosystem analysis and monitoring. However, the potential of terrestrial lidar is utterly reliant on proper characterization and mitigation of its uncertainties, further development of upstream methods and protocols, and refinement and standardization of downstream processing and analysis techniques. We submit that lidar instruments optimized for practicality, coupled with the robust and adaptable conceptual framework of the microstate model may offer vital contributions toward these goals.

Acknowledgments

We gratefully acknowledge assistance from David and Deborah Clark, William Miranda, Leo Campos, Andres Vega, Temilola Fatoyinbo, Linda Deegan, Jack Payette, Jennifer Ly, and the Harrison and Turner families of Treak Cliff Caverns. We also acknowledge financial support from the Oracle Graduate Fellowship, the NASA Harriet Jenkins Graduate Fellowship, the UMass Boston International Research Initiative Seed Grants Program, and NASA grants NNX14AK12G and NNX14AI73G.

Conflict of Interest

None declared.

References

Avitabile, V., M. Herold, G. Heuvelink, S. L. Lewis, O. L. Phillips, G. P. Asner, et al. 2016. An integrated pan-tropical biomass map using multiple reference datasets. *Glob. Change Biol.* **22**,1406–1420.

Béland, M., D. D. Baldocchi, J. L. Widlowski, R. A. Fournier, and M. M. Verstraete. 2014. On seeing the wood from the leaves and the role of voxel size in determining leaf area distribution of forests with terrestrial LiDAR. *Agric. For. Meteorol.* **184**, 82–97.

Bosse, M., R. Zlot, and P. Flick. 2012. Zebedee: design of a spring-mounted 3-D range sensor with application to mobile mapping. *Robotics*, IEEE Transactions on **28**, 1104–1119.

Calders, K., G. Newnham, A. Burt, S. Murphy, P. Raumonen, M. Herold, et al. 2015. Nondestructive estimates of above-ground biomass using terrestrial laser scanning. *Methods Ecol. Evol.* **6**, 198–208.

Clark, D. B., and D. A. Clark. 2000. Landscape-scale variation in forest structure and biomass in a tropical rain forest. *For. Ecol. Manage.* **137**, 185–198.

Cook, B. D., P. V. Bolstad, E. Næsset, R. S. Anderson, S. Garrigues, J. T. Morisette, et al. 2009. Using LiDAR and quickbird data to model plant production and quantify uncertainties associated with wetland detection and land cover generalizations. *Remote Sens. Environ.* **113**, 2366–2379.

Côté, J. F., J. L. Widlowski, R. A. Fournier, and M. M. Verstraete. 2009. The structural and radiative consistency of three-dimensional tree reconstructions from terrestrial lidar. *Remote Sens. Environ.* **113**, 1067–1081.

Côté, J. F., R. A. Fournier, and R. Egli. 2011. An architectural model of trees to estimate forest structural attributes using terrestrial LiDAR. *Environ. Model. Softw.* **26**, 761–777.

Danson, F. M., R. Gaulton, R. P. Armitage, M. Disney, O. Gunawan, P. Lewis, et al. 2014. Developing a dual-wavelength full-waveform terrestrial laser scanner to characterize forest canopy structure. *Agric. For. Meteorol.* **198**, 7–14.

Dassot, M., T. Constant, and M. Fournier. 2011. The use of terrestrial LiDAR technology in forest science: application fields, benefits and challenges. *Ann. For. Sci.* **68**, 959–974.

De Sy, V., M. Herold, F. Achard, G. P. Asner, A. Held, J. Kellndorfer, et al. 2012. Synergies of multiple remote sensing data sources for REDD+ monitoring. *Curr. Opin. Environ. Sustain.* **4**, 696–706.

Douglas, E. S., A. H. Strahler, J. Martel, T. Cook, C. Mendillo, R. Marshall, et al. 2012. DWEL: A Dual-Wavelength Echidna® Lidar for ground-based forest scanning, Proceedings IEEE International Geoscience and Remote Sensing Symposium 2012, Munich, Germany, **2012**, 5.

Douglas, E. S., J. Martel, Z. Li, G. Howe, K. Hewawasam, R. Marshall, et al. 2015. Finding leaves in the forest: the dual-wavelength echidna lidar. *Geosci. Remote Sens. Lett, IEEE* **12**, 776–780.

Dubayah, R. O., and J. B. Drake. 2000. Lidar remote sensing for forestry. *J. Forest.* **98**, 44–46.

Feliciano, E. A., S. Wdowinski, and M. D. Potts. 2014. Assessing mangrove above-ground biomass and structure using terrestrial laser scanning: a case study in the Everglades National Park. *Wetlands* **34**, 955–968.

Hancock, S., R. Essery, T. Reid, J. Carle, R. Baxter, N. Rutter, et al. 2014. Characterising forest gap fraction with terrestrial

lidar and photography: an examination of relative limitations. *Agric. For. Meteorol.* **189**, 105–114.

Hilser, V. J., E. B. García-Moreno, T. G. Oas, G. Kapp, and S. T. Whitten. 2006. A statistical thermodynamic model of the protein ensemble. *Chem. Rev.* **106**, 1545–1558.

Hoffmeister, D. S., A. Zellmann, M. Pastoors, P. Kehl, J. Cantalejo, G. C. Ramos, et al. 2015. The investigation of the Ardales Cave, Spain–3D documentation, topographic analyses, and lighting simulations based on terrestrial laser scanning. *Archaeol. Prospec.* **23**, 75–86.

Hosoi, F., and K. Omasa. 2007. Factors contributing to accuracy in the estimation of the woody canopy leaf area density profile using 3D portable lidar imaging. *J. Exp. Bot.* **58**, 3463–3473.

Howe, G. A., K. Hewawasam, E. S. Douglas, J. Martel, Z. Li, A. Strahler, et al. 2015. Capabilities and performance of DWEL, the dual-wavelength echidna® lidar. *J. Appl. Remote Sens.* **9**, 13. doi:10.1117/1.JRS.9.095979.

Hurtt, G. C., R. Dubayah, J. Drake, P. R. Moorcroft, S. W. Pacala, J. B. Blair, et al. 2004. Beyond potential vegetation: combining lidar data and a height-structured model for carbon studies. *Ecol. Appl.* **14**, 873–883.

Jupp, D. L., D. S. Culvenor, J. L. Lovell, G. J. Newnham, A. H. Strahler, and C. E. Woodcock. 2009. Estimating forest LAI profiles and structural parameters using a ground-based laser called 'Echidna®. *Tree Physiol.* **29**, 171–181.

Kaasalainen, S., A. Krooks, A. Kukko, and H. Kaartinen. 2009. Radiometric calibration of terrestrial laser scanners with external reference targets.*Remote Sens.* **1**, 144–158.

Kaasalainen, S., A. Krooks, J. Liski, P. Raumonen, H. Kaartinen, M. Kaasalainen, et al. 2014. Change detection of tree biomass with terrestrial laser scanning and quantitative structure modelling. *Remote Sens.* **6**, 3906–3922.

Kamal, M., S. Phinn, and K. Johansen. 2015. Object-based approach for multi-scale mangrove composition mapping using multi-resolution image datasets. *Remote Sens.* **7**, 4753–4783.

Kelbe, D., P. Romanczyk, J. van Aardt, and K. Cawse-Nicholson. 2013. Reconstruction of 3D tree stem models from low-cost terrestrial laser scanner data. *SPIE Defense, Secur. Sens.* **8731**, 6.

Kelbe, D., J. van Aardt, P. Romanczyk, M. van Leeuwen, and K. Cawse-Nicholson. 2015. Single-scan stem reconstruction using low-resolution terrestrial laser scanner data. *IEEE J. Sel. Top Appl. Earth Observ. Remote Sens.* **8**, 3414–3427.

Knight, J. M., P. E. Dale, J. Spencer, and L. Griffin. 2009. Exploring LiDAR data for mapping the micro-topography and tidal hydro-dynamics of mangrove systems: an example from southeast Queensland, Australia. *Estuar. Coast. Shelf Sci.* **85**, 593–600.

Krooks, A., S. Kaasalainen, V. Kankare, M. Joensuu, P. Raumonen, and M. Kaasalainen. 2014. Predicting tree structure from tree height using terrestrial laser scanning and quantitative structure models. *Silva Fenn.* **48**, 1125.

Lefsky, M. A., W. B. Cohen, G. G. Parker, and D. J. Harding. 2002. Lidar remote sensing for ecosystem studies lidar, an emerging remote sensing technology that directly measures the three-dimensional distribution of plant canopies, can accurately estimate vegetation structural attributes and should be of particular interest to forest, landscape, and global ecologists. *Bioscience* **52**, 19–30.

Li, Z., D. L. Jupp, A. H. Strahler, C. B. Schaaf, G. Howe, K. Hewawasam, et al. 2016. radiometric calibration of a dual-wavelength, full-waveform terrestrial lidar. *Sensors* **16**, 313.

Lovell, J. L., D. L. Jupp, D. S. Culvenor, and N. C. Coops. 2003. Using airborne and ground-based ranging lidar to measure canopy structure in Australian forests. *Canad. J. Remote Sens.* **29**, 607–622.

Lucas, R. M., A. L. Mitchell, and J. Armston. 2015. Measurement of forest above-ground biomass using active and passive remote sensing at large (subnational to global) scales. *Current Forest. Rep.* **1**, 162–177.

Mohammed Oludare, I., and B. Pradhan. 2016. A decade of modern cave surveying with terrestrial laser scanning: a review of sensors, method and application development. *Int. J. Speleol.* **45**, 8.

Perroy, R. L., B. Bookhagen, G. P. Asner, and O. A. Chadwick. 2010. Comparison of gully erosion estimates using airborne and ground-based LiDAR on Santa Cruz Island, California. *Geomorphology* **118**, 288–300.

Pfeifer, N., and C. Briese. 2007. Geometrical aspects of airborne laser scanning and terrestrial laser scanning. *International Archives of Photogrammetry, Remote Sensing and Spatial Information Sciences* **36** (3/W52), 311–319.

Phinn, S. R. 1998. A framework for selecting appropriate remotely sensed data dimensions for environmental monitoring and management. *Int. J. Remote Sens.* **19**, 3457–3463.

Raumonen, P., M. Kaasalainen, M. Åkerblom, S. Kaasalainen, H. Kaartinen, M. Vastaranta, et al. 2013. Fast automatic precision tree models from terrestrial laser scanner data. *Remote Sens.* **5**, 491–520.

Raumonen, P., E. Casella, K. Calders, S. Murphy, M. Åkerblom, and M. Kaasalainen. 2015. Massive-scale tree modelling from TLS data. *ISPRS Ann. Photogrammetry, Remote Sens. Spat. Inform. Sci.* **2**, 189.

Rincon, R. F., T. Fatoyinbo, G. Sun, K. J. Ranson, M. Perrine, M. Deshapnde, et al. 2011. The ECOSAR P-band Synthetic Aperture Radar. In 2011 IEEE International Geoscience and Remote Sensing Symposium.

Strahler, A. H., D. L. Jupp, C. E. Woodcock, C. B. Schaaf, T. Yao, F. Zhao, et al. 2008. Retrieval of forest structural parameters using a ground-based lidar instrument (Echidna®). *Canad. J. Remote Sens.* **34**(sup2), S426–S440.

Tang, H., M. Brolly, F. Zhao, A. H. Strahler, C. L. Schaaf, S. Ganguly, et al. 2014. Deriving and validating leaf area index (LAI) at multiple spatial scales through lidar remote sensing: a case study in Sierra National Forest, CA. *Remote Sens. Environ.* **143**, 131–141.

Van der Zande, D., W. Hoet, I. Jonckheere, J. van Aardt, and P. Coppin. 2006. Influence of measurement set-up of ground-based LiDAR for derivation of tree structure. *Agric. For. Meteorol.* **141**, 147–160.

Woodhouse, I. H., C. Nichol, P. Sinclair, J. Jack, F. Morsdorf, T. J. Malthus, et al. 2011. A multispectral canopy LiDAR demonstrator project. *IEEE Geosci. Remote Sens. Lett.* **8**, 839–843.

Wulder, M. A., J. C. White, R. F. Nelson, E. Næsset, H. O. Ørka, N. C. Coops, et al. 2012. Lidar sampling for large-area forest characterization: a review. *Remote Sens. Environ.* **121**, 196–209.

Zhao, F., A. H. Strahler, C. L. Schaaf, T. Yao, X. Yang, Z. Wang, et al. 2012. Measuring gap fraction, element clumping index and LAI in Sierra Forest stands using a full-waveform ground-based lidar. *Remote Sens. Environ.* **125**, 73–79.

Zolkos, S. G., S. J. Goetz, and R. Dubayah. 2013. A meta-analysis of terrestrial aboveground biomass estimation using lidar remote sensing. *Remote Sens. Environ.* **128**, 289–298.

Satellite remote sensing to monitor species diversity: potential and pitfalls

Duccio Rocchini[1], Doreen S. Boyd[2], Jean-Baptiste Féret[3], Giles M. Foody[2], Kate S. He[4], Angela Lausch[5], Harini Nagendra[6], Martin Wegmann[7] & Nathalie Pettorelli[8]

[1]Department of Biodiversity and Molecular Ecology, Research and Innovation Centre, Fondazione Edmund Mach, Via E. Mach 1, 38010, S. Michele all'Adige, TN, Italy
[2]School of Geography, University of Nottingham, Nottingham, NG7 2RD, United Kingdom
[3]UMR-TETIS, IRSTEA Montpellier, Maison de la Télédétection, 500 rue JF Breton, 34093, Montpellier Cedex 5, France
[4]Department of Biological Sciences, Murray State University, Murray, Kentucky, 42071, USA
[5]Department of Computational Landscape Ecology, Helmholtz Centre for Environmental Research-UFZ, Permoserstreet 15, D-04318, Leipzig, Germany
[6]Azim Premji University, PES Institute of Technology Campus, Pixel Park, B Block, Electronics City, Hosur Road, Bangalore, 560100, India
[7]Department of Remote Sensing, Remote Sensing and Biodiversity Research Group, University of Wuerzburg, Wuerzburg, Germany
[8]Institute of Zoology, The Zoological Society of London, Regent's Park, London, United Kingdom

Keywords
Alpha-diversity, beta-diversity, biodiversity, distance decay models, remote sensing, spatial ecology

Correspondence
Duccio Rocchini, Department of Biodiversity and Molecular Ecology, Research and Innovation Centre, Fondazione Edmund Mach, Via E. Mach 1, 38010 S. Michele all'Adige (TN), Italy. E-mail: ducciorocchini@gmail.com, duccio. rocchini@fmach.it

Editor: Andrew Skidmore
Associate Editor: Alienor Chauvenet

Abstract

Assessing the level of diversity in plant communities from field-based data is difficult for a number of practical reasons: (1) establishing the number of sampling units to be investigated can be difficult; (2) the choice of sample design can impact on results; and (3) defining the population of concern can be challenging. Satellite remote sensing (SRS) is one of the most cost-effective approaches to identify biodiversity hotspots and predict changes in species composition. This is because, in contrast to field-based methods, it allows for complete spatial coverages of the Earth's surface under study over a short period of time. Furthermore, SRS provides repeated measures, thus making it possible to study temporal changes in biodiversity. Here, we provide a concise review of the potential of satellites to help track changes in plant species diversity, and provide, for the first time, an overview of the potential pitfalls associated with the misuse of satellite imagery to predict species diversity. Our work shows that, while the assessment of alpha-diversity is relatively straightforward, calculation of beta-diversity (variation in species composition between adjacent locations) is challenging, making it difficult to reliably estimate gamma-diversity (total diversity at the landscape or regional level). We conclude that an increased collaboration between the remote sensing and biodiversity communities is needed in order to properly address future challenges and developments.

Introduction

The importance of measuring species diversity as an indicator of ecosystem health has been recognized by major initiatives worldwide (Skidmore et al. 2015), including the Group on Earth Observation (GEO BON, http://www.earthobservations.org/geobon.shtml) initiative, the International Geosphere Biosphere Programme (IGBP, http://www.igbp.net/), the World Climate Research Programme (WCRP, http://wcrp-climate.org/), the Committee on Earth Observation Systems (CEOS) Biodiversity task (http://ceos.org/), among others.

Assessment of biodiversity at local and regional scales has traditionally relied on the assessment of both local diversity (alpha-diversity) and species turnover (beta-diversity); the combination of these two measures leading to an estimate of the whole diversity of an area (gamma-diversity, Whittaker 1972; Lande 1996). A large number of indices have been used to estimate alpha-diversity (e.g. species richness, Simpson, Berger–Parker, Shannon–Wiener,

Table 1. Mostly used indices to measure alpha- and beta-diversity.

Diversity type	Index	Formula	References
Alpha-diversity	Species richness	S	Colwell (2009)
	Simpson index	$I_S = 1/\Sigma p^2$	Simpson (1949)
	Berger–Parker index	$I_{BP} = 1/(p_{max})$	Berger and Parker (1970)
	Shannon–Wiener index	$H' = -\Sigma p \times \ln(p)$	Shannon and Weaver (1948)
	Brillouin index	$I_B = (\ln N! - \Sigma n!)/N$	Maurer and McGill (2011)
	McIntosh index	$I_{Mc} = (N - \sqrt{\Sigma n^2}/N - \sqrt{N})$	McIntosh (1967)
	Pielou evenness	$J' = H'/H'_{max} = H'/\ln(S)$	Pielou (1966)
Beta-diversity (turnover)	Jaccard index	$\beta_j = C/(A + B + C)$	Jaccard (1912)
	Sørensen index	$\beta_{sor} = 2C/(2C + A + B)$	Sørensen (1948)
	Wilson & Shmida index	$\beta_{ws} = (A + B)/(2C + A + B)$	Wilson and Shmida (1984)
	Colwell & Coddington index	$\beta_{cc} = (A + B)/(A + B + C)$	Colwell and Coddington (1994)
	Lennon index	$\beta_l = 2(A - B)/(2C + A + B)$	Lennon et al. (2001)

S = total number of species, n = number of individuals belonging to each species, N = total number of individuals, p = relative proportion of each species, A = exclusive species composition of the sampling unit A, B = exclusive species composition of the sampling unit B, C = intersection of the species composition of sampling units A and B.

Brillouin, McIntosh, Pielou indices, Table 1). Species turnover is generally assessed using information on species compositional distance measures among sampling units, and expressed using a measure such as the Sørensen index or the Jaccard index (Table 1).

Species monitoring in relatively large areas has always been a challenging task for ecologists, mainly because of the intrinsic difficulty in evaluating the completeness of the resulting species' lists and in quantifying sampling effort (e.g. Palmer 1995). Inventorying species over a large region is complicated by the fact that field biologists cannot inspect every individual organism in the region while accounting for changes in species composition over time (Palmer et al. 2002).

Moreover, when sampling species, a number of issues need to be resolved, such as: (1) the number of sampling units to be investigated; (2) the choice of sample design; (3) the identification of the statistical population of concern; and (4) the operational definition of a community to be considered (see Chiarucci 2007, for a review on these issues).

Additionally, ground surveys are time consuming and costly. Moreover, in many biodiversity-rich locations, field survey can be risky due to challenging environmental and socio-political conditions (Hanson et al. 2009).

Field surveys sometimes experience low spatial and thematic accuracy. As an example, Bacaro et al. (2009) demonstrated that species accumulation curves can vary according to the identity of the biologist sampling the area under consideration. Moreover, in a study addressing the causes of species misidentification in vegetation monitoring, Scott and Hallam (2003) found an average misidentification rate of 2.7–25.6% depending on surveyors' expertise and species involved.

Identifying areas likely to have a high level of diversity may help to minimize the amount of time and funds required for setting up efficient monitoring programs, given that increased attention is likely to be given to biodiversity hotspots (e.g. Bacaro et al. 2009).

We acknowledge that criticism exists about the validity of the direct relationship between species richness and turnover versus biodiversity. In light of previous work on the matter, however, (e.g. Hurlbert 1971; Noss 1990; Grime 1998; Dıaz et al. 2003; Fleishman et al. 2006; Petchey and Gaston 2006; Sundstrom et al. 2012; Giorgini et al. 2015), in this article, we consider species richness and turnover as proxies, and useful metrics, for biodiversity estimate at different spatial scales.

Satellite remote sensing (SRS) might prove to be an extremely powerful tool since it allows for coverage of large regions in a short period of time, having the potential to provide a continuous source of information on biodiversity distribution (He et al. 2015). In this period of major environmental change, SRS represents a powerful opportunity for ecologists to gain critical knowledge about the drivers of the spatial and temporal distribution of biodiversity (Rocchini et al. 2005; Pettorelli et al. 2014).

The relationship between spectral variability over space and species diversity might be of great importance for maximizing the inventory of species diversity giving priority to sites which are spectrally more different, hence more diverse in species composition (Rocchini et al. 2005).

However, a number of pitfalls are associated with the use of remote sensing for predicting species diversity, as deriving measures of diversity from a spectral and a spatial signature of environmental features is not trivial.

The aim of this article is to highlight the potential for SRS to support the monitoring of species assemblages, and help predicting patterns in species diversity. We will limit this review to SRS, simply referred in this article to remote sensing. This review will also discuss current pitfalls associated with such approaches; as far as we know, there has been limited discussion on this topic in the literature. We will here primarily focus on plant communities, but the ideas and suggestions raised applied to a number of different other taxa, including butterflies (Kumar et al. 2009) and mammals (Oindo and Skidmore 2002).

Assessing Alpha- and Beta-Diversity from Space

Potential and pitfalls of remote sensing-based estimates of alpha-diversity

Most of the research dealing with remote sensing-based estimates of species diversity has focused on mapping localized biodiversity hotspots (see also Tucker et al. 2004), based on the Spectral Variation Hypothesis (SVH, Gould 2000; Palmer et al. 2002; Rocchini 2007).

The SVH states that the spatial variability in the remotely sensed signal, that is, the spectral heterogeneity, is expected to be related to environmental heterogeneity and could therefore be used as a powerful proxy of species diversity. In other terms, the greater the habitat heterogeneity, the greater the local species diversity within it (Palmer et al. 2002), regardless of the taxonomic group under consideration. Besides random variation in species distribution (Hubbell 2001), a higher heterogeneity of habitats will host a higher number of species each occupying a particular niche (niche difference model, Nekola and White 1999).

Different modeling techniques have been used to unveil a relationship between local species diversity and the level of spatial variation in the spectral signal, ranging from simple univariate models (Gould 2000), to multivariate statistics (Feilhauer and Schmidtlein 2009), neural networks (Foody and Cutler 2003), and Generalized Additive Models (GAMs, Parviainen et al. 2009).

Figure 1 explicitly shows how alpha-diversity is commonly predicted using remote sensing data. Starting from a heterogeneity map based on the satellite sensor image and on field sampling data, a regression model is commonly built and a back transformation is applied to derive a map of species richness variation over space (Fig. 1).

A number of different measures of spectral heterogeneity have been proposed and used to assess ecological heterogeneity and thus species diversity such as: the variance in a neighborhood of the spectral response (Gillespie 2005), the variability in the reflectance values among pixels using the texture of a remotely sensed image (Hernández-Stefanoni et al. 2012), the distance from the spectral centroid, that is, the mean of spectral values in a multiple dimensional system whose axes are represented by each image band (Palmer et al. 2002), and the distance from the spectral centroid in a principal component space, that is, the compacted spectral space where noise related to band collinearity has been removed (Rocchini 2007). All such measures have been shown to be useful to predict species richness at local scale (Table 2).

Moreover, in addition to the use of common spectral indices such as the normalized difference vegetation index (NDVI), some studies have demonstrated an increase in the strength of relationship between species alpha-diversity and remotely sensed spectral heterogeneity when using additional spectral information (e.g. Landsat 7 shortwave IR-band 5, from 155 to 175 nm and band 7, from 209 to 235 nm, see Rocchini (2007) and Nagendra et al. (2010).

In addition to the importance of having the correct measure or spectral band/index to be used for relating spectral and species diversity at local scale (alpha-diversity), different species diversity measures (see Table 1 for an overview of the metrics classically used to assess alpha-diversity) can lead to differences in the type and strength of the relationship between spectral and species diversity. For example, Oldeland et al. (2010); dealing with plant species diversity in African savannas, relied on relative abundances of species, as measured by the Shannon index to quantify the difference in the relative proportion of each species. They demonstrated that accounting for species relative abundances improves the capability of local species diversity estimation with hyperspectral remotely sensed data, with R^2 values obtained up to five times higher than those achieved by only considering species richness (R^2 values of 0.62 and 0.12, respectively). This is mainly because the Shannon index is less affected than species richness by the presence of rare species, which represent a relatively incidental set of species of more 'dispersed' origin (Ricotta et al. 2008).

Recent advances in biodiversity mapping are based on the processing of high spatial resolution imaging spectroscopy and use an original approach to test the validity of SVH for the estimation of alpha-diversity in tropical rainforests (Féret and Asner 2014a). One original aspect of this method is that it takes advantage of both high spatial and spectral resolution to arbitrarily assign a 'spectral species' identity to each individual pixel of the image, using unsupervised clustering. It consecutively performs pixel inventories over all individual surface units of a given size across the image. Féret and Asner (2014a)

LOCAL DIVERSITY (E.G. SPECIES RICHNESS) ALPHA DIVERSITY

SPECIES TURNOVER BETA DIVERSITY

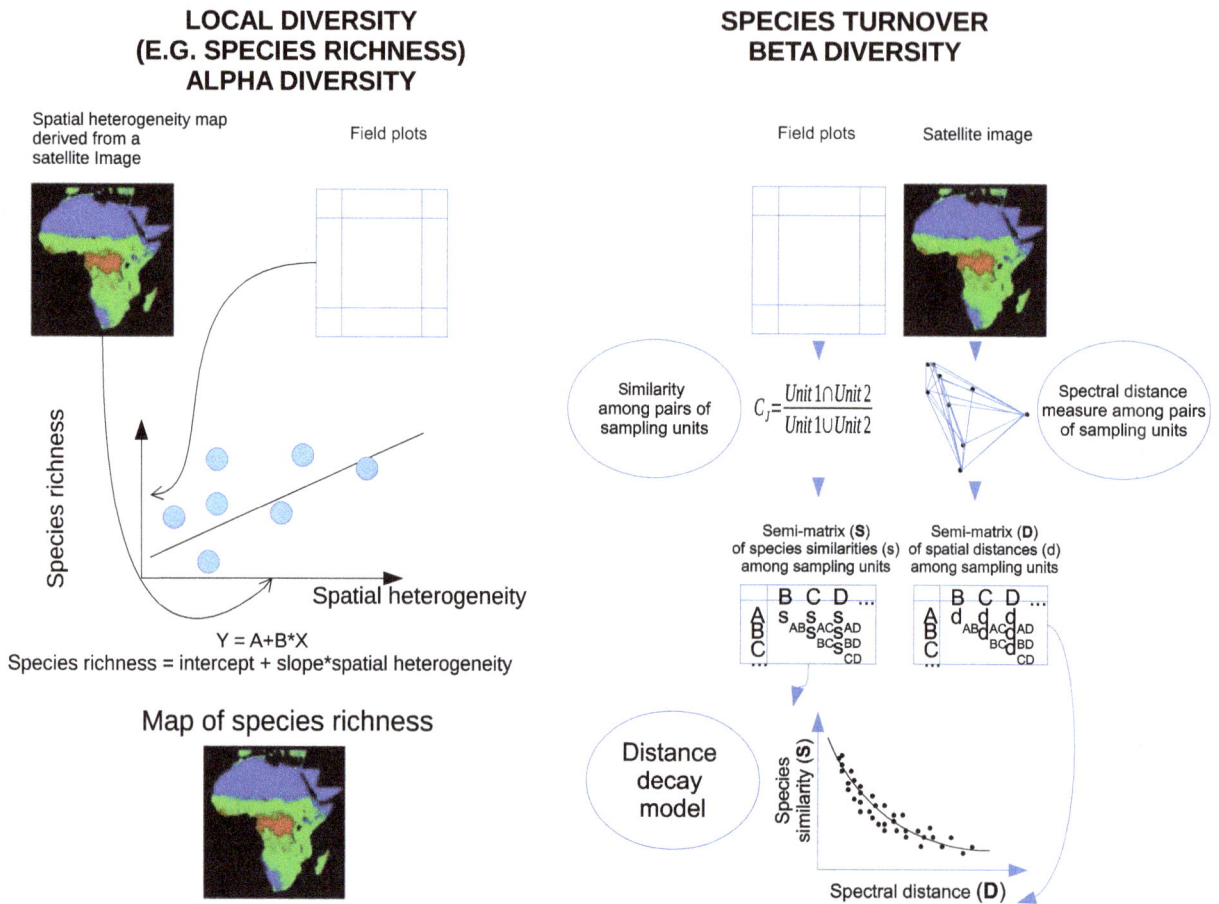

Figure 1. Alpha-diversity (left side) is commonly predicted starting from remote sensing data, calculating a heterogeneity map based on local heterogeneity estimators applied to the image and building a regression model with field-based species diversity calculated with one of the methods reported in Table 1. Beta-diversity is commonly assessed relying on differences among areas in terms of spectral signal values (Table 2) and the relationship with field based measures of species distance (Table 1). In this case an example of a distance decay model is shown where species similarity decays with increasing spectral variability among areas. The satellite image shown in this figure is used as a symbol of a raster file. Refer to the main text for additional explanations.

applied a size of 1 ha for individual surface units, but it can be adapted depending on the spatial resolution of the image and expected patterns of biodiversity. This method is based on the hypothesis that species or groups of species can be identified across the landscape based on their spectral signature (Clark et al. 2005; Féret and Asner 2013), with each actual species/group of species showing lower within-group spectral variability than among-groups variability. Consequently, even without supervision, pixels from the same actual species/group of species statistically tend to naturally converge toward the same cluster. Deriving classic diversity metrics (e.g. Shannon index, Simpson index, etc.) based on histograms of spectral species offers appreciable properties compared to one-dimensional spectral distance metrics such as the ones discussed earlier. In fact, this approach can also be less sensitive to the inherent nonuniform distribution of

species groups in the spectral space, or to pixels corresponding to undesired surfaces (shadow, water, soil), which may artificially increase spectral variability in high spatial resolution images. This is a well documented and common issue (Nagendra and Rocchini 2008). Such methods currently lack validation based on remote sensing, due to the unavailability of satellites combining high spatial and spectral resolution, and high signal to noise characteristics. However, several possibilities are foreseen for the near future, and adjustments can already be anticipated to fit specificities of different types of biomes.

Tropical environments may be particularly dependent on high spectral resolution information to discriminate at least between broad groups of species because of the extremely high number of species and the subtle differences in spectral signature measured among species (Asner and Martin 2009). Future satellite programs such

Table 2. Advances in biodiversity assessment using remote sensing.

Topic	Approach	Habitat, location and source data	References
Alpha-diversity assessment	Univariate regression models using vegetation indices	Tropical dry forests, Florida, US, Landsat Enhanced Thematic Mapper Plus (ETM+) data	Gillespie (2005)
	Univariate regression models using single bands as predictors	Wetlands, Italy, Quick-Bird data	Rocchini et al. (2007)
	Univariate regression models using Principal Components as predictors	Wetlands, Italy, Quick-Bird data	Rocchini et al. (2004)
	LOcally WEighted Smoothing Surfaces (LOWESS) regression models, testing images with different spatial and spectral resolution	Tropical dry forests, India, Landsat Enhanced Thematic Mapper Plus (ETM+)/IKONOS data	Nagendra et al. (2010)
	Multiple nonparametric regression models (Generalized Additive Models, GAM) including remote sensing variables	Boreal forests, Finland, Landsat Enhanced Thematic Mapper Plus (ETM+) data	Parviainen et al. (2009)
	Multivariate regression models including remote sensing data as covariates	Walnut fruit forests, Kyrgyzstan/Central Asia, ASTER data	Feilhauer and Schmidtlein (2009)
	Regression models testing images with different spatial and spectral resolution	Tropical dry forests, India, Landsat Enhanced Thematic Mapper Plus (ETM+)/IKONOS data	Nagendra et al. (2010)
	Neural networks for predicting species richness and abundance	Bornean tropical rainforests, Malaysia, Asia, Landsat Thematic Mapper (TM) data	Foody and Cutler (2003)
Beta-diversity assessment	Mantel-based correlation between species compositional turnover and spectral distance	Worldwide, WWF Ecoregion database, MODIS data	He and Zhang (2009)
	Multivariate approaches (based on detrended correspondence analysis, DCA) to find beta-diversity gradients	Walnut fruit forests, Kyrgyzstan, Central Asia, ASTER data	Feilhauer and Schmidtlein (2009)
	Quantile regression applied to species beta-diversity estimate (spectral distance decay)	Tropical dry forests, India, Landsat ETM+ data	Rocchini and Cade (2008)
	Beta-diversity mapping based on explicit distance maps	Tropical forests, Amazon, Peru, data from the Carnegie Airborne Observatory (CAO) airborne taxonomic mapping system (AToMS)	Féret and Asner (2014a)

Rows are ordered based on a complexity criterion, that is, following an increase in the complexity of the approaches being used.

as HyspIRI (USA), EnMAP (Germany) and HYPXIM (France) may provide the necessary amount of spectral information to allow the implementation of this method. The fusion with higher spatial resolution sensors and appropriate methods (such as image sharpening) will also contribute to improved biodiversity mapping in complex environments. However, many difficulties have to be overcome in such environments, starting with high-quality atmospheric correction and efficient fusion methods of multiple sensors.

In temperate to boreal environments, spectral information may not be as crucial as it is in tropical environments. First, the moderate species richness may require less spectral information for accurate species discrimination; second, the strong diversity of temporal–seasonal dynamics among species and vegetation types provides particularly helpful information for discrimination, as the temporal variations of spectral properties can be related to differences in phenology and physiology. Multispectral satellite sensors with high to very high spatial resolution and short revisit period, such as Sentinel-2, Vens, and other high spatial resolution multispectral sensors may be good candidates for biodiversity mapping based on spectro-temporal variations.

Potential and pitfalls of remote sensing estimates of beta-diversity

While alpha-diversity is related to local variability, species turnover (beta-diversity) is a crucial parameter when trying to identify high biodiversity areas (Baselga 2013). In fact, for a given level of local species richness, high beta-diversity indeed leads to high global diversity of the area. This is one of the basic rules underpinning the concept of irreplaceability of protected areas (e.g. Wegmann et al. 2014).

In general, beta-diversity is assessed by plotting the compositional similarity among sites measured in the field versus their spatial distance (but see Table 1 for an overview of the metrics classically used to assess beta-diversity). The higher the slope of the resulting curve, the higher the beta-diversity of the area. In other terms, the higher the decay in the similarity of species among sites the higher the turnover in terms of species composition.

Therefore, it is expected that species turnover should increase with increasing spatial extent. The curvilinear nature of this relationship, however, means in practice that the validity of extrapolation will depend on the sampling effort, that is, the extent of field knowledge (Ferrier et al. 2007).

In some cases spatial distance/dispersal ability might not be the only driver of species turnover, which seems to be more strictly related to environmental conditions (Tuomisto et al. 2003). Hence, models have been built to relate species and spectral turnover to explain their potential relationship and its causes (Rocchini et al. 2009; Fig. 1).

Tuomisto et al. (2003), studying plant diversity in Amazonia, found that spatial distance accounted only for a small fraction of variance in species similarity, while environmental variation, measured by both soil properties and spectral distance in a Landsat TM image, accounted for a much larger one. When using spatial distances, distance decay does not necessarily account for environmental heterogeneity (Palmer 2005), especially in heavily fragmented landscapes. Thus, the use of spectral distances for summarizing beta-diversity patterns may be more reliable as this method explicitly takes environmental heterogeneity into account instead of mere spatial distances among sites (Fig. 1). Therefore, it is expected that the higher the spectral distance among sites, the higher their difference in terms of environmental niche, thus leading to higher beta-diversity. This has been demonstrated at a number of spatial scales and in several habitat types, ranging from local scaled studies in Mediterranean forests (Rocchini and Cade 2008), Amazonian tropical forests (Tuomisto et al. 2003), Western Ghats (India) tropical forests (Krishnaswamy et al. 2009), tropical dry forests (Rocchini et al. 2009), North and South Carolina (U.S.) lowlands, and floodplains (He et al. 2009), to worldwide assessments on the relationship between biodiversity and productivity (He and Zhang 2009).

A straightforward method for measuring beta-diversity is to calculate the differences between pairs of plots in terms of their species composition using one out of the many possible (dis)similarity coefficients proposed in the ecological literature (Legendre and Legendre 1998; Koleff et al. 2003; Baselga 2013), and assess the spectral turnover variability derived remotely from the variation in species composition among sites.

This has been mainly related to the spectral distance decay models in which species similarity decays once spectral distance d increases, using all pairwise distances $\frac{123N}$ once a total number N of plots is considered, based on an *a priori* defined statistical sampling design (Fig. 1).

A potential pitfall is that the relationship between beta-diversity and habitat heterogeneity is also rarely linear, even when appropriate log transforms of environmental variables are made (Ferrier et al. 2007) because of variations in the rate of species turnover along an environmental gradient. New statistical approaches need to be developed to deal with such challenges (e.g. generalized dissimilarity modeling, Ferrier et al. 2007).

Predicting and mapping beta-diversity using remotely sensed images acquired over large areas may become a computationally intensive task when it is based on distance matrices, compared to the 'raw data' approach ('distance' vs. 'raw data' approaches as described by Legendre et al. 2005; see also Rocchini et al. 2010).

Another difficulty with mapping beta-diversity is the need to use appropriate visualization strategies in order to produce spatially explicit maps respecting the continuous nature of changes in species composition (Penner et al. 2011). On this particular point, prediction maps of species composition based on supervised classification has obvious drawbacks in terms of (1) class definition which requires exhaustive description of all classes, and risk of confusion increasing with the number of classes, and (2) crisp borders which do not correspond to the gradual nature of changes in species composition, and directional turnovers along continuous environmental/physical gradients.

A solution proposed by Thessler et al. (2005), and addressing these two issues consists in combining ordination methods derived from field observations with clustering methods applied to remotely sensed data, in order to obtain a uniform prediction of species composition over an entire satellite acquisition. The solution proposed by Thessler et al. (2005) still requires important field observation in order to run the method and build the models, and there is no guarantee that the field sampling accounted for all major species communities to identify. However, this method significantly decreased the amount of work. The authors estimated that mapping all Ecuadorian Amazonia (70 000 km^2) would require 5 years with *c.* 20 people.

The approach of Thessler et al. (2005) was extended to purely remotely sensed data thanks to high spatial resolution imaging spectroscopy (Baldeck and Asner 2013; Baldeck et al. 2014; Féret and Asner 2014a,b). These studies are based on the preliminary unsupervised clustering of spectral data, assigning each pixel to a 'spectral species' as described in the previous section. After spectral clustering, the image is divided into homogeneous elementary surface units, and a dissimilarity metric (Bray–Curtis dissimilarity, Bray and Curtis 1957) is then used to compute pairwise dissimilarity between each pair of surface units. Finally, the resulting dissimilarity matrix is processed using nonmetric multidimensional scaling to project elementary units in a three-dimensional Euclidean space, allowing the creation of a colored map in the standard Red–Green–Blue referential system. This colored map expresses changes in species composition with changes in color tone. This method proved to perform well with different vegetation types. In savanna landscapes, both preliminary definition of spectral species based on supervised and unsupervised classification of tree species were compared (Baldeck and Asner 2013). Supervised classification using support vector machine algorithms resulted in more accurate estimation of pairwise Bray–Curtis dissimilarity than k-mean unsupervised classification. However, such possibilities can be considered only for landscapes with moderate species richness, due to the confusion caused by increased species richness. In tropical environments, exhaustive supervised classification of individual tree crowns is unrealistic, which leads to selecting unsupervised classification for the assignment of spectral species to pixels. Therefore, Féret and Asner (2014a) developed a fully unsupervised method to process hyperspectral images acquired over various sites in Peruvian Amazonian rainforest. They successfully mapped spatial variations in species composition and Shannon diversity index for various sites in Peruvian Amazonian rainforest using a preliminary spectral species mapping derived from repetitive k-means clustering. This method was compared to various other methods relying on SVH and proposed in the literature, and dramatically outperformed indicators such as variations in NDVI and mean distance from centroid. In a second study, Féret and Asner (2014b) analyzed variations in both alpha- and beta-diversity related to changes in microtopography derived from a digital elevation model obtained with airborne LiDAR (Light Detection And Ranging) acquisitions. Therefore, they proposed a way to take advantage of the combination of imaging spectroscopy and LiDAR acquisitions in order to map biodiversity and relate the spatial variations in species composition to environmental and physical factors (Fig. 2). Therefore, novel approaches integrating multisensors acquisitions can help to improve understanding of the various environmental, physical, climatic, and human factors influencing biodiversity, by monitoring spatial and temporal variations in species composition.

Adding a further confounding factor, the relationship of beta-diversity with environmental heterogeneity is also scale dependent, perhaps even more than alpha-diversity. Areas of ecological transition, where the factors influencing patterns of biodiversity distribution change at different spatial scales, represent therefore a particular challenge for field monitoring. Yet, it is in precisely such areas where remote sensing may be especially helpful, enabling swift and easy computation of proxies of vegetation heterogeneity at different spatial scales, to generate hypotheses about scales at which such ecological transitions may be taking place: this can then be tested using appropriately designed field datasets. For instance, Mairota et al. (2015) found differences in models of the association between remotely sensed values and biodiversity across scales, with plant diversity being most appropriately measured at the patch scale, while bird and insect diversity showed stronger associations with remotely sensed variables at the landscape and plot level, respectively.

Additional Limitations Associated with Remote Sensing-Based Approaches for Assessing Alpha- and Beta-Diversity

Spectral information can be a good proxy of diversity estimate; however, care must be taken in using only remotely sensed variables without considering additional multiscale drivers like climate, soil types, topographic variables, and biotic interactions.

A potential pitfall in the use of remotely sensed data for species diversity estimation is related to spatial scale. Finding a perfect match between remotely sensed imagery and species diversity sampling units is difficult. Obviously, pixels should ideally be smaller than the sampling units, at least when calculating local spectral heterogeneity for local species diversity estimates. Nonetheless, as previously stated, when pixels with a very low dimension (e.g. a ground spatial distance, of 1–5 m) are used (high spatial resolution), shadows may create a higher spatial heterogeneity among spectra leading to higher noise rather than information content (Nagendra and Rocchini 2008; Stickler and Southworth 2008).

On the other hand, a lower spatial resolution may hamper catching the actual heterogeneity due to information smoothing processes which can hinder the detection of fine-grained patterns. Quoting Turner et al. (2003) 'the challenge for the researcher is to ensure that the scale of the imagery matches that of the species richness data and that both are scaled appropriately for the theory being tested'. An inappropriate match of satellite spatial resolution and the grain size of field data could hide actual spatial heterogeneity with subpixel variability remaining undetected (Small 2004; Rocchini 2007).

Figure 2. A lowland Amazonian area shown using: (a) a natural color composite image from the Carnegie Airborne Observatory (CAO) visible-to-shortwave infrared (VSWIR) imaging spectrometer; (b) alpha-diversity (Shannon index); and (c) beta-diversity based on Bray–Curtis dissimilarity. A larger Bray–Curtis dissimilarity between two plots corresponds to larger differences in color in the RGB space between the two corresponding pixels). Reproduced from Féret and Asner (2014a) with kind permission from the Ecological Society of America.

Remotely sensed data sets may also vary in suitability for diversity estimation. For example, issues such as the radiometric resolution of the sensor often get ignored, but two sensors with differing radiometric resolution may yield different estimates for the same site if all other sensor properties are equal. Similarly, sensors of differing spectral resolution may yield different diversity estimates.

Concerning temporal fluctuations of species diversity, an interesting aspect has been raised by Oindo and Skidmore (2002) who posed the attention to the interannual variability in NDVI in explaining species diversity patterns (considering both vascular plants and mammal species). The best predictor was found to be the interannual integrated NDVI, including both its average (negative polynomial relationship with species richness) and its coefficient of variation (linear relationship). From the 'temporal' point of view, remote sensing is a valuable tool since it offers the capability of extracting multitemporal univariate or multivariate statistics as predictors instead

of relying on single-date predictors of species diversity. The same holds for intraannual variability as demonstrated by He et al. (2009) who found the NDVI variability calculated for March to be range of NDVI-based measures, mainly because of the phenological changes of the vegetation under study.

The use of spatial heterogeneity in the spectral signal as a proxy of species diversity also has its limitations, particularly in the conservation and management of biodiversity. Simple measures of species diversity in biology, and habitat diversity in landscape ecology, have been criticized because diversity contains no information on the actual species composition of a community or the habitat composition of a landscape (Luoto et al. 2005). Habitat diversity estimated by spectral heterogeneity is a landscape summary measure that does not take into account the uniqueness or potential ecological importance of different habitats. Furthermore, there are situations where increasing habitat diversity may contradict management objectives with regard to threatened species that require large and homogeneous habitat patches of a specific type.

Despite the 'heterogeneity pitfall', on the other hand, remotely sensed spectral heterogeneity information offers an inexpensive means to derive spatially complete environmental information for large areas in a consistent and regular manner. For this reason, spectral heterogeneity may provide a valuable 'first filter' estimate for the location of species hotspots and the prediction of spatial patterns of biodiversity and their change over space, overall since remote sensing offers straightforward multiscale measurements and analyses at different scales will lead to a more effective biodiversity assessment.

Conclusion

Landscape ecology can provide a useful framework for improving the potential of remote sensing for predicting and monitoring species diversity, allowing the consideration of environmental gradients and spatial discontinuities, through, for example, the use of patch-matrix-corridor models (Turner 1989) of biodiversity distribution. Traditional approaches relating remote sensing to species diversity indeed consider alpha- or beta-diversity to be primarily dependent of environmental variations along defined gradients, without sufficiently taking into account how environmental discontinuities at patch edges and variability in spatial configuration (e.g. patch size, shape, connectivity) can impact species distributions (Anderson et al. 2009).

Remote sensing-based analyses also need to be conducted at multiple spatial scales using approaches such as texture analysis at different window sizes, moving windows, and/or pixel aggregation, to assess the scale most suitable for biodiversity monitoring of specific taxa, in specific contexts (Mairota et al. 2015). Field sampling protocols need to be modified accordingly. Increased use of hierarchical nested field sampling approaches is most needed, collecting field data on species diversity at nested plot (pixel within an environmental gradient), patch (larger environmentally distinct unit bounded by discontinuities that separate it from other patches) and landscape (corresponding to a single image window, or a region of interest such as a protected area) levels (Nagendra and Gadgil 1999).

This review has shown the high potential of remote sensing in biodiversity research as well as the challenges underpinning the development of this interdisciplinary field of research. Further sensitivity studies on environmental parameters derived from remote sensing for biodiversity mapping need to be undertaken to understand the pitfalls and impacts of different data collection processes and models. Such information is crucial for a continuous

global biodiversity analysis and an improved understanding of our current global challenges.

Quantifying and monitoring global biodiversity using remote sensing-based techniques will require increasingly complex data analyses. These can only be implemented in the future with a completely new orientation of Big Data analysis using Linked Open Data (LOD) approaches as well as the rapidly growing Open Database Initiative – Freebase (Lausch et al. 2015). LOD availability is constantly evolving, also under the umbrella of Volunteering Geographic Information (Fonte et al. 2015), whereby all large datasets that are freely available on the Internet can be related through semantic networks. The basic advantage of LOD models compared to traditional data-mining models is that, due to already existing semantic links, it is not up to the analyst to decide the criteria to link data in the model (Bizer et al. 2009).

Following the research needs suggested in our article, new and unknown patterns of biodiversity, insights, and model-based forecasts might be developed based on a more robust use of remote sensing.

Acknowledgments

DR was partially funded by: (i) the EU BON (Building the European Biodiversity Observation Network) project, funded by the European Union under the 7th Framework programme, Contract No. 308454 and (ii) by the ERA-Net BiodivERsA, with the national funders ANR, BelSPO and DFG, part of the 2012–2013 BiodivERsA call for research proposals. This work benefited from support from EU COST Action TD1202 'Mapping and the Citizen Sensor'.

References

Anderson, B., B. Arroyo, Y. Collingham, B. Etheridge, J. Fernandez-De-Simon, S. Gillings, et al. 2009. Using distribution models to test alternative hypotheses about a species' environmental limits and recovery prospects. Biol. Conserv. 142, 488–499.

Asner, G. P., and R. E. Martin. 2009. Airborne spectranomics: mapping canopy chemical and taxonomic diversity in tropical forests. Front. Ecol. Environ. 7, 269–276.

Bacaro, G., E. Baragatti, and A. Chiarucci. 2009. Using taxonomic data to assess and monitor biodiversity: are the tribes still fighting? J. Environ. Monit. 11, 798–801.

Baldeck, C. A., and G. P. Asner. 2013. Estimating vegetation beta diversity from airborne imaging spectroscopy and unsupervised clustering. Remote Sens. 5, 2057–2071.

Baldeck, C. A., M. S. Colgan, J.-B. Féret, S. R. Levick, R. E. Martin, and G. P. Asner. 2014.Landscape-scale variation in plant community composition of an African savanna from airborne species mapping. Ecol. Appl. 24, 84–93.

Baselga, A. 2013. Multiple site dissimilarity quantifies compositional heterogeneity among several sites, while average pairwise dissimilarity may be misleading. *Ecography*, **36**, 124–128.

Berger, W. H., and F. L. Parker. 1970. Diversity of planktonic foraminifera in deep-sea sediments. *Science* **168**, 1345–1347.

Bizer, C., T. Heath, and T. Berners-Lee. 2009. Linked data – The story so far. *Int. J. Semant. Web Inf. Syst.* **5**, 1–22.

Bray, J. R., and J. T. Curtis. 1957. An ordination of the upland forest communities of southern Wisconsin. *Ecol. Monogr.* **27**, 325–349.

Chiarucci, A. 2007. To sample or not to sample? That is the question. for the vegetation scientist. *Folia Geobot.* **42**, 209–216.

Clark, M., D. Roberts, and D. Clark. 2005. Hyperspectral discrimination of tropical rain forest tree species at leaf to crown scales. *Remote Sens. Environ.* **96**, 375–389.

Colwell, R. K. 2009. Biodiversity: concepts, patterns and measurement. Pp. 257–263 in S. A. Levin, ed, *The Princeton guide to ecology*. Princeton University Press, Princeton, NJ.

Colwell, R. K., and J. A. Coddington. 1994. Estimating terrestrial biodiversity through extrapolation. *Philos. Trans R. Soc. Lond. B: Biol. Sci.* **345**, 101–118.

Dıaz, S., A. J. Symstad, F. III Stuart Chapin, D. A. Wardle, and L. F. Huenneke. 2003. Functional diversity revealed by removal experiments. *Trends Ecol. Evol.*, **18**, 140–146.

Féret, J.-B., and G. P. Asner. 2013. Tree species discrimination in tropical forests using airborne imaging spectroscopy. *IEEE Trans. Geosci. Remote Sens.* **51**, 73–84.

Féret, J.-B., and G. P. Asner. 2014a. Mapping tropical forest canopy diversity using high-fidelity imaging spectroscopy. *Ecol. Appl.* **24**, 1289–1296.

Féret, J.-B.and, G. P. Asner.2014b.Microtopographic controls on lowland Amazonian canopy diversity from imaging spectroscopy. *Ecol. Appl.* **24**, 1297–1310.

Feilhauer, H., and S. Schmidtlein. 2009. Mapping continuous fields of forest alpha and beta diversity. *Appl. Veg. Sci.* **12**, 429–439.

Ferrier, S., G. Manion, J. Elith, and K. Richardson. 2007. Using generalized dissimilarity modelling to analyse and predict patterns of beta diversity in regional biodiversity assessment. *Divers. Distrib.* **13**, 252–264.

Fleishman, E., R. Noss, and B. Noon. 2006. Utility and limitations of species richness metrics for conservation planning. *Ecol. Ind.* **6**: 543–553.

Fonte, C. C., L. Bastin, L. See, G. Foody, and F. Lupia. 2015. Usability of VGI for validation of land cover maps. *Int. J. Geogr. Inf. Sci.* **29**, 1269–1291.

Foody, G. M., and M. E. J. Cutler. 2003. Tree biodiversity in protected and logged Bornean tropical rain forests and its measurement by satellite remote sensing. *J. Biogeogr.* **30**, 1053–1066.

Gillespie, T. 2005. Predicting woody-plant species richness in tropical dry forests: a case study from South Florida, U.S.A. *Ecol. Appl.* **15**, 27–37.

Giorgini, D., P. Giordani, G. Casazza, V. Amici, M. G. Mariotti, and A. Chiarucci. 2015. Woody species diversity as predictor of vascular plant species diversity in forest ecosystems. *For. Ecol. Manage.* **345**, 50–55.

Gould, W. 2000. Remote Sensing of vegetation, plant species richness, and regional biodiversity hot spots. *Ecol. Appl.* **10**, 1861–1870.

Grime, J. P. 1998. Benefits of plant diversity to ecosystems: immediate, filter and founder effects. *J. Ecol.* **86**, 902–910.

Hanson, T., T. M. Brooks, D. Fonseca, A. B. Gustavo, M. Hoffmann, J. F. Lamoreux, et al. 2009. Warfare in biodiversity hotspots. *Conserv. Biol.* **23**, 578–587.

He, K., and J. Zhang. 2009. Testing the correlation between beta diversity and differences in productivity among global ecoregions, biomes, and biogeographical realms. *Ecol. Inform.* **4**, 93–98.

He, K. S., J. Zhang, and Q. Zhang. 2009. Linking variability in species composition and MODIS NDVI based on beta diversity measurements. *Acta Oecol.* **35**, 14–21.

He, K. S., B. A. Bradley, A. F. Cord, D. Rocchini, M.-N. Tuanmu, S. Schmidtlein, et al., 2015. Will remote sensing shape the next generation of species distribution models? *Remote Sens. Ecol. Conserv.* **1**, 4–18.

Hernández-Stefanoni, J. L., J. A. Gallardo-Cruz, J. A. Meave, D. Rocchini, J. Bello-Pineda, and J. Omar L′opez-Mart′ınez. 2012. Modeling alpha- and beta-diversity in a tropical forest from remotely sensed and spatial data. *Int. J. Appl. Earth Obs. Geoinf.* **19**, 359–368.

Hubbell, S. P. 2001. *The Unified Neutral Theory of Biodiversity and Biogeography*. Princeton University Press, Princeton, USA.

Hurlbert, S. H. 1971. The nonconcept of species diversity: a critique and alternative parameters. *Ecology* **52**, 577–586.

Jaccard, P. 1912. The distribution of the flora in the alpine zone. *New Phytol.* **11**, 37–50.

Koleff, P., K. Gaston, and J. Lennon. 2003. Measuring beta diversity for presence-absence data. *J. Anim. Ecol.* **72**, 367–382.

Krishnaswamy, J., K. S. Bawa, K. N. Ganeshaiah, and M. C. Kiran. 2009. Quantifying and mapping biodiversity and ecosystem services: utility of a multi-season NDVI based Mahalanobis distance surrogate. *Remote Sens. Environ.* **113**, 857–867.

Kumar, S., S. Simonson, and T. Stohlgren. 2009. Effects of spatial heterogeneity on butterfly species richness in Rocky Mountain National Park, CO, U.S.A. *Biodivers. Conserv.* **18**, 739–763.

Lande, R. 1996. Statistics and partitioning of species diversity, and similarity among multiple communities. *Oikos* **76**, 5–13.

Lausch, A., A. Schmidt, and L. Tischendorf. 2015. Data mining and linked open data – New perspectives for data analysis in environmental research. *Ecol. Model.* **295**, 5–17. Use of ecological indicators in models.

Legendre, P., and L. Legendre. 1998. *Numerical ecology*, 2nd ed. Elsevier Science, Amsterdam, NL.

Legendre, P., D. Borcard, and P. R. Peres-Neto. 2005. Analyzing beta diversity: partitioning the spatial variation of community composition data. *Ecol. Monogr.* **75**, 435–450.

Lennon, J. J., P. Koleff, J. J. D. GreenwooD, and K. J Gaston. 2001. The geographical structure of British bird distributions: diversity, spatial turnover and scale. *J. Anim. Ecol.* **70**, 966–979.

Luoto, M., J. Pöyry, R. K. Heikkinen, and K. Saarinen. 2005. Uncertainty of bioclimate envelope models based on the geographical distribution of species. *Glob. Ecol. Biogeogr.* **14**, 575–584.

Mairota, P., B. Cafarelli, R. Labadessa, F. Lovergine, C. Tarantino, R. M. Lucas, et al. 2015. Very high resolution Earth observation features for monitoring plant and animal community structure across multiple spatial scales in protected areas. *Int. J. Appl. Earth Obs. Geoinf.* **37**, 100–105.

Maurer, B. A., and B. J. McGill. 2011. Measurement of species diversity *in* A. E. Magurran and B. J. McGill, eds. *Biological diversity: frontiers in measurement and assessment*. Oxford University Press.

McIntosh, R. P. 1967. An index of diversity and the relation of certain concepts to diversity. *Ecology* **48**, 392–404.

Nagendra, H., and M. Gadgil. 1999. Biodiversity assessment at multiple scales: linking remotely sensed data with field information. *Proc. Natl Acad. Sci. USA* **96**, 9154–9158.

Nagendra, H., and D. Rocchini. 2008. High resolution satellite imagery for tropical biodiversity studies: the devil is in the detail. *Biodivers. Conserv.* **17**, 3431–3442.

Nagendra, H., D. Rocchini, R. Ghate, B. Sharma, and S. Pareeth. 2010. Assessing plant diversity in a dry tropical forest: comparing the utility of Landsat and Ikonos Satellite Images. *Remote Sens.* **2**, 478–496.

Nekola, J., and P. White. 1999. The distance decay of similarity in biogeography and ecology. *J. Biogeogr.* **26**, 867–878.

Noss, R. F. 1990. Indicators for monitoring biodiversity: a hierarchical approach. *Conserv. Biol.* **4**, 355–364.

Oindo, B. O., and A. K. Skidmore. 2002. Interannual variability of NDVI and species richness in Kenya. *Int. J. Remote Sens.* **23**, 285–298.

Oldeland, J., W. Dorigo, L. Lieckfeld, A. Lucieer, and N. Jürgens. 2010. Combining vegetation indices, constrained ordination and fuzzy classification for mapping semi-natural vegetation units from hyperspectral imagery. *Remote Sens. Environ.* **114**, 1155–1166.

Palmer, M. W. 1995. How should one count species? *Nat. Areas J.* **15**, 124–135.

Palmer, M. W. 2005. Distance decay in an old-growth neotropical forest. *J. Veg. Sci.* **16**, 161–166.

Palmer, M. W., P. Earls, B. W Hoagland, P. S White, and T. Wohlgemuth. 2002. Quantitative tools for perfecting species lists. *Environmetrics* **13**, 121–137.

Parviainen, M., M. Luoto, and R. Heikkinen. 2009. The role of local and landscape level measures of greenness in modelling boreal plant species richness. *Ecol. Model.* **220**, 2690–2701.

Penner, J., M. Wegmann, A. Hillers, M. Schmidt, and M.-O. Rödel. 2011. A hotspot revisited – a biogeographical analysis of West African amphibians. *Divers. Distrib.* **17**, 1077–1088.

Petchey, O., and K. Gaston. 2006. Functional diversity: back to basics and looking forward. *Ecol. Lett.* **9**, 741–758.

Pettorelli, N., W. F. Laurance, T. G. O'Brien, M. Wegmann, H. Nagendra, and W. Turner. 2014. Satellite remote sensing for applied ecologists: opportunities and challenges. *J. Appl. Ecol.* **51**, 839–848.

Pielou, E. C. 1966. The measurement of diversity in different types of biological collections. *J. Theor. Biol.* **13**, 131–144.

Ricotta, C., S. Godefroid, and L. Celesti-Grapow. 2008. Common species have lower taxonomic diversity: evidence from the urban floras of Brussels and Rome. *Divers. Distrib.* **14**, 530.

Rocchini, D. 2007. Effects of spatial and spectral resolution in estimating ecosystem alpha-diversity by satellite imagery. *Remote Sens. Environ.* **111**, 423–434.

Rocchini, D., and B. S. Cade. 2008. Quantile regression applied to spectral distance decay. *IEEE Geosci. Remote Sens. Lett.* **5**, 640–643.

Rocchini, D., A. Chiarucci, and S. Loiselle. 2004. Testing the spectral variation hypothesis by using satellite multispectral images. *Acta Oecol.* **26**, 117.

Rocchini, D., S. Andreini Butini, and A. Chiarucci. 2005. Maximizing plant species inventory efficiency by means of remotely sensed spectral distances. *Glob. Ecol. Biogeogr.* **14**, 431–437.

Rocchini, D., C. Ricotta, and A. Chiarucci. 2007. Using satellite imagery to assess plant species richness: the role of multispectral systems. *Appl. Veg. Sci.* **10**, 325–331.

Rocchini, D., H. Nagendra, R. Ghate, and B. Cade. 2009. Spectral distance decay: assessing species beta-diversity by quantile regression. *Photogramm. Eng. Remote Sens.* **75**, 1225–1230.

Rocchini, D., N. Balkenhol, G. A. Carter, G. M. Foody, T. W. Gillespie, K. S. He, et al. 2010. Remotely sensed spectral heterogeneity as a proxy of species diversity: recent advances and open challenges. *Ecol. Inform.* **5**, 318–329.

Scott, W. A., and C. Hallam. 2003. Assessing species misidentification rates through quality assurance of vegetation monitoring. *Plant Ecol.* **165**, 101–115.

Shannon, C. E., and W. Weaver. 1948. A mathematical theory of communication. *Bell Syst. Tech. J.* **27**, 623–656.

Simpson, E. H. 1949. Measurement of diversity. *Nature* **163**, 688.

Skidmore, A. K., N. Pettorelli, N. C. Coops, G. N. Geller, M. Hansen, R. Lucas, et al. 2015. Environmental science: agree on biodiversity metrics to track from space. *Nature* **523**, 403–405.

Small, C. 2004. The Landsat ETM + spectral mixing space. *Remote Sens. Environ.* **93**, 1–7.

Sørensen, T. 1948. A method of establishing groups of equal amplitude in plant sociology based on similarity of species and its application to analyses of the vegetation on Danish commons. *Kongelige Danske Videnskabernes Selskab* **5**, 1–34.

Stickler, C., and J. Southworth. 2008. Application of multi-scale spatial and spectral analysis for predicting primate occurrence and habitat associations in Kibale National Park, Uganda. *Remote Sens.Environ.* **112**, 2170.

Sundstrom, S. M., C. R. Allen, and C. Barichievy. 2012. Species, functional groups, and thresholds in ecological resilience. *Conserv. Biol.* **26**, 305–314.

Thessler, S., K. Ruokolainen, H. Tuomisto, andE. Tomppo. 2005. Mapping gradual landscape-scale floristic changes in Amazonian primary rain forests by combining ordination and remote sensing. *Glob. Ecol. Biogeogr.* **14**, 315–325.

Tucker, C., D. M. Grant, and J. D. Dykstra. 2004. NASA's global orthorectified landsat data set. *Photogramm. Eng. Remote Sens.* **70**, 313–322.

Tuomisto, H., A. Poulsen, K. Ruokolainen, R. Moran, C. Quintana, J. Celi, et al. 2003. Linking floristic patterns with soil heterogeneity and satellite imagery in Ecuadorian Amazonia. *Ecol. Appl.* **13**, 352–371.

Turner, M. 1989. Landscape ecology: the effect of pattern on process. *Annu. Rev. Ecol. Syst.* **20**, 171–197.

Turner, W., S. Spector, N. Gardiner, M. Fladeland, E. Sterling, and M. Steininger. 2003. Remote sensing for biodiversity science and conservation. *Trends Ecol. Evol.* **18**, 306.

Wegmann, M., L. Santini, B. Leutner, K. Sa, D. Rocchini, M. Bevanda, et al. 2014. Role of African protected areas in maintaining connectivity for large mammals. *Phil. Trans. R. Soc. B.* **369**, 20130193.

Whittaker, R. H. 1972. Evolution and measurement of species diversity. *Taxon*, **21**, 213–251.

Wilson, M. V., and A. Shmida. 1984. Measuring beta diversity with presence–absence data. *J. Ecol.* **72**, 1055–1064.

A simple remote sensing based information system for monitoring sites of conservation importance

Zoltan Szantoi[1], Andreas Brink[1], Graeme Buchanan[2], Lucy Bastin[1], Andrea Lupi[1], Dario Simonetti[1], Philippe Mayaux[3], Stephen Peedell[1] & James Davy[1]

[1]European Commission, Joint Research Centre, Institute for Environment and Sustainability, Ispra, Italy
[2]RSPB Centre for Conservation Science, RSPB, Edinburgh, UK
[3]Climate Change, Environment, Natural Resources, Water Unit, Directorate General for International Cooperation and Development, European Commission, Brussels, Belgium

Keywords
Information system, land cover change, Landsat, monitoring, protected areas, remote sensing

Correspondence
Zoltan Szantoi, Land Resource Management Unit, Joint Research Centre, European Commission, TP440, Ispra 21027, Italy.
E-mail: zoltan.szantoi@jrc.ec.europa.eu, zoltan. szantoi@remote-sensing-biodiversity.org

Funding Information
This study has been supported by funding through the Biodiversity and Protected Area Management (BIOPAMA) joint initiative of the European Union and the African, Caribbean and Pacific Group of States secretariat.

Editor: Harini Nagendra
Associate Editor: Duccio Rocchini

Abstract

Monitoring is essential for conservation of sites, but capacity to undertake it in the field is often limited. Data collected by remote sensing has been identified as a partial solution to this problem, and is becoming a feasible option, since increasing quantities of satellite data in particular are becoming available to conservationists. When suitably classified, satellite imagery can be used to delineate land cover types such as forest, and to identify any changes over time. However, the conservation community lacks (a) a simple tool appropriate to the needs for monitoring change in all types of land cover (e.g. not just forest), and (b) an easily accessible information system which allows for simple land cover change analysis and data sharing to reduce duplication of effort. To meet these needs, we developed a web-based information system which allows users to assess land cover dynamics in and around protected areas (or other sites of conservation importance) from multi-temporal medium resolution satellite imagery. The system is based around an open access toolbox that pre-processes and classifies Landsat-type imagery, and then allows users to interactively verify the classification. These data are then open for others to utilize through the online information system. We first explain imagery processing and data accessibility features, and then demonstrate the toolbox and the value of user verification using a case study on Nakuru National Park, Kenya. Monitoring and detection of disturbances can support implementation of effective protection, assist the work of park managers and conservation scientists, and thus contribute to conservation planning, priority assessment and potentially to meeting monitoring needs for Aichi target 11.

Introduction

The protected area network is the cornerstone of site-based conservation, and it is specifically named in Aichi Target 11 of the CBD, which requires world governments to conserve 17% of land through "protected areas and other effective area-based conservation measures" (CBD, 2010). Target 11 also calls for these sites to be effectively and equitably managed (CBD, 2010). Site managers need to know what is happening on sites if they are to respond to current or potential threats to sites. There are standardized methods for *in situ* monitoring of sites, including management effectiveness – PAME (Coad et al. 2013) and Important Bird and Biodiversity Area monitoring (BirdLife International, 2006; Mwangi et al. 2010). However, accepted methods for monitoring land cover and assessing park integrity are lacking for many areas of the globe. Standardization of monitoring would allow data to be collated at national, regional and international levels, allowing progress toward targets, such as CDB Target 11, to be monitored globally.

The loss of natural habitat, especially deforestation and conversion for agriculture, are perhaps the largest threats to biodiversity (Pimm et al. 1995, 2014), so solutions need to be found that allow for easy monitoring of habitat loss across conservation sites. Remote sensing, and satellite data in particular, has been identified as a useful tool for conservation in tracking land cover (e.g. Turner et al. 2003; Buchanan et al. 2009; Leidner et al. 2013). However, classifying and processing the data can be complex, and the need for a simple solution for protected area monitoring was highlighted recently by Rose et al. (2015).

There are already a number of tools and data sources that are available to the conservation community for making assessments of land cover. Indeed, several global products related to land cover state and change, developed by using medium spatial resolution satellite imagery, have been published (e.g. Global Forest Cover (Hansen et al. 2013), Global Land cover (Gong et al. 2012), CLASlite – (http://claslite.carnegiescience.edu/en/) and the Protected Area Archive (http://asterweb.jpl.nasa.gov/paa/). However, these products either focus on one land cover type or they cover limited time periods. In other cases, accuracies are not always acceptable or consistent with other studies (Achard et al. 2014). CLASlite and Global Forest Cover help to lower the entry barrier for mapping and assessing vegetation from satellite imagery, but they focus exclusively on loss and regrowth of tree cover. The conservation and biodiversity community requires accurate data on the dynamics of all vegetation types (grassland, herbaceous vegetation, shrubs, etc.) (Rose et al. 2015), and not just forest. Consequently, the toolbox we present here is a step forward in conservation monitoring.

Here, we introduce and demonstrate an open access, remote sensing based toolbox for site monitoring and a biodiversity and protected areas information system. The toolbox meets many of the needs identified by conservationists (Buchanan et al. 2015; Rose et al. 2015), and allows individuals with little experience of the use of remote sensing to undertake dedicated assessments of land cover and use, and their change, at the site scale. The use of a standardized assessment of land cover/use and type allows data to be collated to produce comparable estimates of change across sites. This allows analysis of the effectiveness of actions, but more importantly, allows consistent statistics to be calculated at the local and regional scales. The toolbox, which can be used offline, requires little capacity or training to pre-process and classify satellite imagery and validate the thematic map results, and is designed specifically for non-(geospatial)-expert conservationists, and PA managers. The biodiversity and protected areas information system integrates various protected area and biodiversity data including the toolbox's thematic land cover/use maps, using open source web services. We illustrate the use of the toolbox with a case study on Lake Nakuru National Park, Kenya before aggregating the results from 10 protected areas in East Africa in order to describe broad patterns of land cover change and show the accessibility of the processed data through the information system.

Materials and Methods

Toolbox – data, pre-processing and image classification

The satellite IMagery ProCessing Toolbox (v1.2b), IMPACT (http://forobs.jrc.ec.europa.eu/products/software/) was developed with pre and post image processing capabilities and combines automated processing chains with minimal user interaction (Fig. 1). The toolbox utilizes data from the USGS Landsat program which, by stretching back to the 1980s (Turner et al. 2013, 2015), provides a readily accessible retrospective baseline land cover.

In the toolbox, an automated pre-processing chain converts digital numbers to top-of-atmosphere reflectance, undertakes clouds and cloud shadows masking, and performs image normalization. The next step is an automated unsupervised classification, using an empirical knowledge-based decision tree approach based on spectral band reflectance characteristics and spectral indices as described in Szantoi and Simonetti (2013) (Fig. 1). The procedure is based on a minimal mapping unit (MMU) of 5 ha, and multi-date image segmentation that assigns individual pixels into objects for each year for which an image has been selected. These objects are then automatically classified using an automated knowledge-based classification algorithm, where the algorithm groups individual pixels within a segment (by year) into a land cover class based on their occurrence frequencies.

Six major land cover types are mapped and an additional 'cloud/shadow' class is used. The applicable land cover classes are: tree cover (over 70% canopy cover and tree height over 5 m), tree cover mosaic (between 30% and 70% canopy and tree height over 5 m), other wooded land (less than 30% of canopy cover, less than 5 m of canopy height, shrubs included), other land cover (non-woody land cover, includes herbaceous vegetation and grass), bare and burnt (a mostly temporary class, depends on seasonality) and permanent water.

We pre-defined the land cover classes based on previous large-scale land cover monitoring studies such as Achard et al. (2014) so that it would fit the various geographical locations to be mapped. The use of a stan-

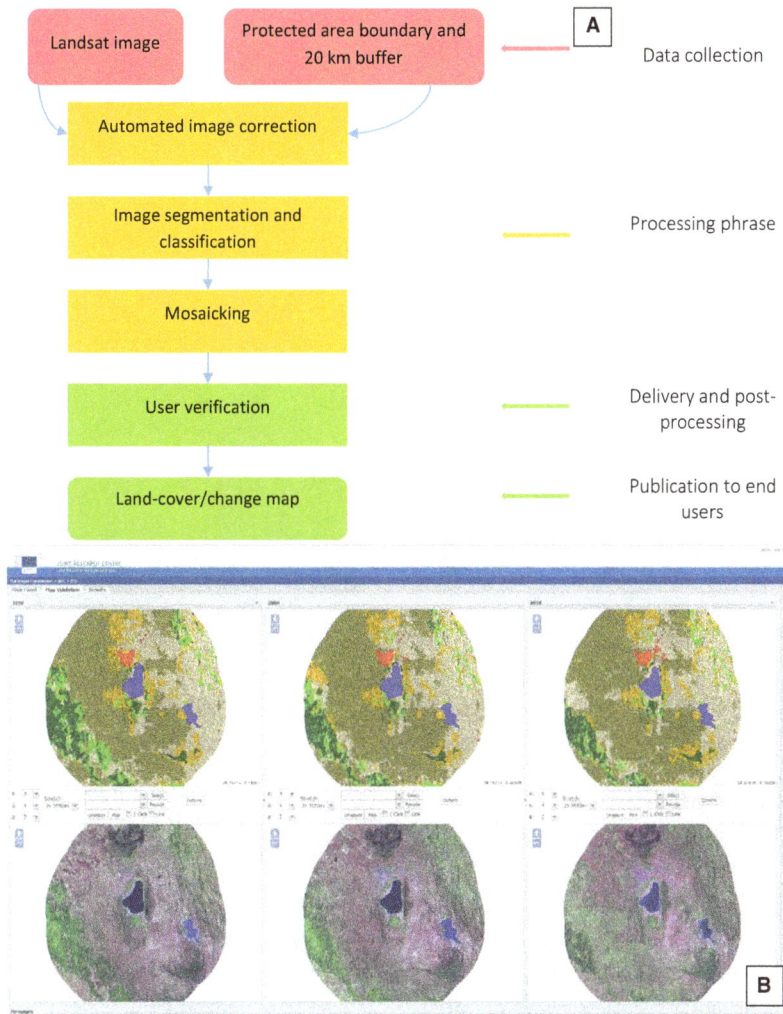

Figure 1. Schematic representation of the data processing flow in IMPACT. Pink layers represent inputs, orange are the automated steps and green indicates steps where user input is needed (A). The interface of the map refinement and verification tool (B).

dard algorithm means the land cover classes should be comparable between sites, allowing amalgamation of results for national and regional scale analyses. These land cover types are broad, but they are appropriate to capture the dynamics which have been identified as being the major land cover related threats to biodiversity, namely conversion and degradation of natural areas (i.e. forest loss) and expansion of agriculture.

Land use classes, most importantly agriculture and human settlements/urban, are not included in the automatic classification since based on their spectral signatures they are generally part of the 'other land cover' and the 'bare and burnt' land cover classes. Inclusion of these land uses in the image classification algorithm would be inaccurate due to the spectral similarities. However, if the user is interested in discriminating such land use classes, there is a 'verification' option within the toolbox where

refinement and verification of the automated classification output is possible (Figs. 1 and 2). This map refinement and verification procedure allows users to derive reliable and highly accurate maps and land cover change statistics, and to include agriculture and human settlement classes.

The user input (verification) phase allows the user to (a) review the land cover maps generated by the automated system and (b) revise and modify (correct) them interactively if mis-classification errors are present using higher spatial resolution imagery (Fig. 1) through various other data sources (e.g. Google Earth™ (https://earthgoogle.org), OpenNebula (http://opennebula.org/) or Global Risk Assessment Services (https://www.gras-system.org). The unverified maps must be used with caution as temporal seasonality effects are present in many cases. These could lead to misclassification errors, and hence to inaccurate land cover maps and change statistics.

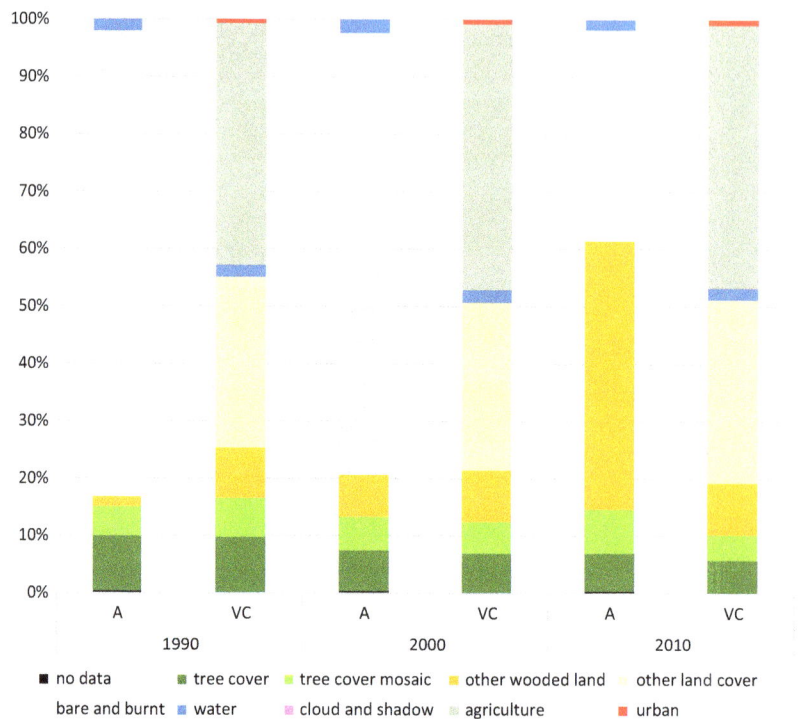

Figure 2. Land cover and land use statistics for Nakuru National Park (and 20 km buffer zone) before and after map refinement and verification. (A – automated, VC – verified and refined).

Monitoring land cover on sites – the protected areas land cover information system

The BIOPAMA Regional Reference Information System (RRIS) is an online biodiversity and protected areas information system hosted by the Joint Research Centre of the European Commission, bringing together relevant information to support decision making for the protection and management of protected areas (http://rris.biopama.org/content/About-RRIS). The protected areas land cover information system represents one module within the RRIS (http://rris.biopama.org/lcc). Processed imagery and validated land cover data are currently available for 772 and 23 protected areas (PAs) in Sub-Saharan Africa, respectively. These protected areas were selected based on their IUCN categories (I–IV) and on some other ranks (e.g. Ramsar wetlands, UNESCO sites). The boundaries were obtained from the World Database on Protected Areas (WDPA) as of March, 2015 (IUCN & UNEP-WCMC, 2015). As the verification is an ongoing procedure, validated land cover data sets are being added to the system in a continuous manner. The protected areas land cover information system under the BIOPAMA RRIS allows users to search any protected area in Sub-Saharan Africa and at a later stage the Caribbean and Pacific regions by name, WDPA ID, and country. In addition, users can select the option to view only validated land cover data or to include unvalidated land cover maps in their selection. Landsat image mosaics

for up to three analyzed time periods (currently representing the decades 1990, 2000, 2010) are available for each PA and their 20 km surrounding buffer zones. Users can specify the color band combination using custom Red-Green-Blue values of the multispectral bands or a predefined standard for viewing satellite imagery (false color and natural color) to aid visual interpretation. Users also have the option to toggle auto scaling on the satellite imagery, which attempts to generate a cohesive uniform color pallette for the satellite images that represent the protected area. The resulting maps are displayed simultaneously with synchronized pan and zoom capabilities to allow the user to easily compare differences between decades.

Refinement and verification of the thematic maps

Those protected areas whose land cover maps have been verified are published online and can be viewed, analyzed and downloaded by any registered/anonymous user. The data consists of labeled polygons with a classification and a final category for each year. The unverified data have also been made freely available. However, since the automated classification requires user input in order to verify labels (see Fig. 1), unverified data are not supplied as a true record of land cover – instead it is foreseen that users will download this data to perform their own verifications (e.g. by using the IMPACT Toolbox presented here) and contribute them back to the information system.

Results

Land cover/use change on Lake Nakuru

To demonstrate the use of the toolbox we use Lake Nakuru, Kenya (0°22′0″ S 36°4′60″ E; WDPA ID 762; http://rris.biopama.org/lcc/762) as a case study, not least because this site illustrates the need for the user verification procedure. Following the steps outlined above, and using satellite images from 1990, 2000 and 2010, land cover and land cover change were assessed. The automated image classification algorithm was able to discriminate between most of the major land covers, based on the visual assessment (Fig. 2). Moreover, an accuracy assessment using 429 randomly generated points based on the corrected thematic map (ground truth) and compared to the automated classifier generated thematic map revealed that the overall accuracy of the latter map was over 60% for each investigated year (1990 – 65.73%, 2000 – 60.60% and 2010 – 67.13%).

The verification (and refinement) suggested that the automated algorithm was less able to identify the 'other wooded land', 'other land cover' and 'bare and burnt' mixed classes, confusing these with large agricultural and urban land use classes. Without a refinement and verification step, the agricultural land use would have been classified as simply 'other land cover'. The decadal land cover maps, subsequent to refinement and verification, show a clear trend in the expansion of agriculture and human settlements (i.e. urban) at the expense of tree cover (close forest) and tree cover mosaic (open forest) over the decades (Fig. 3). A gradual loss of tree cover in the south/southwest area of the study site is apparent, and tree cover/tree cover mosaic were replaced by agriculture. Change can be noticed in the northeastern part, where the transition from tree cover and tree cover mosaic classes to other land, other wooded land and agriculture

occurred. The city of Nakuru has expanded at the expense of wooded areas.

Broader patterns of land cover change

In addition to Lake Nakuru, we assessed land cover change on nine other protected areas in East Africa (Table 1). All showed broadly similar patterns in their land cover change dynamics, with human settlements and agricultural land use increasing in extent in the buffers while the land cover within the protected areas remained stable (Fig. 4). Udzungwa Mountains National Park (19297) was particularly notable for the clear increase in the extent of agriculture and urban (with a concomitant decrease in forest cover). This might indicate that this park is a concern compared to, for example Nechisar National Park (2278) and Mahale Mountains National Park (7521) which remain relatively intact, both within the park and in the buffer surrounding the park.

Discussion and Conclusions

Monitoring is a key part of the conservation process, which allows problems to be identified and solutions developed, and also allows the effectiveness of actions to be assessed. Traditionally, monitoring has been based on *in situ* field assessments, but this can incur considerable costs, especially where sites are remote, inaccessible or extensive. The conservation community has identified an as-yet unmet need to apply remote sensing data in monitoring sites of conservation importance (Buchanan et al. 2015; Rose et al. 2015). Our tool, which can be applied globally, allows assessment of land cover change to be made from anywhere in the world. It meets the criteria that have previously been suggested for a remote sensing based system for monitoring land cover change on sites (Buchanan et al. 2009) in that it is free to use, requiring

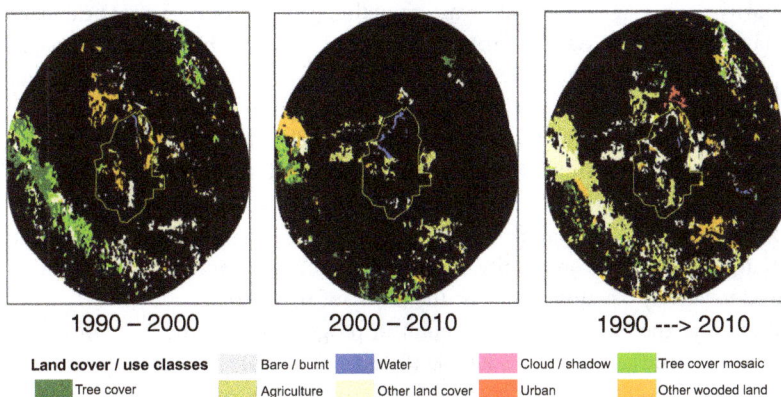

Figure 3. Change detection results of the verified Nakuru National Park and its vicinity (20 km). The change maps for 1990–2000 and 2000–2010 show the original class which has changed during the particular period, whereas 1990→2010 shows the actual (new) land cover class for the year 2010.

Table 1. Protected areas analyzed for land cover/land use change.

WDPA ID	Park name	Country	IUCN category	Centroid (Long/Lat)
653	Simien Mountains National Park	Ethiopia	II	38.17, 13.16
753	Marsabit National Park	Kenya	II	37.95, −2.33
760	Mount Elgon National Park	Kenya	II	34.69, 1.09
762	Lake Nakuru National Park	Kenya	II	36.09, −0.39
764	Ol Donyo Sabuk National Park	Kenya	II	37.26, −1.14
2278	Nechisar National Park	Ethiopia	II	37.89, 5.99
2279	Abijatta-Shalla Lakes National Park	Ethiopia	II	38.52, 7.54
7521	Mahale Mountains National Park	United Republic of Tanzania	II	29.90, −6.21
19297	Udzungwa Mountains National Park	United Republic of Tanzania	II	36.66, −7.79
19726	Malka Mari National Park	Kenya	II	40.76, 4.18

just a login. It utilizes free data (an essential feature if it is to be sustainable) and it requires little expertise in remote sensing, since the built-in logical processing chain utilizes metadata supplied with images to undertake pre-processing. We expect that it will have a major impact in the field of conservation.

This tool adds to an expanding number of methods available to the conservation community for tracking land cover. Many of these tools are simple and easy to use, making them ideal for the conservation community which may have limited remote sensing capacity. However, many of the tools are confined to a consideration of forest (e.g. Global Forest Cover or CLASlite) and not other land cover (habitat) types. Given that the majority of species are dependent on forest, these tools have obvious value for monitoring change in sites of conservation importance, up to a global scale (e.g. Tracewski et al. 2016). However, a consideration of other land cover types is useful for long term monitoring and planning, in order to identify what is replacing the lost forest.

Application of the land cover change tool to monitor Nakuru National Park in Kenya indicated that the land cover within and around the park had stayed relatively stable over the past 30 years. This might indicate that the park was being effective at conserving the natural land cover within its boundaries. The extent of natural land

cover in the other parks was broadly stable too, but there was a notable increase in agriculture, and decrease in tree cover, in the 20 km buffers around the parks. Previous studies have found that broadly, protected areas reduce the rate of loss of forest (Geldmann et al. 2013) and all land cover (Beresford et al. 2013). Forest loss and agricultural encroachment are recognized as major threats to biodiversity (Pimm et al. 1995, 2014), and the loss of trees and increase in agricultural land around the parks indicates that pressures on the parks themselves might be increasing. Conservation responses could be informed by these data, allowing the authorities to maintain the integrity of the parks under greatest threat.

The image processing toolbox has evolved from code developed to help in remote sensing assessments of forest loss (Achard et al. 2014). Consequently, as it was originally developed for remote sensing applications, it will also be of use to the remote sensing community. In particular, some of the features which were previously standalone applications (e.g. Baatz's segmentation algorithm http://www.terralib.org/html/v360/TePDIBaatz_8hpp-source.html) are fully integrated in the box's processing chain.

The toolbox is complemented by an information system that allows sharing of results. The toolbox, while requiring some web access, can be used offline and is suitable for situations where remote sensing expertise and capacity to track changes on sites are limited. It can be used to make assessments at the local scale (i.e. by park managers or community site management groups) or for regional level studies, and it covers multiple land cover types.

The toolbox specifically was developed for individuals in the conservation community who have little experience of remote sensing and image pre-processing. The workflow allows such individuals to utilize images which have not had higher levels of processing, opening up a new pool of potential data sources to them.

We presented examples of the toolbox being used with Landsat data, which are available retrospectively, back until the 1980s. This enables conservationists to access a back catalogue of images and make retrospective baselines for monitoring; something not possible from field data. As the imagery selection and pre-processing is highly automated, the inclusion of new medium spatial resolution imagery into the workflow is foreseen, including Landsat 8 OLI and Sentinel 2ab multispectral imagery from 2015 onwards. Continued availability of these data has been highlighted as essential for the continued (and expanded) use of remote sensing by the conservation community (Turner et al. 2015), and will certainly be important for the utility of the RRIS.

At this midpoint in progress toward the 2020 Aichi targets for the CBD (Secades et al. 2014), we envisage that the toolbox could make a contribution toward measuring pro-

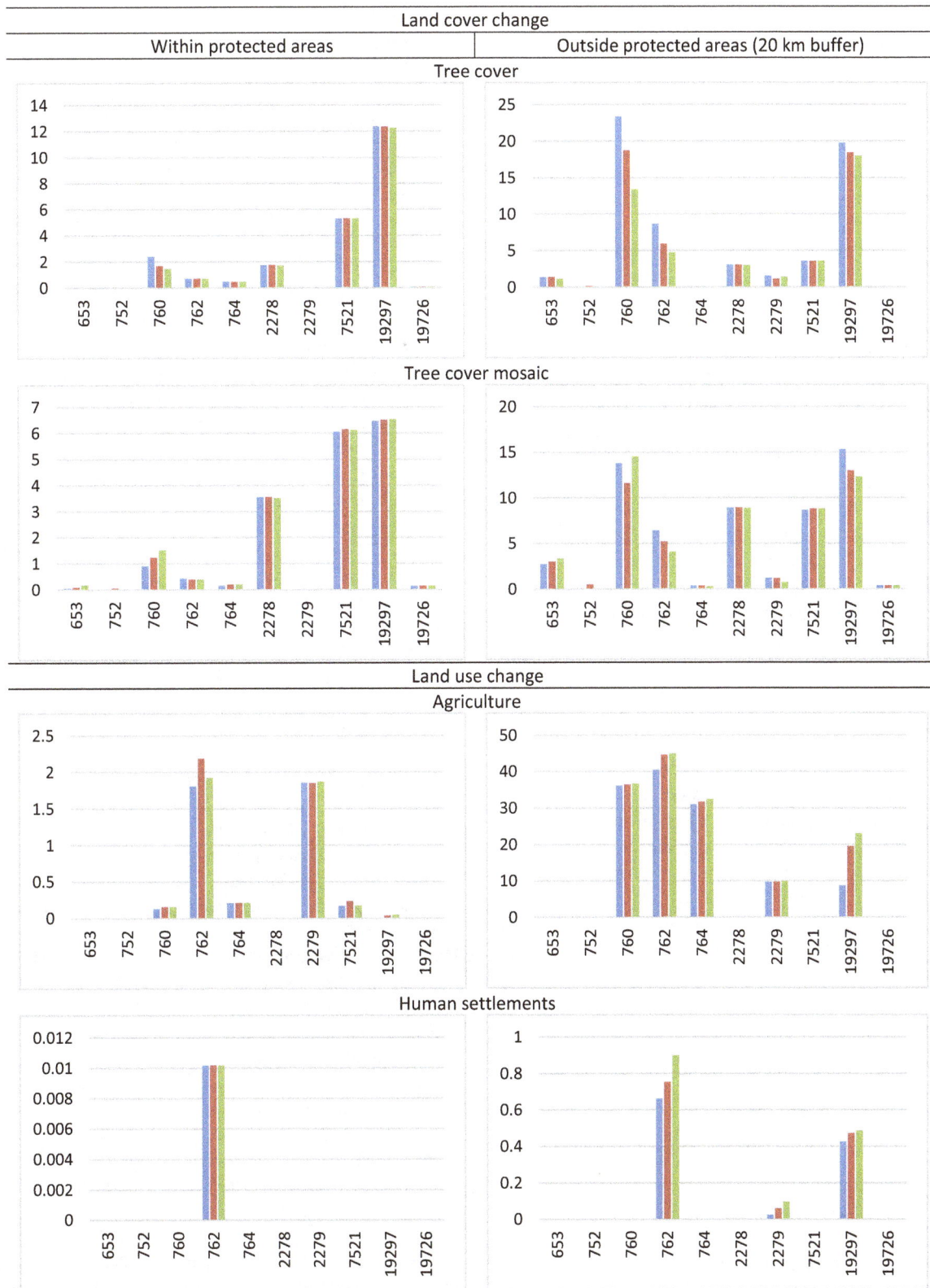

Figure 4. Land cover and land use change dynamics in selected validated protected areas and their buffer zones, expressed in (%) compared to the covered area, in 1990 (blue), 2000 (red) and 2010 (green). Reference numbers (x-axis) are the WDPA IDs for each protected area (see Table 1 for more details).

gress to the 2020 CBD targets (O'Connor et al. 2015). The toolbox could also make a contribution toward measuring the effectiveness of protected areas, something that is needed to measure progress toward Aichi Target 11. Progress toward this target has so far been measured in terms of the extent of coverage of protected areas (e.g. Butchart et al. 2015), rather than effectiveness. The effectiveness of protection in general has been studied (e.g. Geldmann et al. 2013), but the effectiveness at the country level remains unassessed. Analysis of individual protected areas according to a common legend, followed by the upload of these data to a shared web resource will enable collation of data on how effective individual protected areas are at reducing detrimental land cover change, and how effective the entire network of PAs is at halting land cover change, making progress toward measuring Aichi target 11 across all land cover types.

In addition to target 11, the toolbox could contribute directly to target 12, which relates to species conservation. In particular, if species have distributions that are restricted to one or two protected areas or other sites (AZE and IBAs are identified on such criteria), the toolbox could measure change in suitable land cover within these sites as a surrogate for population change for species. Large-scale sampling could also make a contribution toward measuring progress toward target 5 on reduction in the loss of natural habitats. Application of the toolbox for sampling, developing the approach of, for example Brink and Eva (2009) would allow land cover change assessments to be made efficiently across extensive areas which might not otherwise be assessed.

The tool will perhaps be of greatest use in the biodiversity-rich, but conservation-capacity poor, tropics. Analysis could be carried out close to the source of application (i.e. by the staff of a protected area themselves), but given that resources are not available to allow this in every park, analysis could be carried out for multiple parks or sites at a national or regional level. It will always be essential that the data are returned to the point of use, to allow information to be used rapidly by those managing a park on the ground.

Conflict of Interest

The authors declare no conflict of interest.

References

Achard, F., R. Beuchle, P. Mayaux, H.-J. Stibig, C. Bodart, A. Brink, et al. 2014. Determination of tropical deforestation rates and related carbon losses from 1990 to 2010. *Glob. Change Biol.* **20**:2540–2554.

Beresford, A. E., G. W. Eshiamwata, P. F. Donald, A. Balmford, B. Bertzky, A. B. Brink, et al. 2013. Protection reduces loss of natural land cover at sites of conservation importance across Africa. *PLoS ONE* **8**:e65370.

BirdLife International. 2006. *Monitoring important bird areas: a global framework.* BirdLife International, Cambridge, UK.

Brink, A. B., and H. D. Eva. 2009. Monitoring 25 years of land cover change dynamics in Africa: a sample based remote sensing approach. *Appl. Geogr.* **29**:501–512.

Buchanan, G. M., A. Nelson, P. Mayaux, A. Hartley, and P. F. Donald. 2009. Delivering a global, terrestrial, biodiversity observation system through remote sensing. *Conserv. Biol.* **23**:499–502.

Buchanan, G. M., A. B. Brink, A. K. Leidner, R. Rose, and M. Wegmann. 2015. Advancing terrestrial conservation through remote sensing. *Ecol. Inform.* **30**:318–321.

Butchart, S. H. M., M. Clarke, R. J. Smith, R. E. Sykes, J. P. W. Scharlemann, M. Harfoot, et al. 2015. Shortfalls and solutions for meeting national and global conservation area targets. *Conserv. Lett.* **8**:329–337.

CBD. 2010. *Global biodiversity outlook 3.* Secretariat of the Convention on Biological Diversity, Montreal, Quebec, Canada.

Coad, L., F. Leverington, N. D. Burgess, I. C. Cuadros, J. Geldmann, T. R. Marthews, et al. 2013. Progress towards the CBD protected area management effectiveness targets. *Parks* **19**:13–24.

Geldmann, J., M. Barnes, L. Coad, I. D. Craigie, M. Hockings, and N. D. Burgess. 2013. Effectiveness of terrestrial protected areas in reducing habitat loss and population declines. *Biol. Conserv.* **161**:230–238.

Gong, P., J. Wang, L. Yu, Y. Zhao, Y. Zhao, L. Liang, et al. 2012. Finer resolution observation and monitoring of global land cover: first mapping results with Landsat TM and ETM+ data. *Int. J. Remote Sens.* **34**:2607–2654.

Hansen, M. C., P. V. Potapov, R. Moore, M. Hancher, S. A. Turubanova, A. Tyukavina, et al. 2013. High-resolution global maps of 21st-century forest cover change. *Science* **342**:850–853.

IUCN & UNEP-WCMC. 2015. Available at http://www.unep-wcmc.org/policies/wdpa-data-licence.

Leidner, A., A. Brink, and Z. Szantoi. 2013. Leveraging remote sensing for conservation decision making. *EOS* **94**:508.

Mwangi, M. A. K., S. H. M. Butchart, F. B. Munyekenye, L. A. Bennun, M. I. Evans, L. D. C. Fishpool, et al. 2010. Tracking trends in key sites for biodiversity: a case study using Important Bird Areas in Kenya. *Bird Conserv. Int.* **20**:215–230.

O'Connor, B., C. Secades, J. Penner, R. Sonnenschein, A. Skidmore, N. D. Burgess, et al. 2015. Earth observation as a tool for tracking progress towards the Aichi Biodiversity Targets. *Remote Sens. Ecol. Conserv.* **1**:19–28.

Pimm, S. L., G. J. Russell, J. L. Gittleman, and T. M. Brooks. 1995. The future of biodiversity. *Science* **269**:347–350.

Pimm, S. L., C. N. Jenkins, R. Abell, T. M. Brooks, J. L. Gittleman, L. N. Joppa, et al. 2014. The biodiversity of species and their rates of extinction, distribution, and protection. *Science* **344**:1246752.

Rose, R. A., D. Byler, J. R. Eastman, E. Fleishman, G. Geller, S. Goetz, et al. 2015. Ten ways remote sensing can contribute to conservation. *Conserv. Biol.* **29**:350–359.

Secades, C., B. O'Connor, C. Brown, and M. Walpole. 2014. *Earth observation for biodiversity monitoring: a review of current approaches and future opportunities for tracking progress towards the Aichi biodiversity targets.* Secretariat of the Convention on Biological Diversity, Montréal, Canada.

Szantoi, Z., and D. Simonetti. 2013. Fast and robust topographic correction method for medium resolution satellite imagery using a stratified approach. *IEEE J. Select. Topics Appl. Earth Obs. Remote Sens.* **6**:1921–1933.

Tracewski, L., S. H. M. Butchart, M. Evans, L. D. C. Fishpool, and G.M. Buchanan 2016. Patterns of twenty-first century forest loss across a global network of important sites for biodiversity. *Remote Sens. Ecol. Conserv.*, 1–8.

Turner, W., S. Spector, N. Gardiner, M. Fladeland, E. Sterling, and M. Steininger. 2003. Remote sensing for biodiversity science and conservation. *Trends Ecol. Evol.* **18**:306–314.

Turner, W., G. Buchanan, C. Rondinini, J. L. Dwyer, M. Herold, L. P. Koh, et al. 2013. Satellites: make data freely accessible. *Nature* **498**:37.

Turner, W., C. Rondinini, N. Pettorelli, B. Mora, A. K. Leidner, Z. Szantoi, et al. 2015. Free and open-access satellite data are key to biodiversity conservation. *Biol. Conserv.* **182**:173–176.

13

Sea turtle nesting patterns in Florida vis-à-vis satellite-derived measures of artificial lighting

Zachary A. Weishampel, Wan-Hwa Cheng & John F. Weishampel

Department of Biology, University of Central Florida, Orlando, Florida, USA

Keywords

Artificial light, DMSP, light pollution, marine turtles, nest surveys, simultaneous autoregressive modeling

Correspondence

John F. Weishampel, Department of Biology, University of Central Florida, Orlando, FL 32816 USA.
E-mail: John.Weishampel@ucf.edu

Editor: Rob Williams
Associate Editor: Helen de Klerk

Abstract

Light pollution contributes to the degradation and reduction of habitat for wildlife. Nocturnally nesting and hatching sea turtle species are particularly sensitive to artificial light near nesting beaches. At local scales (0.01–0.1 km), artificial light has been experimentally shown to deter nesting females and disorient hatchlings. This study used satellite-based remote sensing to assess broad scale (~1–100s km) effects of artificial light on nesting patterns of loggerhead (*Caretta caretta*), leatherback (*Dermochelys coriacea*) and green turtles (*Chelonia mydas*) along the Florida coastline. Annual artificial nightlight data from 1992 to 2012 acquired by the Defense Meteorological Satellite Program (DMSP) were compared to an extensive nesting dataset for 368, ~1 km beach segments from this same 21-year period. Relationships between nest densities and artificial lighting were derived using simultaneous autoregressive models to adjust for the presence of spatial autocorrelation. Though coastal urbanization increased in Florida during this period, nearly two-thirds of the surveyed beaches exhibited decreasing light levels ($N = 249$); only a small fraction of the beaches showed significant increases ($N = 52$). Nest densities for all three sea turtle species were negatively influenced by artificial light at neighborhood scales (<100 km); however, only loggerhead and green turtle nest densities were influenced by artificial light levels at the individual beach scale (~1 km). Satellite monitoring shows promise for light management of extensive or remote areas. As the spectral, spatial, and temporal resolutions of the satellite data are coarse, ground measurements are suggested to confirm that artificial light levels on beaches during the nesting season correspond to the annual nightlight measures.

Introduction

As humans appropriate the terrestrial surface of the earth, areas of total natural darkness are becoming increasingly rare (Bogard 2014). Though perhaps not as feared as chemical forms of pollution, light pollution (excessive and/or misdirected light) may have significant negative impacts on wildlife (e.g., Hölker et al. 2010). Artificial light disrupts animal (amphibian, bird, fish, mammal, reptile and invertebrate) physiology and behavior, which affects their ability to orient, communicate, forage and reproduce (Longcore and Rich 2004; Rich and Longcore 2006). Organisms that live on earth's surface have evolved within the constraints of regular diel and seasonal light cycles that have been interrupted by the glow created by artificial light. Light pollution is fundamentally a form of habitat degradation—a primary cause of biodiversity loss (Tilman et al. 1994).

The two sea turtle families (Cheloniidae and Dermochelyidae) have persisted for >100 million years (Naro-Maciel et al. 2008). The seven species that comprise these families are listed as Vulnerable (1), Endangered (2), Critically Endangered (3) and Data Deficient (1) by the International Union for Conservation of Nature (IUCN 2015). As a slowly maturing, long-lived taxon, sea turtles have had only a few generations to adapt to the widespread appearance of artificial lighting along the coast since the patent of the electric light bulb in 1880. Sea turtles occupy land for a minuscule fraction of their lives: when they incubate as eggs then crawl to the ocean as hatchlings and, if adult female, when they return to their natal beaches to nest. Both of these life-history-defining activities are primarily nocturnal for six of these species and the influence of artificial light on these behaviors are well documented (see Witherington and Martin 2003 for a summary).

When sea finding after emergence, hatchlings often become mis- or disoriented in the presence of artificial light (Peters and Verhoeven 1994; Salmon et al. 1995; Tuxbury and Salmon 2005). Artificial light from behind the beach, interferes with the normal, visual cues of starlight or moonlight reflecting off the water (Salmon 2006). In Florida, this has been projected to lead to thousands of hatchling deaths annually as they wander away from the ocean (Salmon 2003). For adult females, experimental field studies (at 10–100 m scales) on Florida and Costa Rica beaches showed the presence of mercury vapor lights (multiple spectral peaks between 400 and 700 nm) inhibited loggerhead and green turtle nesting, respectively (Witherington 1992). This negative effect was not apparent with low pressure sodium lamps (spectral peak at 590 nm; Witherington 1992). Deterring gravid females leads to denser nest congregations on darker beaches, which may increase depredation rates and divert hatchlings from more suitable dispersal-related nearshore features (Hamann et al. 2011).

Remote sensing is used to map extensive and inaccessible areas and over time can be used to monitor habitat change (Pettorelli et al. 2014). For sea turtles, airborne and satellite data have been used to map beach morphology (e.g., Long et al. 2011; Yamamoto et al. 2012), sea surface temperature (e.g., Hawkes et al. 2007; Weishampel et al. 2010), sargassum movements (e.g., Witherington et al. 2012; Hu et al. 2015), and recently artificial lighting (e.g., Magyar 2008; Kamrowski et al. 2012, 2014a). A series of Australian studies showed trends in satellite-derived artificial lighting on nesting beaches of five sea turtle species around the continent (Kamrowski et al. 2012, 2014a). These studies focused on population management units of the different species that include stretches of coastline (>100 km) and represent important nesting beaches, but did not directly relate artificial light levels to annual nest numbers on specific beaches. However, an eastern Mediterranean study summed loggerhead and green turtle nest counts over a 19-year period and found that nest density negatively related to two satellite-derived measures of artificial light intensity (from 2003 to 2007) on 190, 1-km beaches along the coastline of Israel (Mazor et al. 2013).

Because of systematic nest surveying efforts over the last 26 years across >320 km of Florida beaches and a 21-year archive of satellite-derived nightlight imagery, we can directly compare year-to-year nesting and artificial light patterns. Where the maximum combined nest density in the Israeli study was <10 nests km^{-1} $year^{-1}$, several monitored Florida beaches have densities that exceed 700 loggerhead, 100 green turtle, and 10 leatherback nests km^{-1} $year^{-1}$. This study's primary objectives were to: (1) use satellite data to identify coastal light pollution trends on key Florida nesting beaches from 1992 to 2012 and with nest survey records (2) assess broad scale (1–100s km) artificial lighting effects on nocturnal nesting patterns of the three predominant sea turtle species in Florida.

Materials and Methods

Study area and species

Approximately 90% of all sea turtle nesting in the continental United States occurs in Florida (FWC 2015a). The three main sea turtles that nest in Florida in decreasing nest numbers are loggerheads (Caretta caretta; family Cheloniidae), green turtles (Chelonia mydas; family Cheloniidae), and leatherbacks (Dermochelys coriacea; family Dermochelyidae). Florida possesses the highest number of loggerhead nests in the Western Hemisphere, the highest number of nesting green turtles in the United States, and is the only continental state where leatherbacks regularly nest (FWC 2015a). In 1996 the IUCN designated loggerheads as endangered; the U. S. Fish and Wildlife Service (USFWS) lists Florida populations as threatened. These populations have been fluctuating over the last 30 years (Witherington et al. 2009). Florida loggerheads nest between April and September laying 4–7 nests (~115 eggs per nest) and return approximately to the same beach area every 2–4 years (Ceriani et al. 2015). In 2004 the IUCN categorized Green turtles as endangered. Though listed as endangered by the USFWS, Florida green turtle populations have been exponentially increasing (Chaloupka et al. 2008) since the 1980s; they nest at 2–4-year intervals between June and September and lay ~3–5 nests (~135 eggs per nest) each season. In 2013 the IUCN listed leatherbacks as vulnerable; the USFWS lists the Florida populations as endangered. Florida leatherback populations have been steadily increasing (Stewart et al. 2011); they nest between March and August laying 5–7 nests per season (~80 eggs per nest), returning every 2–3 years.

In terms of physiological responses to light, differences among species are somewhat unclear. Based on the anatomical, optical eye geometries, though universally not well-adapted for vision in dim light conditions, leatherbacks are more sensitive to low light intensities than loggerheads and green turtles (Fritsches and Warrant 2013). Loggerhead and green turtles have similar visual pigment spectral peak locations, but the magnitudes of the peaks differ between the species. Though possessing similar visual pigments, leatherbacks have a higher sensitivity at shorter wavelengths (Crognale et al. 2008) reflecting their lifestyle that involves deeper dives than loggerheads and green turtles. Moreover, adult leatherbacks also have a slower temporal, visual resolution than loggerheads and green turtles, suggesting adaptations for dimmer (deeper) water (Southwood et al. 2008).

Sea turtle nesting survey data

To conserve these species as mandated by the 1973 Endangered Species Act, the Florida Fish and Wildlife Conservation Commission (FWC) collaborates with the USFWS to monitor nesting patterns across the state. In 1989, the FWC established the Index Nesting Beach Survey (INBS). The INBS (FWC 2015b) follows standard protocols where FWC-trained surveyors and volunteers traverse specific beaches every morning during nesting season (15 May to 31 August) and record nesting by species based on track and digging characteristics. Witherington et al. (2009) divided the Florida coast into eight survey regions, Northeast, Central Northeast, Central East, Central Southeast, Southeast, South, Southwest, and Northwest (Fig. 1). Though these geographic regions may possess similar environmental factors (e.g., ocean currents, sea surface temperatures), they do not correspond to genetic subdivisions of the sea turtle populations as

shown for loggerheads which are estimated to have seven different subunits for Florida (Shamblin et al. 2011). A total of 27 beaches comprising 368, ~1 km segments, from six survey regions have been monitored consistently, using INBS standards since the program's inception. Five more beaches (126, ~1 km segments) from the Gulf coast have been added since 1992; however, these were not included in this analysis. The 368 beaches analyzed here do not include any from the South (which represents the Everglades and Florida Keys) and Northwest (Florida panhandle) survey regions.

We acquired the INBS sea turtle nesting data from the FWC Fish and Wildlife Research Institute for 1992–2012 to coincide with the satellite data described below. The FWC data consisted of 7 728 nesting tallies per species (368 beach segments × 21 years). We calculated annual nest density for the state, each region, and each INBS beach segment for each species (Figs. 2 and 3) using segment lengths from the FWC-provided ArcGIS shapefiles. In a similar approach as Witherington et al. (2009), to better visualize the data and simplify spatial analyses, we coarsely estimated distances along the coastline (not considering fractal properties; Jiang and Plotnick 1998) from the midpoint of each INBS segment in ArcGIS and spatially transformed the nesting (and artificial light data) from latitude–longitude coordinates into 1-dimensional, 953.2-km long transects comprised of the 368 sample points from Ft. Clinch (NE 0 km) to Sanibel Island (SW 953.2 km) going clockwise around the state (Fig. 3). This

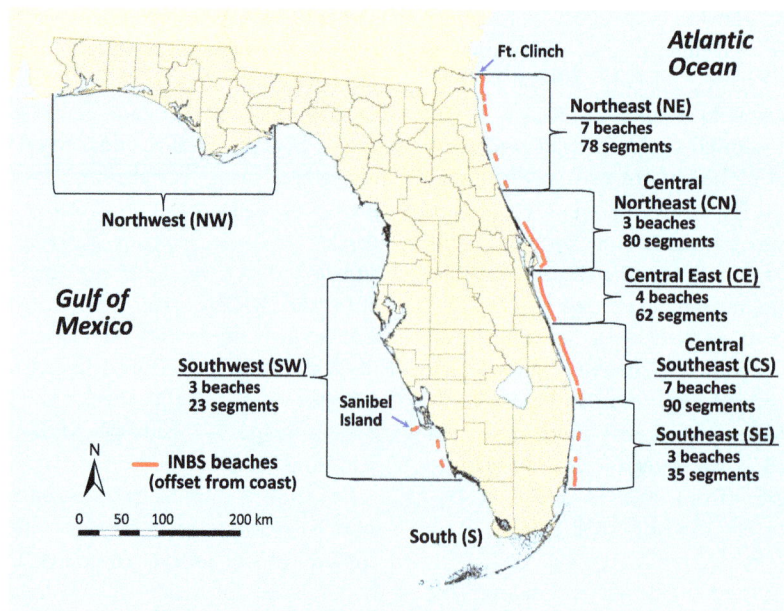

Figure 1. Map of study area along the Florida coast. Nesting regions with numbers of beaches and beach segments that were analyzed are designated.

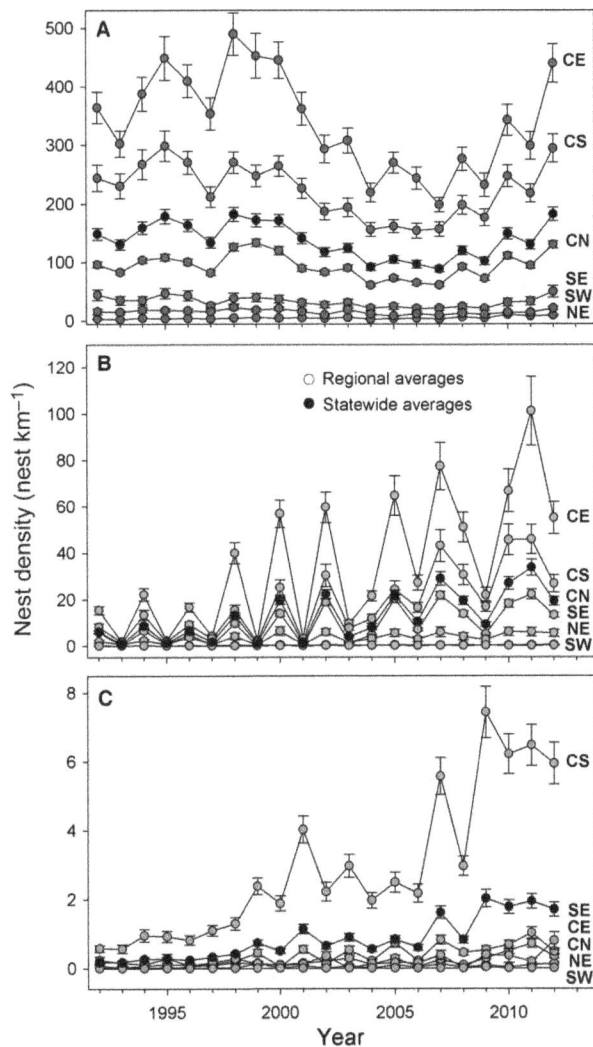

Figure 2. Dynamics of (A) loggerhead, (B) green turtle, and (C) leatherback nest densities for Florida nesting regions and overall average (black markers) from 1992 to 2012. Extensions above and below markers represent ±1 standard error.

transformation ignores the coastline along bays with narrow inlets and produces minor distortions as the distance light travels differs from the distance a sea turtle swims from one beach to the next.

Artificial lighting data

The Defense Meteorological Satellite Program (DMSP) Operational Linescan System (OLS) monitors weather for the United States Air Force. Visible and infrared sensors (440–940 nm) collect 3 000 km swaths of data, covering the earth twice daily. To detect clouds at night, the visible signal is intensified. This enables light detection at the planet surface (see Elvidge 2014 for examples using nighttime lights data). Nighttime, cloud-free data (without

solar glare, moonlight, and auroral effects) are mapped into a latitude-longitude grid with an equatorial pixel size of ~1 km. The pixel size around Florida is closer to 800 m. We used the stable lights product, which removes ephemeral events such as fires and background noise (NOAA 2015a). We downloaded DMSP imagery data from 1992 to 2012 from the NOAA National Geophysical Data Center (now the National Centers for Environmental Information) using the recommended six satellite sources for the years that Kamrowski et al. (2014a) used for their 19-year study. Because there is no on-board calibration, sensitivity may change over the lifespan of the satellite as well as differ among satellites. To compensate for this, we used the second-order regression models developed by Elvidge et al. (2014) to intercalibrate the time series of night light data. We rescaled the annual 6-bit data from 0 to 100 with 0 representing total darkness and 100 representing complete light saturation for the period of detection.

Figure 4A shows the average light value for each pixel from 1992 to 2012. Urban areas, for example, Jacksonville, Miami, and Tampa-St. Petersburg, are large red amorphous features representing maximum light (saturated) values. We used the *insectlinerst* command in the Geospatial Modelling Environment (Beyer 2014) software package to calculate the yearly artificial light value for each beach segment weighted by the portion of the segment in a DMSP raster. As done with the nesting data, we calculated the change in light over the 21-year period by region (Fig. 5) and examined the average light levels by INBS segment (Fig. 6).

Data analysis and statistical modeling

Trends in artificial light were calculated based on the slope of the linear regressions of the light levels for each INBS beach segment over the 21-year period. To determine the influence of artificial light (X) and other possible contributing factors on nest density (Y), we followed a multi-model selection approach. But as was found with nesting data on a ~40-km INBS subsample (Weishampel et al. 2003) and is common with species distribution data (Kissling and Carl 2008; Miller 2012), we anticipated that nesting, as well as artificial light patterns, would not be spatially independent, that is, they would exhibit spatial autocorrelation (SAC). This assumption that spatially closer values will be more related than spatially distant values is, in part, caused by site fidelity and philopatry of nesting sea turtles and by the non-random location and sprawling growth patterns of cities. To measure the extent of SAC across different distances (i.e., correlograms) between INBS locations, we calculated Moran's I using the software program GS+ (Gamma Design Software

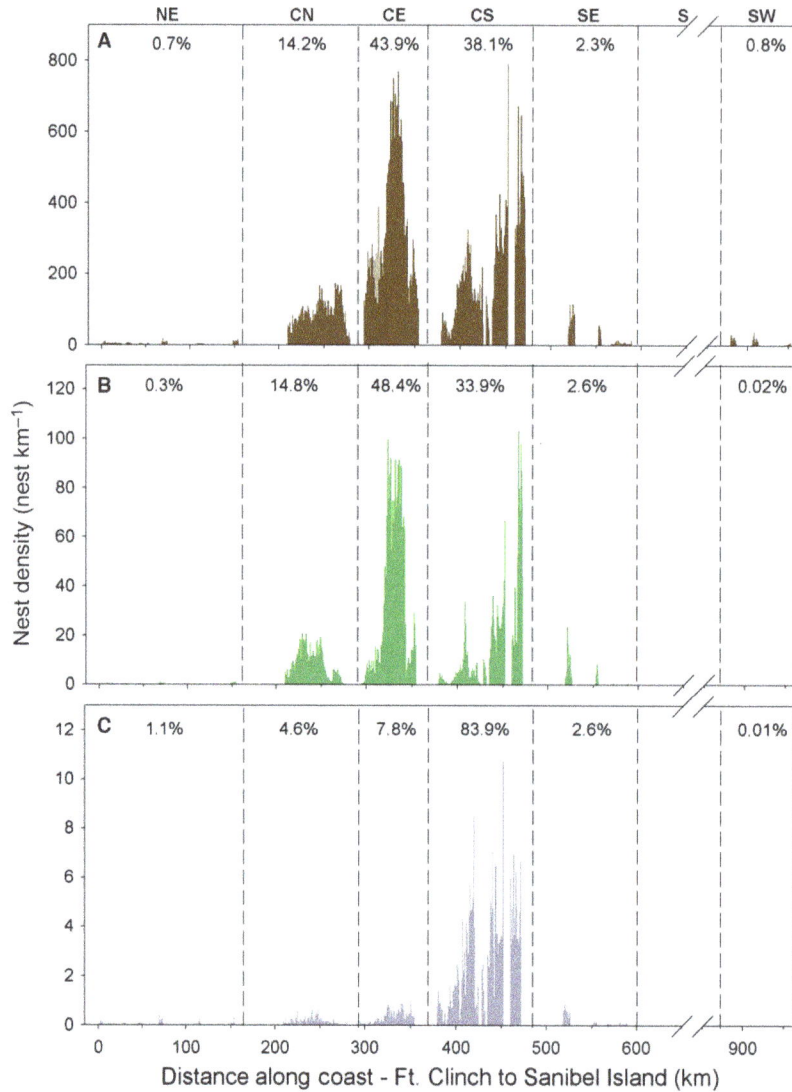

Figure 3. Average (A) loggerhead, (B) green turtle, and (C) leatherback nest densities for INBS nesting beaches from 1992 to 2012 going clockwise around the Florida peninsula. Percentages are based on average nest numbers for each region.

2014) of the independent (nest density) and predictor (artificial light) variables.

Because the presence of SAC violates assumptions of sample independence, we used simultaneous (or spatial) autoregressive (SAR) models that account for the influence of SAC in the models. Following analyses of simulated SAC data using a variety of spatial regression approaches (Dorman et al. 2007; Kissling and Carl 2008; Beale et al. 2010), we used a spatial error model (SAR$_{err}$). The SAR$_{err}$ model expands on the basic generalized least squares model ($Y = X\beta + \varepsilon$; where β is a vector of the slopes of predictors X and ε is the error term) with the term $\lambda W\mu$, where λ is the spatial autoregression coefficient, W is the spatial weights matrix, and μ is the spatially dependent error term (Dorman et al. 2007). Because

independent variables (e.g., artificial light) can also be autocorrelated, at scales different from the dependent variable, we used a mixed SAR$_{err}$ model or spatial Durbin error model (*errorsarlm* command with an "emixed" error type in the *spdep* package in R; Bivand 2015) which adds spatially lagged predictors. As nest density for a given year at a specific INBS beach location may relate to regional or statewide fluctuations, which are most conspicuous with green turtles (Fig. 2), these were also included as predictors (Table 1). Resulting SAR$_{err}$ models comprised of different predictors were compared using the Akaike Information Criterion (AIC). An analysis of model residuals was performed to assess where the regression models over- and under-predicted nest densities and to compare species nesting patterns. The effectiveness of the SAR$_{err}$

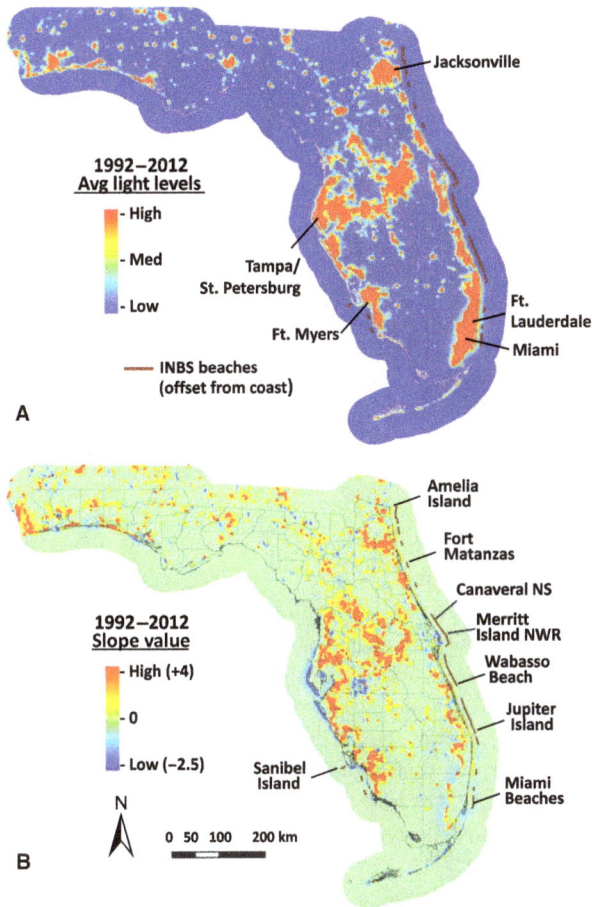

Figure 4. (A) Per-pixel averages and (B) trends of DMSP recorded night lights across Florida from 1992 to 2012. Coastal urban areas and beaches mentioned in the text are identified. The brown segments adjacent to the coast are offset locations of the 368 INBS beaches.

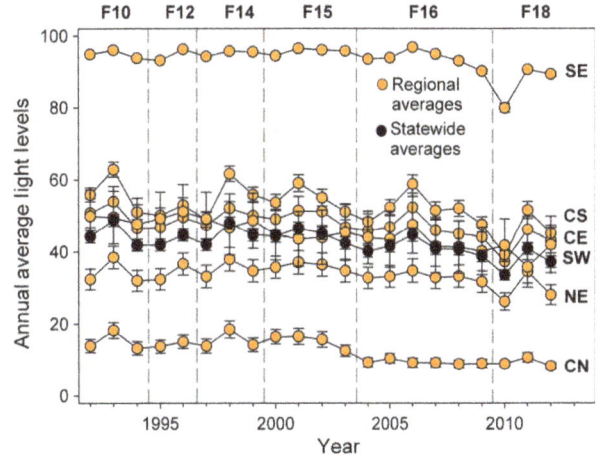

Figure 5. Dynamics of artificial light averages by nesting regions and overall average (black markers) from 1992 to 2012. F-numbers refer to different satellites. Extensions above and below markers represent ±1 standard error.

models were assessed by comparing autocorrelation patterns (using Moran's I) of the models' residuals to those of their corresponding generalized least squares models.

Results

Trends in light levels on surveyed beaches

Though most urban areas grew over this 21-year period as shown by the positive slopes (light increases) around the cities (Fig. 4B), trends in artificial light along the nesting beaches were variable, but mostly showed decreases (Fig. 7). Some beach communities seemingly had no artificial light mitigation policies allowing light to increase (e.g., Amelia Island, Wabasso Beach, Jupiter Island); others very conscientiously reduced artificial light (e.g., Ft. Matanzas, Merritt Island, Sanibel Island); while others basically maintained the status quo which was typi-

cally relatively dark (e.g., Canaveral National Seashore) or exceedingly bright (e.g., Miami Beach).

Spatial autocorrelation of coastal artificial light levels and nest densities

Artificial light was significantly positively autocorrelated at scales <20 km and ~50 km (Fig. 8), using a Moran's I value of ±0.2 as a rough cutoff of significance ($P < 0.05$). This indicates that variables exhibit neighborhood effects in these ranges and are not spatially independent. Nest densities for all three species and artificial light exhibited strong SAC patterns. Loggerhead, green turtle, and leatherback nest densities were significantly positively autocorrelated at <40, <20, and <50 km scales, respectively (Fig. 9). Thus, using a mixed SAR$_{err}$ model that considers autocorrelation associated with both independent variables (artificial light) and dependent (sea turtle nest density) as reflected by the model error term was appropriate.

Simultaneous autoregressive (SAR) model selection

Prior to multiple regression analyses, sea turtle nest densities and artificial light values were log-transformed to obtain normal distributions. Based on the lowest AIC value, the most informative (best fit and most parsimonious) SAR$_{err}$ models (highlighted in dark gray in Table 2) for all three sea turtle species included artificial light on the nesting beach (*Artificial_Light*), overall annual nest density by species (*Statewide_Density*), and the spatially autocorrelated term (with a lag distance up to 100 km) for artificial

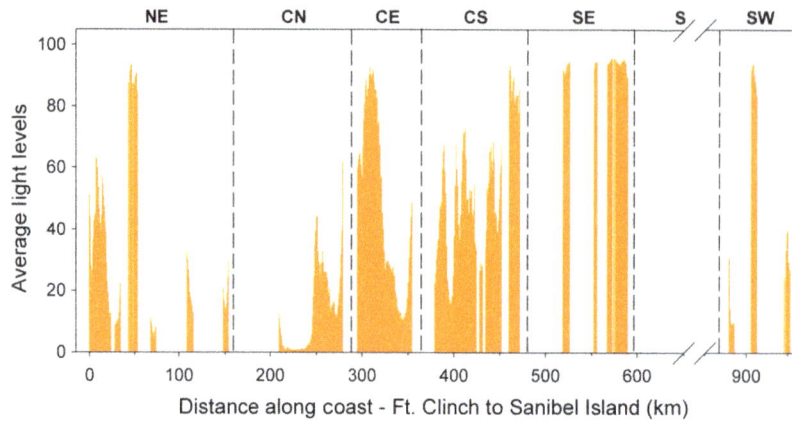

Figure 6. Average artificial light levels (1992–2012) for each INBS beach segment going clockwise around the Florida peninsula.

light ($lag_{100}Artificial\_Light$). The second most informative models (highlighted in light gray in Table 2), which were also consistent for the three species, dropped the $lag_{100}Artificial\_Light$ term. The remaining eight models had much higher AIC values as did others with smaller (50 km) and larger (150 km) lag distances (not shown).

For all species, the spatial auto-regressive (neighborhood effect) terms associated with the model error (λ, which is on a -1 to $+1$ scale) were all highly ($P < 0.001$) significant and the resulting SAR models were also significantly different (as shown by the Wald statistic) and more

Table 1. List of simple and spatially-autocorrelated (lag_{100}) predictor variables used to estimate nest density patterns separately for loggerhead, leatherback, and green turtles on the INBS beaches from 1992 to 2012.

Predictor variable	Description
Artificial_Light	Annual DMSP intensity values of pixels that intersected with INBS beach segments weighted by the proportion of the length of the segment that occurred in each pixel
Regional_Density	Annual average nest density on INBS beach segments for each of the six regions surveyed consistently from 1992 to 2012
Statewide_Density	Annual average nest density for all INBS beach segments
$lag_{100}Artificial\_Light$	Annual levels of spatially autocorrelated light levels that occur within 100 km of a particular INBS beach segment
$lag_{100}Regional\_Density$	Annual levels of spatially autocorrelated regional density levels that occur within 100 km of a particular beach segment

The SAR$_{err}$ model by default includes the spatially autocorrelated error that may be associated with the dependent variable (i.e., nest density).

informative (as shown by the lower AIC) than the corresponding ordinary least squares (OLS) models (Table 3). Though the best models as determined by AIC were the same, the significance of the individual predictors differed by species. To summarize:

- The two spatially autocorrelated variables ($lag_{100}Artificial\_Light$ and the λ coefficient associated with the error terms) were significant ($P < 0.001$) for all species. The $lag_{100}Artificial\_Light$ predictor was negatively related to nest density at the beach segment scale for all species.
- Artificial_Light values at the INBS beach segment locations were significant and negatively related to nest densities for loggerhead ($P < 0.001$) and green turtles ($P = 0.007$), but did not show a relationship with leatherback nest densities.
- Statewide_Density values were significant ($P = 0.007$) and negatively related only to green turtle nest density at the beach segment scale. This may relate to their widely fluctuating, somewhat biennial nesting patterns found in Florida (Fig. 2).

Analysis of SAR$_{err}$ model residuals

A comparison of autocorrelation of the residuals between linear models and the SAR$_{err}$ models showed that the levels were consistently lower for the SAR$_{err}$ models (Table 3). Residuals of the most informative SAR$_{err}$ models (Table 2) indicated how well the regression models predicted nest densities for the 368 beach segments. For loggerheads and green turtles, residuals were within ± 6 nests per km (Figs. 10A and B), which show the models to be accurate given their high nest densities (Figs. 3A and B). For leatherbacks, however, residuals were within ± 3 nests per km (Fig. 10C), which is not as notable, given their relatively low densities (Fig. 3C). This may

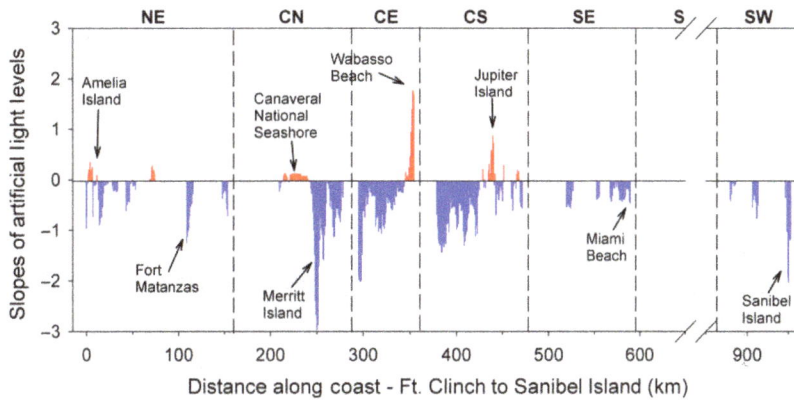

Figure 7. Artificial light trends on INBS beach segments from 1992 to 2012. Red and blue bars designate increasing and decreasing light levels, respectively.

reflect their lower densities and more fluctuating nesting patterns (Fig. 9C).

The regression models fairly consistently over-predicted nest densities for well-lit heavily urbanized areas (e.g., Atlantic-Jacksonville and Miami Beaches) and beaches near inlets (e.g., Wabasso Beach, Ft. Pierce Inlet, St. Lucie Inlet, Sanibel Island) while under-predicting others (e.g., South Brevard, Hutchinson Island, Jupiter Island, Juno Beach, MacArthur State Park). This suggests additional environmental (e.g., human activities, ocean currents) or biological (e.g., historic philopatry) variables are influencing the nesting patterns.

Though all general nesting patterns were similar, based on an analysis of the residuals, loggerhead and green turtle nesting, with respect to artificial light, behaved more alike than compared to those of leatherbacks (Fig. 11). This was not unexpected given the resemblance of average nesting patterns, especially between loggerhead and green turtles across the INBS segments (Fig. 3).

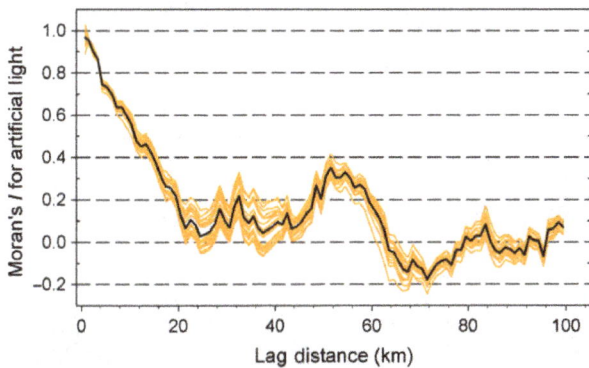

Figure 8. Spatial autocorrelation patterns of artificial light on INBS beaches. Orange lines are the annual levels and the black line is the average from 1992 to 2012.

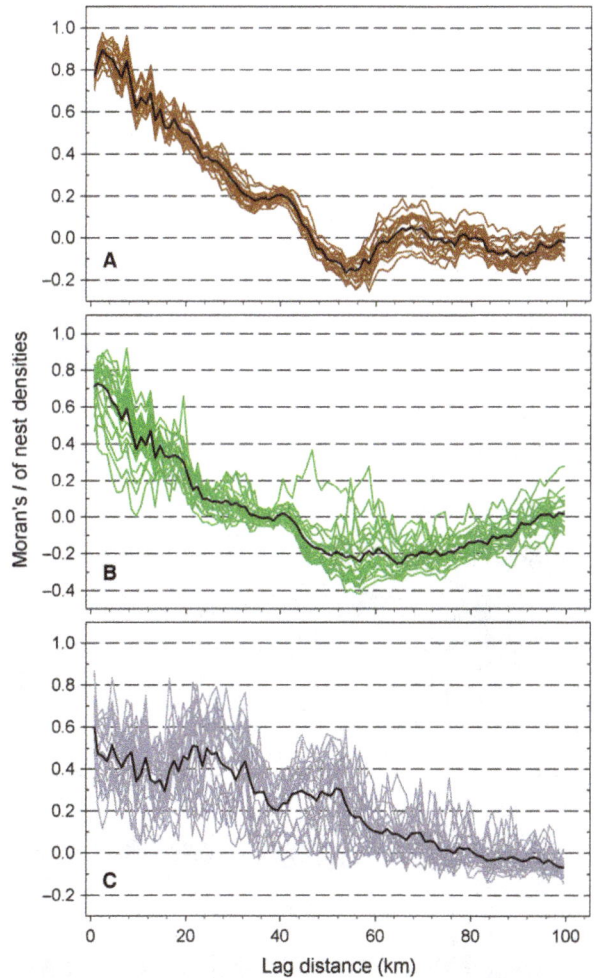

Figure 9. Spatial autocorrelation patterns of (A) loggerhead, (B) green turtle, and C) leatherback nest densities on INBS beaches. Colored lines are the annuals levels and the black lines are the averages from 1992 to 2012.

Table 2. Summary of SAR_err model selection to predict nest densities for the three species at the 368 INBS beaches for the 1992–2012 period.

Predictor variables (X)	Total parameters estimated	Loggerhead AIC	Loggerhead ΔAIC	Green turtle AIC	Green turtle ΔAIC	Leatherback AIC	Leatherback ΔAIC
Artificial_Light	4	23 246	5 214	23 383	5 523	11 389	2 870.4
Artificial_Light + Statewide_Density	5	18 123	91	17 976	116	8 534	15.4
Artificial_Light + Regional_Density	5	19 830	1 798	21 723	3 863	9 440.4	921.8
Artificial_Light + Statewide_Density + Regional_Density	6	19 815	1 783	21 723	3 863	9 442.3	923.7
Artificial_Light + lag_{100}Artificial_Light	5	23 243	5 211	23 360	5 500	11 347	2 828.4
Artificial_Light + Statewide_Density + lag_{100}Artificial_Light	6	18 032	0	17 860	0	8 518.6	0
Artificial_Light + Regional_Density + lag_{100}Artificial_Light + lag_{100}Regional_Density	7	19 614	1 582	21 564	3 704	9379.3	860.7
Artificial_Light + Statewide_Density + Regional_Density + lag_{100}Artificial_Light + lag_{100}Regional_Density	8	19 598	1 566	21 560	3 700	9372.9	854.3
Regional_Density + lag_{100}Regional_Density	5	20 056	2 024	21 862	4 002	9386.3	867.6
Regional_Density + Statewide_Density + lag_{100}Regional_Density	6	20 042	2 010	21 861	4 001	9376.2	857.6

Lower AIC values indicate the best fit and most parsimonious models (dark gray is the lowest, followed by the lighter gray). AIC, Akaike information criterion.

Table 3. Parameters from the most informative SAR_err models shown in Table 2 that predict (A) loggerhead, (B) green turtle, and (C) leatherback nest densities.

A. Predictors (X) for loggerhead nest density	Coefficient (B) estimates	Standard error	z-value	Pr(>|z|)
(Intercept)	2.109	1.032	2.043	0.041
Artificial_Light	−0.090	0.009	−9.723	<.001
Statewide_Density	0.006	0.007	0.857	0.392
lag_{100}Artificial_Light	−0.371	0.038	−9.807	<.001
λ	0.961	0.003	306.85	<.001
Wald statistic	94 157; P-value < 0.001			
AIC for SAR_err:AIC for OLS	18 032:30 671			
Avg Moran's I of residuals <100 km for SAR_err:OLS	0.11:0.56			

B. Predictors (X) for green turtle nest density	Coefficient (B) estimates	Standard error	z-value	Pr(>|z|)
(Intercept)	1.270	0.335	3.793	<.001
Artificial_Light	−0.025	0.009	−2.691	0.007
Statewide_Density	−0.057	0.020	−2.896	0.003
lag_{100}Artificial_Light	−0.406	0.037	−10.921	<.001
λ	0.957	0.003	278.52	<.001
Wald statistic	77 572; P-value < 0.001			
AIC for SAR_err:AIC for OLS	17 860:29 912			
Avg Moran's I of residuals <100 km for SAR_err:OLS	0.03:0.27			

C. Predictors (X) for leatherback nest density	Coefficient (B) estimates	Standard error	z-value	Pr(>|z|)
(Intercept)	0.266	0.105	2.537	0.011
Artificial_Light	0.001	0.005	0.129	0.897
Statewide_Density	−0.014	0.085	−0.170	0.865
lag_{100}Artificial_Light	−0.078	0.018	−4.242	<0.001
λ	0.909	0.007	132.88	<0.001
Wald statistic	17 658; P-value < 0.001			
AIC for SAR_err:AIC for OLS	8518.6:13 604			
Avg Moran's I of residuals <100 km for SAR_err:OLS	0.11:0.24			

Parameters and tests highlighted in gray represent statistically significant relationships (P < 0.05).

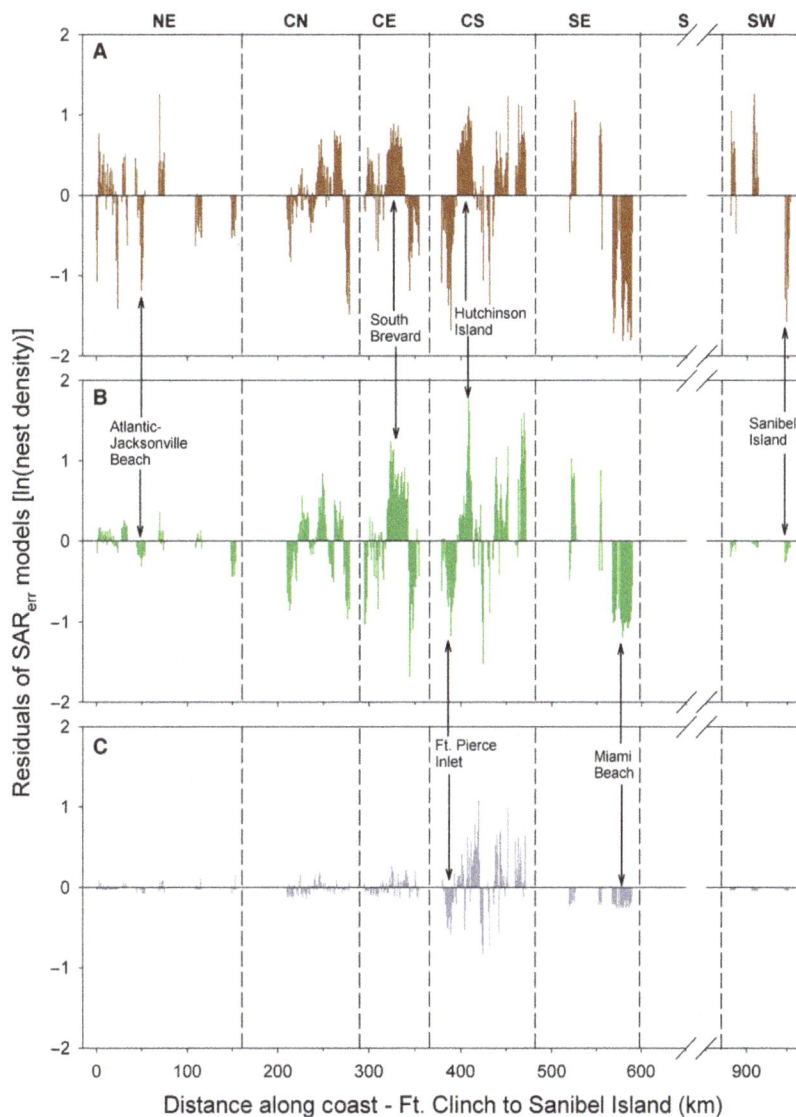

Figure 10. Distribution of SAR$_{err}$ model residuals from INBS beaches for (A) loggerhead, (B) green turtle, and (C) leatherback nest densities. Positive and negative values correspond to model under- and over-predictions, respectively.

Discussion

Over the 21-year period from 1992 to 2012, Florida's human population grew by more than 5.7 million (U. S. Census Bureau 2015). Following general United States trends, much of this growth occurred along the coast (NOAA 2015b). Thus, it was assumed that most nesting beaches would exhibit an increase in artificial light associated with increasing urbanization. However, 67.7% (N = 249) of INBS beaches showed a decrease in satellite-derived artificial light, 14.1 (N = 52) showed an increase, and the remaining 18.2% (N = 67) showed little change. Some U.S. National Parks (and National Seashores like Canaveral National Seashore; Fig. 4) with active light

management strategies have been shown to provide a refuge from light pollution (Manning et al. 2015).

As found in the Mediterranean study (Mazor et al. 2013), the presence of artificial light was negatively associated with nest numbers. Unlike Mazor et al. (2013), SAC associated with nest densities and artificial light played an important role in the regression models. The artificial light level for the pixels that included the nesting beaches was found to be a significant predictor of loggerhead and green turtle nest densities and at broader scales for nest densities of all three species (Table 3). Leatherbacks were less impacted than loggerheads and green turtles by artificial lighting, a trait which was also noted by Walker (2010). Long-range effects associated with sky glow from

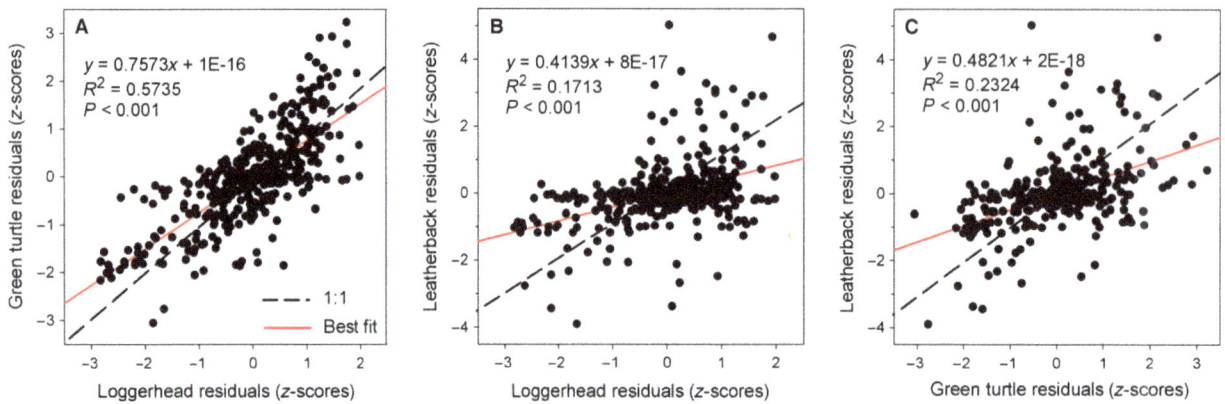

Figure 11. Two-species comparisons, that is, (A) loggerheads and green turtles, (B) loggerheads and leatherbacks, and (C) green turtles and leatherbacks, of SAR_{err} model residuals for the INBS beaches. Model residuals were converted to z-scores to account for the differences in nest densities.

sources as far as 18 km away have been shown to influence nesting flatback (*Natator depressus*) turtles (Hodge et al. 2007). Thus, light from nearby urban areas, not at the specific nesting beach, may be a factor in determining the spatial patterns of nesting.

The satellite-derived artificial light measures undoubtedly conflate other potential factors associated with urbanization (e.g., noise, chemical pollution, per cent of adjacent impervious surfaces, nighttime beach activity; Southwood et al. 2008; Taylor and Cozens 2010) that could also influence nesting. In addition to light, Mazor et al. (2013) found human population density and infrastructure to be negatively related to nest numbers. Though satellite nightlight data have been correlated to human population density in coastal zones (e.g., Small and Nicholls 2003), obtaining annual human density numbers (and other urbanization metrics) for areas surrounding specific Florida beaches is difficult as census (and land use) surveys are less frequent. It is likely that the model over-prediction for Miami Beaches may reflect additional urban factors that were not included.

Even though the SAR_{err} models developed in this study did not incorporate a litany of potential contributing factors (e.g., sea surface temperature, ocean currents, beach renourishment, shoreline armoring) in addition to satellite-derived artificial light, they nonetheless provided fairly accurate predictions of nest density as shown by the relatively small residuals (Fig. 10).

Undoubtedly some of the model error relates to the spatial, temporal, and spectral resolution limitations of the DMSP OLS sensors and available data. A fundamental question that remains is to what extent does the satellite capture the light environment that is being experienced by the nesting female? The 800 m resolution of the DMSP pixel is much coarser (probably by an order of magni-

tude) than the width of most Florida nesting beaches (Reece et al. 2013). Hence, light from sources not directly exposed on the beach are included. Furthermore, beaches may be shaded by vegetation, buildings, dunes, escarpments or simply protected from direct light because of the beach slope. Thus, additional metrics that incorporate beach morphology and adjacent 3-dimensional structures (e.g., condominiums and trees) perhaps using LiDAR (Long et al. 2011) could be developed to improve predictions. An examination of the artificial light trends and model residuals may show beaches that should be examined to determine what management policies (e.g., light ordinances, beach use restrictions, dune construction features, tree or other light blocking structures) may be contributing to higher and lower than expected nest densities.

As the Stable Light DMSP product is an annual measure, it includes time periods outside the nesting seasons of the turtles. Artificial light activities on beaches may differ during different times of the year. Some local municipalities have ordinances that restrict lighting on the beaches during nesting season (FWC 2015c). Contrarily, human beach activity may increase during the summer months during the prime nesting season of loggerhead and green turtles. If the temporal resolution of the satellite data were finer, these seasonal changes corresponding to nesting periods could be more precisely captured. To perhaps support a future study, monthly nightlight data from the more light sensitive, finer spatial resolution Visible Infrared Imaging Radiometer Suite (VIIRS) are currently being distributed, but the archive only begins in 2014. Also some light ordinances permit "turtle friendly" light wavelengths (i.e., red light). However, DMSP OLS data are not divided into spectral bands that would allow differentiation of turtle friendly versus unfriendly wavelengths. To further investigate the results, a ground truth effort using light-sensitive instruments that directly compares ambient nightlight on the beach (e.g., Pendoley et al. 2012; Verutes et al. 2014; Constant 2015) to the DMSP OLS values is warranted.

In Florida, the general increases in sea turtle nesting for all three species (Fig. 2) suggest that artificial lighting may not be critically impairing these populations or the adopted artificial light mitigation policies may be successful which may be reflected by the general decreasing light trends on the INBS beaches (Fig. 7). However, nesting females represent only one side of the equation. Artificial lighting effects on hatchlings may not be manifest in nest density measures and may not be apparent until decades later. This study identified pixels that subsume nesting beaches where there are a variety of artificial light levels as well as areas where light has been increasing or decreasing. To determine if there is an impact, hatchling dis and misorientation and mortality rates should be assessed on a cross-section of these beaches representing different satellite-derived light levels. Futhermore, flatback hatchlings have been shown to be influenced by sky-glow that originates from 15 km away (Kamrowski et al. 2014b); thus, Florida hatchlings may also be affected by distance light.

The long-term sea turtle nest monitoring effort by FWC permitted an extensive, year-to-year analysis of broad-scale sea turtles nesting patterns in relation to artificial light for these important Western Hemisphere breeding grounds. This study, along with its predecessors (Kamrowski et al. 2012, 2014a; Mazor et al. 2013), shows promise that satellite measures of artificial light can be used to help guide not only statewide, but global management of light pollution for sea turtles including the other three predominantly nocturnally nesting species, i.e. flatback, hawksbill (*Eretmochelys imbricata*), and olive ridley (*Lepidochelys olivacea*) as well as other light-sensitive species.

Acknowledgments

We are grateful for the nesting data provided by the Florida Fish and Wildlife Conservation Commission's Florida Marine Research Institute and the artificial nightlight satellite data provided by NOAA made available through the U.S. Department of Defense. The research benefited from suggestions provided by Beth Brost, Simona Ceriani, Robert Hardy, Tomo Hirama, and Anne Meylan from the FWC, members of the UCF Geospatial Analysis and Modeling of Ecological Systems (GAMES) Lab and attendees at the International Sea Turtle Symposium in Malaman, Mugla-Turkey and the International Symposium on Research/Conservation and Future Prospects of Sea Turtles in Taiwan. We also thank Anthony Weishampel for assistance with the R programming and statistical analyses and two anonymous reviewers for helpful suggestions.

References

Beale, C. M., J. J. Lennon, J. M. Yearsley, M. J. Brewer, and D. A. Elston. 2010. Regression analysis of spatial data. *Ecol. Lett.* **13**, 246–264.

Beyer, H. L. 2014. Geospatial Modelling Environment v. 0.7.2. Available at http://www.spatialecology.com (accessed 15 August 2014).

Bivand, P. 2015. Package 'spdep' spatial dependence: weighting schemes, statistics and models. Available at https://cran.r-project.org/web/packages/spdep/spdep.pdf (accessed 3 April 2015).

Bogard, P. 2014. *End of night: searching for natural darkness in an age of artificial light.* Back Bay Books, New York.

Ceriani, S. A., J. D. Roth, A. D. Tucker, D. R. Evans, D. S. Addison, C. R. Sasso, et al. 2015. Carry-over effects and foraging ground dynamics of a major loggerhead breeding aggregation. *Mar. Biol.* **162**, 1955–1968.

Chaloupka, M., K. A. Bjorndal, G. H. Balazs, A. B. Bolton, L. M. Ehrhart, C. J. Limpus, et al. 2008. Encouraging outlook for recovery of a once severely exploited marine mega herbivore. *Glob. Ecol. Biogeogr.* **17**, 297–304.

Constant, N. 2015. Geospatial assessment of artificial lighting impacts on sea turtles in Tortuguero, Costa Rica. Master's Thesis. Nicholas School of the Environment, Duke University. 65 pp.

Crognale, M. A., S. A. Eckert, D. H. Levenson, and C. A. Harms. 2008. Leatherback sea turtle *Dermochelys coriacea* vision capacities and potential reduction of bycatch by pelagic longine fisheries. *Endange. Species Res.* **5**, 249–256.

Dorman, C. F., J. M. McPherson, M. B. Araújo, R. Bivand, J. Bolliger, G. Carl, et al. 2007. Methods to account for spatial autocorrelation in the analysis of species distribution data: a review. *Ecography* **30**, 609–628.

Elvidge, C. D. 2014. Special issue "remote sensing with nighttime lights". *Remote Sens.*. Available at http://www.mdpi.com/journal/remotesensing/special_issues/nightimelights (accessed 20 July 2015).

Elvidge, C. D., F. Hsu, K. E. Baugh, and T. Ghosh. 2014. National trends in satellite-observed lighting 1992–2012. Pp. 97–119. in Q. Wend, ed. *Global urban monitoring and assessment through earth observation.* CRC Press, Boca Raton, FL.

Fritsches, K. A., and E. J. Warrant. 2013. Vision. Pp. 31–58. in J. Wyneken, K. J. Lohmann, and J. A. Musick, eds. *The biology of sea turtles, Volume III.* CRC Press, Boca Raton, FL.

FWC. 2015a. Sea turtle nesting. Available at http://myfwc.com/research/wildlife/sea-turtles/nesting/ (accessed 11 November 2015).

FWC. 2015b. Sea Turtle Monitoring (the SNBS and INBS Programs). Available at http://myfwc.com/research/wildlife/sea-turtles/nesting/sea-turtle-monitoring/ (accessed 20 July 2015).

FWC. 2015c. Sea Turtle Protection Ordinances. Available at http://myfwc.com/conservation/you-conserve/lighting/ordinances/ (accessed 20 August 2015)

Gamma Design Software. 2014. GS+. Available at https://www.gammadesign.com/ (accessed 10 September 2014).

Hamann, M., A. Grech, E. Wolanski, and J. Lambrechts. 2011. Modelling the fate of marine turtle hatchlings. *Ecol. Model.* **222**, 1515–1521.

Hawkes, L. A., A. C. Broderick, M. S. Coyn, M. H. Godfrey, and B. J. Godley. 2007. Only some like it hot— quantifying the environmental niche of the loggerhead sea turtle. *Divers. Distrib.* **14**, 447–457.

Hodge, W., C. J. Limpus, and P. Smissen. 2007. Queensland turtle conservation project: hummock Hill Island nesting turtle study december 2006. Technical and Data Report, 6 pp. Environmental Protection Agency, Queensland.

Hölker, F., C. Wolter, E. K. Perkin, and K. Tockner. 2010. Light pollution as a biodiversity threat. *Trends Ecol. Evol.* **25**, 681–682.

Hu, C., L. Feng, R. F. Hardy, and E. J. Hochberg. 2015. Spectral and spatial requirements of remote measurements of pelagic Sargassum macroalgae. *Remote Sens. Environ.* **167**, 229–246.

IUCN. 2015. The IUCN red list of threatened species. Available at http://www.iucnredlist.org/ (accessed 20 July 2015).

Jiang, J., and R. E. Plotnick. 1998. Fractal analysis of the complexity of United States coastlines. *Math. Geol.* **30**, 535–546.

Kamrowski, R. L., C. Limpus, J. Moloney, and M. Hamann. 2012. Coastal light pollution and marine turtles: assessing the magnitude of the problem. *Endanger. Species Res.* **19**, 85–98.

Kamrowski, R. L., C. Limpus, R. Jones, S. Anderson, and M. Hamann. 2014a. Temporal changes in artificial light exposure of marine turtle nesting areas. *Glob. Change Biol.* **20**, 2437–2449.

Kamrowski, R. L., C. Limpus, K. Pendoley, and M. Harmann. 2014b. Influence of industrial light pollution on the sea-finding behavior of flatback turtle hatchlings. *Wildl. Res.* **41**, 421–434.

Kissling, W. D., and G. Carl. 2008. Spatial autocorrelation and the selection of simultaneous autoregressive models. *Glob. Ecol. Biogeogr.* **17**, 59–71.

Long, T. M., J. Angelo, and J. F. Weishampel. 2011. LiDAR-derived measures of hurricane- and restoration-generated beach morphodynamics in relation to sea turtle nesting behavior. *Int. J. Remote Sens.* **32**, 231–241.

Longcore, T., and C. Rich. 2004. Ecological light pollution. *Front. Ecol. Environ.* **2**, 191–198.

Magyar, T. 2008. The impact of artificial lights and anthropogenic noise on loggerheads (*Caretta caretta*) and green turtles (*Chelonia mydas*) assessed at index nesting beaches in Turkey and Mexico. Ph.D. Thesis. Universitäts-und-Landesbibliothek Bonn, Germany.

Manning, R., E. Rovelstad, C. Moore, J. Hallo, and B. Smith. 2015. Indicators and standards of quality for viewing the night sky in the national parks. *PARKScience* **32**. Available at http://www.nature.nps.gov (accessed 4 September 2015).

Mazor, T., N. Levin, H. P. Possingham, Y. Levy, D. Rocchini, A. J. Richardson, et al. 2013. Can satellite-based night lights be used for conservation? The case of nesting sea turtles in the Mediterranean. *Biol. Conserv.* **159**, 63–72.

Miller, J. A. 2012. Species distribution models: spatial autocorrelation and non-stationarity. *Prog. Phys. Geogr.* **36**, 681–692.

Naro-Maciel, E., M. Le, N. N. FitzSimmons, and G. Amato. 2008. Evolutionary relationships of marine turtles: a molecular phylogeny based on nuclear and mitochondrial genes. *Mol. Phylogenet. Evol.* **49**, 659–662.

NOAA. 2015a. Version 4 DMSP-OLS Nighttime Lights Time Series. NOAA National Centers for Environmental Information. Available at http://www.ngdc.noaa.gov/eog/dmsp/downloadV4composites.html (accessed 20 July 2015).

NOAA. 2015b. State of the Coast: The U.S. population living at the coast. Available at http://stateofthecoast.noaa.gov/population/welcome.html (accessed 20 August 2015).

Pendoley, K. L., A. Verveer, A. Kahlon, J. Savage, and R. T. Ryan. 2012. A novel technique for monitoring light pollution. *Soc. Petrol. Eng.* SPE-158034-MS.

Peters, A., and K. J. F. Verhoeven. 1994. Impact of artificial lighting on the seaward orientation of hatchling loggerhead turtles. *J. Herpetol.* **28**, 112–114.

Pettorelli, N., K. Safi, and W. Turner. 2014. Satellite remote sensing, biodiversity research and conservation of the future. *Phil. Trans. R. Soc. B* **369**, 20130190.

Reece, J. S., D. Passeri, L. Ehrhart, S. C. Hagen, A. Hays, C. Long, et al. 2013. Sea level rise, land use, and climate change influence the distribution of loggerhead turtle nests at the largest USA rookery (Melbourne Beach, Florida). *Mar. Ecol. Prog. Ser.* **493**, 259–274.

Rich, C., and T. Longcore. 2006. *Ecological consequences of Artificial Night Lighting.* Island Press.

Salmon, M. 2003. Artificial night lighting and sea turtles. *Biologist* **50**, 163–168.

Salmon, M. 2006. Protecting sea turtles from artificial lighting at Florida's oceanic beaches. Pp. 141–168. *in* C. Rich, and T. Longcore, eds. *Ecological consequences of artificial night lighting.* Island Press, Washington, DC.

Salmon, M., M. Garro Tolbert, D. Pender Painter, M. Goff, and R. Reiners. 1995. Behavior of loggerhead sea turtles on an urban beach II. Hatchling orientation. *J. Herpetol.* **29**, 568–576.

Shamblin, B. M., M. G. Dodd, D. A. Bagley, L. M. Ehrhart, A. D. Tucker, C. Johnson, et al. 2011. Genetic structure of the southeastern United States loggerhead turtle nesting aggregation: evidence of additional structure within the peninsular Florida recovery unit. *Mar. Biol.* **156**, 571–587.

Small, C., and R. J. Nicholls. 2003. A global analysis of human settlement in coastal zones. *J. Coastal Res.* **19**, 584–599.

Southwood, A., K. Fritsches, R. Brill, and Y. Swimmer. 2008. Sound, chemical, and light detection in sea turtles and pelagic fishes: sensory-based approaches to bycatch reduction in longline fisheries. *Endanger. Species Res.* **4**, 225–238.

Stewart, K., M. Sims, A. Meylan, B. Witherington, B. Brost, and L. B. Crowder. 2011. Leatherback nests increasing significantly in Florida, USA: trends assessed over 30 years using multilevel modeling. *Ecol. Appl.* **21**, 263–273.

Taylor, H., and J. Cozens. 2010. The effects of tourism, beachfront development and increased light pollution on nesting Loggerhead turtles *Caretta caretta* (Linnaeus, 1758) on Sal, Cape Verde Islands. *Zoologia Caboverdiana* **1**, 100–111.

Tilman, D., R. M. May, C. L. Lehman, and M. A. Nowak. 1994. Habitat destruction and the extinction debt. *Nature* **371**, 65–66.

Tuxbury, S. M., and M. Salmon. 2005. Competitive interactions between artificial lighting and natural cues during seafinding by hatchling marine turtles. *Biol. Conserv.* **121**, 311–316.

U. S. Census Bureau. 2015. Available at http://www.census.gov (accessed 8 November 2015).

Verutes, G. M., C. Huang, E. Rodríguez Estrella, and K. Loyd. 2014. Exploring scenarios of light pollution from coastal development reaching sea turtle nesting beaches near Cabo Pulmo, Mexico. *Global Ecol. Conser.* **2**, 170–180.

Walker, G. 2010. The influence of morphology and artificial light on terrestrial transit of the leatherback turtle (*Dermochelys coriacea*). Master's Thesis. Faculty of Biomedical & Life Sciences, University of Glasgow. 107 pp.

Weishampel, J. F., D. A. Bagley, L. M. Ehrhart, and B. L. Rodenbeck. 2003. Spatiotemporal patterns of annual sea turtle nesting behaviors along an East Central Florida beach. *Biol. Conserv.* **110**, 295–303.

Weishampel, J. F., D. A. Bagley, L. M. Ehrhart, and A. C. Weishampel. 2010. Nesting phenologies of two sympatric sea turtle species related to sea surface temperatures. *Endanger. Species Res.* **12**, 41–47.

Witherington, B. E. 1992. Behavioral responses of nesting sea turtles to artificial lighting. *Herpetologica* **48**, 31–39.

Witherington, B. E., and R. E. Martin. 2003. Understanding, assessing, and resolving light-pollution problems on sea turtle nesting beaches. FWC FMRI Technical Report TR-2.

Witherington, B., P. Kubilis, B. Brost, and A. Meylan. 2009. Decreasing annual nest counts in a globally important loggerhead sea turtle population. *Ecol. Appl.* **19**, 30–54.

Witherington, B., S. Hirama, and R. Hardy. 2012. Young sea turtles of the pelagic *Sargassum*-dominated drift community: habitat use, population density, and threats. *Mar. Ecol. Prog. Ser.* **463**, 1–22.

Yamamoto, K. H., R. L. Powell, S. Anderson, and P. C. Sutton. 2012. Using LiDAR to quantify topographic and bathymetric details for sea turtle nesting beaches in Florida. *Remote Sens. Environ.* **125**, 125–133.

14

More than counting pixels – perspectives on the importance of remote sensing training in ecology and conservation

Asja Bernd[1,2], Daniela Braun[3], Antonia Ortmann[4], Yrneh Z. Ulloa-Torrealba[2], Christian Wohlfart[5] & Alexandra Bell[6]

[1]EcoDev/ALARM, Yangon, Myanmar
[2]Department of Biogeography, University of Bayreuth, Bayreuth, Germany
[3]Remote Sensing Laboratories, Department of Geography, University of Zurich, Zurich, Switzerland
[4]Food and Agriculture Organization of the United Nations, Rome, Italy
[5]Company for Remote Sensing and Environmental Research (SLU), German Aerospace Center (DLR), Oberpfaffenhofen, Germany
[6]University of Cambridge Conservation Research Institute (UCCRI) and Ecosystems and Global Change Group, University of Cambridge, Cambridge, UK

Keywords
Academic education, conservation, GIS, online survey, remote sensing, training

Correspondence
Alexandra Bell, Ecosystems and Global Change Group, Department of Plant Sciences, University of Cambridge, Downing Street, Cambridge CB2 3EA, UK. E-mail: alexandra.bell@remote-sensing-conservation.org

Funding Information
This publication was funded by the German Research Foundation (DFG) and the University of Bayreuth in the funding programme.

Editor: Duccio Rocchini
Associate Editor: Martin Wegmann

Abstract

As remote sensing (RS) applications and resources continue to expand, their importance for ecology and conservation increases – and so does the need for effective and successful training of professionals working in those fields. Methodological and applied courses often form part of university curricula, but their practical and long-term benefits only become clear afterwards. Having recently received such training in an interdisciplinary master's programme, we provide our perspectives on our shared education. Through an online survey we include experiences of students and professionals in different fields. Most participants perceive their RS education as useful for their career, but express a need for more training at university level. Hands-on projects are considered the most effective learning method. Besides methodological knowledge, soft skills are clear gains, including problem solving, self-learning and finding individual solutions, and the ability to work in interdisciplinary teams. The largest identified gaps in current RS training concern the application regarding policy making, methodology and conservation. To successfully prepare students for a career, study programmes need to provide RS courses based on state-of-the-art methods, including programming, and interdisciplinary projects linking research and practice supported by a sound technical background.

Introduction

Remote sensing (RS) as well as Geographical Information Systems (GIS) have proven to be highly valuable tools in both ecology (Pettorelli et al. 2011; Skidmore et al. 2011; Anderson and Gaston 2013) and conservation science (Scott et al. 1993; Rodrigues et al. 2004; Pettorelli et al. 2014). Their applicability includes habitat assessments (McDermid et al. 2009; Kuenzer et al. 2011), mapping of ecosystem processes and functioning (Cabello et al. 2012) as well as services (De Araujo Barbosa et al. 2015), animal movement analysis (Kays et al. 2015), conservation planning (Nagendra et al. 2013) and future global earth observations monitoring the state of ecosystems and biodiversity (O'Connor et al. 2015; Skidmore et al. 2015). For research and practice using RS data, GIS are crucial,

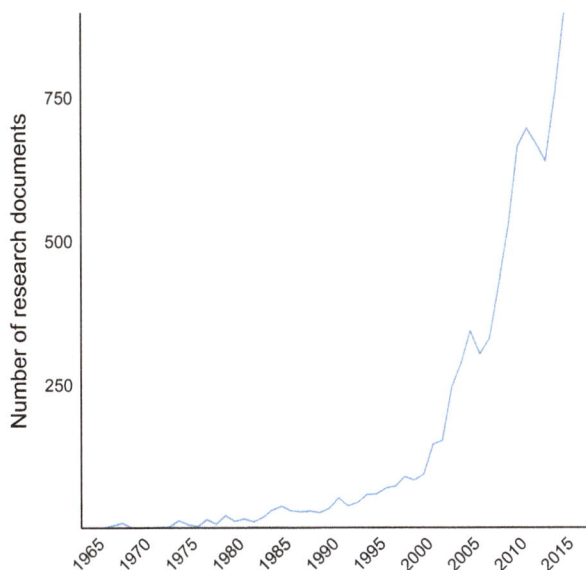

Figure 1. The number of scientific texts published from 1960 to 2015 on remote sensing for ecology or conservation, based on a search of the database SCOPUS (see Appendix S1 for details).

providing powerful tools to create knowledge and communicate it effectively for policy and decision making.

The growing relevance and rapid development of remote sensing in relation to conservation and ecology within academia can be demonstrated through the change in number of scientific publications over the years. The search string *"remote sensing" AND [conservation OR ecology]* yielded 8147 results in the database SCOPUS starting in the year 1966, 90% of which have been published since 1998, and 53% within the last 5 years (Fig. 1, and Appendix S1 for details).

These quickly developing research fields require trained graduates who are able to apply these tools meaningfully to support decision making and answer relevant questions. High-quality education enables students with technical knowledge and the ability to apply RS results in different contexts, such as policy, academia, conservation or governmental agencies. While methodological courses are often part of university study programmes, their practical and long-term benefits only become clear afterwards.

Following shared education, we provide our perspectives on graduate and postgraduate training on RS, including GIS, and how it is useful in our diverse working environments in- and outside academia. We first explore the obtained skills, both in terms of sound technical and methodological knowledge, as well as soft skills. Secondly, we outline gaps where training in RS and other kind of spatial data analysis could still be improved, and where academic education should place a higher emphasis on. We underline our own experience with an online survey that we conducted aiming at broadening and

supporting our insights through the experience of other professionals and students who use RS and GIS in relation to RS for their research and work. Finally, we look towards the future and highlight the opportunities we see for the development of RS training in ecology and conservation.

Survey

As our shared education shapes our experience, we include the perspectives of professionals and students from different backgrounds through an online survey. This survey was distributed through closed mailing lists of universities and conservation groups, namely those for the study programme Global Change Ecology (University of Bayreuth), University of Cambridge Conservation Research Institute, and the training courses AniMove, Spatio-Temporal Data Analysis using Free and Open Source Software (Spatial Ecology, UK) and the Module on International Nature Conservation (German Federal Agency for Nature Conservation), with a total of 73 participants. Despite the limited reach, responses stemmed from all six continents (Fig. 2A), although with a heavy focus on the Global North. Most participants received their training in the UK (28), Germany (17) and US (10), with fewer from many other countries. Respondents are mostly working in research-oriented environments (61.1%), primarily as PhD students (31.2%) and research assistants (18.2%) (Fig. 2B). Early and mature researchers (graduate students, 9.1% and professors, 2.6%) are not highly represented. Respondents are also engaged as consultants (16.9%) and project managers (7.8%) as well as other fields (14.3%). In accordance with this, participants work mostly in universities (38.5%), followed by NGOs (25.3%), research institutes and international organizations (both 16.5%) (Fig. 2C). Only small fractions are employed by companies (2.2%) or government agencies (1.1%). The majority of respondents (90.4%) currently apply the knowledge and skills learnt during their education and training in RS in their research or work in a conservation context. This is consistent with the field of specialization sampled, where keywords like "ecology" and "conservation" were the most prominent, followed by others, such as "biodiversity", "GIS", "remote sensing", "biology" and, to a lesser extent, "landscape", "mapping", "wildlife", "ecological" and "marine" (Fig. 3). Further fields show low numbers. These keywords illustrate in which contexts the community sampled applies RS tools. We included GIS as a means to work with RS data in the survey; when mentioning it, we do not refer to GIS in general, but specifically to its applicability for remote sensing-based research and practice.

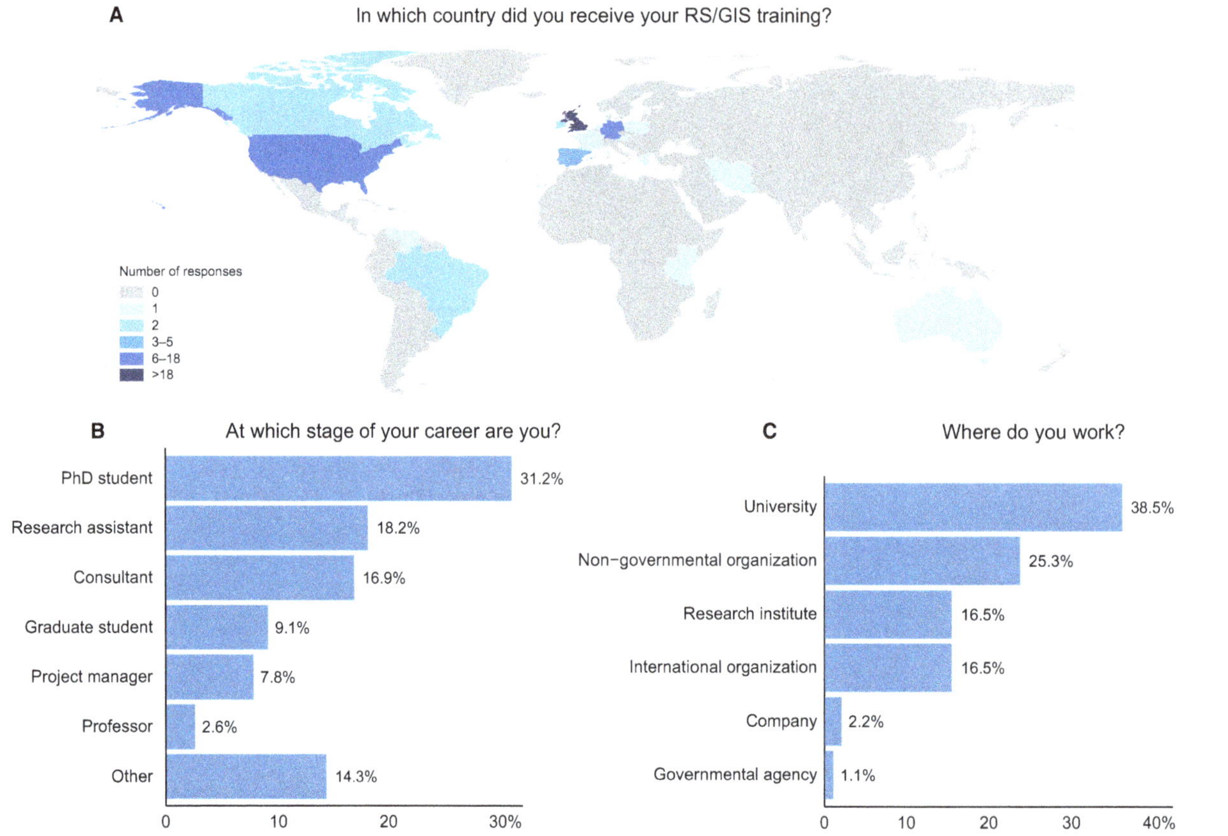

A   In which country did you receive your RS/GIS training?

B   At which stage of your career are you?

C   Where do you work?

Figure 2. (A) Countries in which respondents received their remote sensing (RS)/Geographical Information Systems (GIS) education (multiple answers possible), (B) professional background of the survey participants and (C) workplaces of the survey participants.

What is your field of specialisation?

Figure 3. Fields of work by frequency of occurrence (font size).

Methodological Learnings

Introductory courses in RS convey knowledge about the basic theory of the field. They focus on terminology and data types, different sensors acquiring airborne and space-borne imagery, as well as how to analyse the data and the different products available. These courses are usually integrated in the methods section of the university curriculum. To this end, this theory-driven part of the education is centred on the methodological applications, while also demonstrating the huge potential of RS for diverse applications ranging from animal conservation to land management and urban planning.

Basing courses on the use of free and open-source software is cost-efficient and crucial for showing its values and importance of encouraging its use. It enables users to control each step of their work, increases transparency (Steiniger and Hunter 2013), fosters joint development, collaboration and the establishment of communities (Steiniger and Hay 2009). Moreover, it also allows people with limited resources (i.e., lower capacity to afford costly licenses) to be at the forefront of innovation and on an equal footing in terms of software utilization for their

A

What teaching elements were crucial for you while learning RS/GIS?

Academic seminars (hands–on projects)	20.5%
Web–based forums	16.0%
Academic lectures (theory)	13.2%
Academic tutorials	13.2%
Web–based tutorials (online courses/videos)	13.2%
Books	7.8%
Emailing lists (e.g. google groups)	7.3%
Summer schools	4.1%
Internships/work experience	1.4%
Trial and error	0.5%
Other	2.7%

B

Which soft skills did your training provide?

Capacity of solving problems on your own	25.3%
Capacity to self–learning and finding resources for this goal	25.3%
Networking	9.3%
Capacity of working in interdisciplinary teams	9.3%
Team work	8.8%
Capacity of working in multicultural environments	8.8%
Project management	7.7%
Team management	3.8%
None of the above	1.6%

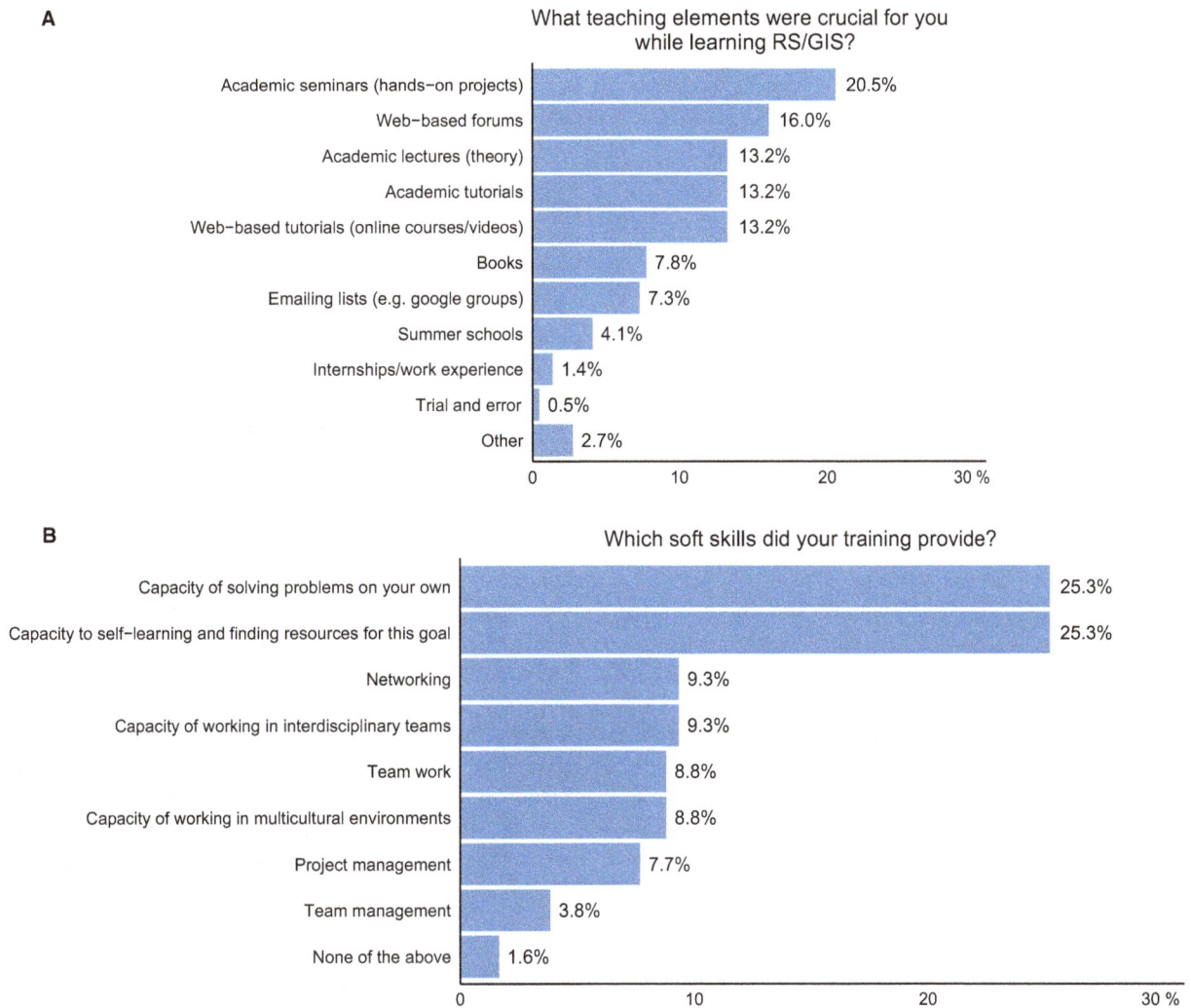

Figure 4. (A) Importance of teaching elements as perceived by interviewee (multiple answers possible) and (B) soft skills obtained during the training (multiple answers possible).

work (Kogut and Metiu 2001), if the necessary computational resources are available.

Besides a sound knowledge in theory, hands-on projects have proven to be most helpful to complement RS training and education in our experience. Courses are more valuable when applying methodological skills in a specific research setting and can especially benefit from the collaboration of lecturers from different disciplines, resulting, for example in the use of RS data for ecosystem service analysis, or species distribution modelling. Furthermore, hands-on projects are more likely to involve understanding, modifying and creating large datasets, honing crucial skills in handling them. This is reflected in the higher importance of university seminars compared to lectures in our survey results (Fig. 4A).

Although RS is known to reduce the amount of data collected in the field, often ground data are necessary to calibrate and validate RS-derived analyses (Fuller et al. 1998; Ceccato et al. 2002; Congalton and Green 2008; Müller and Brandl 2009). Thus, being able to plan and conduct fieldwork to generate accurate RS products is very important. Such practical examples can be implemented in field classes. This allows discussing appropriate data needed and their acquisition (e.g. type and number of plots) in a study site to answer research questions. Conducting field campaigns to collect data such as vegetation or land cover information is valuable to gain understanding of on-site conditions and how they may or may not be reflected in remotely sensed information. This experience can enable to responsibly weigh the benefits and limits of RS research in later real-life projects.

How did you gain your RS/GIS
knowledge and experience?

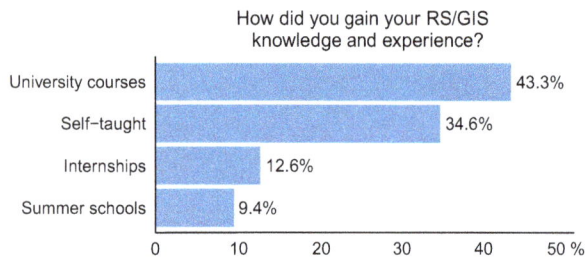

Figure 5. Remote sensing (RS)/Geographical Information Systems (GIS) education history of survey participants.

Learning, however, is not limited to the university. Opportunities to gather knowledge outside specific study programmes already exist, such as during scientific summer (or winter) schools, which are already part of some university curricula. Especially in the field of RS, there are many available. Attending these courses provides multiple benefits in our experience: it allows students to focus on specific topics of interest, also those outside their university curriculum; it brings together scientists, practitioners and students from different fields, which fosters interdisciplinary collaboration and networking, and it allows for a broader perspective of applications and methods.

The results of the survey confirm the application of all the above-mentioned learning strategies of RS and GIS. The learning process is mainly achieved through university courses (43.3%) and self-teaching (34.6%), whereas internships (12.6%) and science summer schools (9.4%) are less common knowledge sources (Fig. 5). However, the perception which of these methods are considered crucial during the training is slightly different: academic seminars (20.5%), web-based forums and tutorials (16.0% and 13.2%, respectively), lectures (13.2%) and academic tutorials (13.2%) are regarded as most important, followed by books (7.8%), emailing lists (7.3%) and summer schools (4.1%). Experiences gained independently (e.g. trial and error, 0.5%) and in working environments (e.g. internships, 1.4%) were rarely regarded as crucial learning sources (Fig. 4A). Importantly though, the opportunities to participate in schools or internships can depend on the financial situation of students and may not be accessible to all.

Soft Skills Gained Through RS Training

Besides learning the theory and application, further relevant skills are acquired and developed during the time of training and complement it (Fig. 4B). Among these are capacity for problem-solving, confidence in the ability to

learn, networking, project management as well as interdisciplinary and intercultural skills.

Solution-oriented thinking and problem-solving skills are fundamental when working with RS data, and are considered some of the most important components of "critical and structured thinking" in general (Schulz 2008). This pattern is reflected in the survey: the capacity of solving problems on one's own (25.3%) and to self-learn and find resources (25.3%) are the most common soft skills gained from RS and GIS education. In comparison, teamwork, project and team management, capacity of working in multicultural and multidisciplinary teams and networking, among others, were considered secondary skills. These competences are highly valuable generally, not only in academia. Study programmes fostering them are therefore crucial in preparing students for successful careers (Schulz 2008; Heckman and Kautz 2012).

The many options of RS in terms of data and methods can make problem solving complex. The availability of approaches requires creativity from students to find their own way and discuss options during group projects. This capacity of developing an appropriate solution to the respective question in an original and empirical way is fostered during RS training in individual projects. Both as groups and individuals, students have to learn this. Resources we found highly useful can be setting up lessons for solving problems, such as programming issues, exchanging information in topic-related emailing lists, searching in the vast numbers of online forums for solutions (e.g. www.gis.stackexchange.com) and educating themselves through online tutorials (e.g. www.r-bloggers.com, also videos on YouTube and other channels), manuals (R: www.cran.r-project.org/manuals.html, GRASS: http://grass.osgeo.org/documentation/manuals/, QGIS: http://qgis.org/en/docs/index.html) and books available through libraries or for purchase.

Another crucial aspect is the confidence in one's own ability to learn. When RS training adequately increases in difficulty over time and includes big datasets and real world data and questions, it helps students develop trust in their own ability to grow and find solutions to new challenges, even if they may seem daunting at first.

More generally, working with people from diverse backgrounds provides advantages in RS training and is considered as comparatively important in the survey. Interdisciplinary and international environments generate an enriching learning experience. Students can benefit from a mixture of diverse perspectives, ideas, solutions and expertise. Being able to learn from each other, and from students with more knowledge in certain fields, allows seeing a range of possibilities and finding a comfortable environment to ask questions no matter how basic.

Two important general learnings are project planning and management. These are practiced as most of the course assignments are set up as projects – developed and conducted individually or in small groups. Setting up a project requires all necessary steps from literature search, identifying a research question, to planning and dividing tasks, all through discussions with fellow students, which is valuable and important (Schulz 2008; Heckman and Kautz 2012).

Lastly, even if the next career steps are not directly in the field of RS, the training provides the knowledge to estimate on the usefulness and feasibility of RS applications for projects. This is of particular advantage for conservation projects where knowledge about RS is limited on site. Although RS may seem like an ideal solution to technical, educational and institutional impediments hindering research on the ground (Jha and Chowdary 2007), the opportunities and limitations need to be understood, and its applicability evaluated. The skills acquired during RS training foster the ability to assess planned projects and provide insights for people from different backgrounds.

Usefulness of RS Training in Different Contexts

While these soft skills are crucial in many work settings irrespective of the field, the methodological knowledge also provides clear advantages. 83.6% of the survey respondents consider their RS and GIS skills as helpful or very helpful for their career, for 74.0% it plays a role for their research and 71.2% perceive their RS and GIS knowledge helpful for their research (Fig. 6B). For 69.9% of the participants their RS and GIS skills have been useful in conservation applications, however, only for 31.5% the same can be said in a policy context.

As the ways of distributing the survey likely created a bias towards people working in academia, the results indicate a higher usefulness of RS and GIS skills in research than in the other fields. The low perception and application of RS and GIS in a policy context could partially stem from the comparative novelty of both as tools in this field and from a lacking focus on the effective communication of results during training.

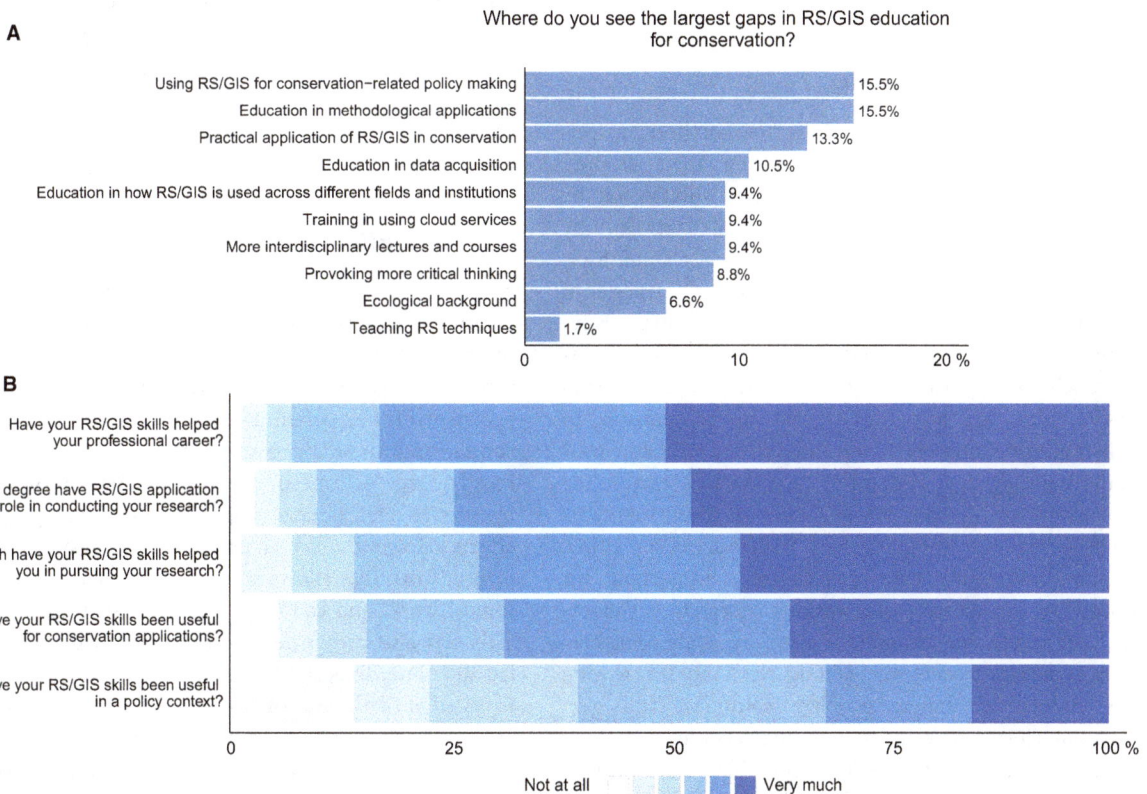

Figure 6. (A) Largest perceived gaps in remote sensing (RS)/Geographical Information Systems (GIS) education for conservation (multiple answers possible) and (B) usefulness and relevance of RS/GIS training for policy, research and practice.

Professionals with a comprehensive and interdisciplinary RS training can fill this gap and foster the application in such areas.

Gaps and Needs

The wish list for effective academic training is infinite, and not everything can be covered in graduate or postgraduate training. However, there are fundamental issues that academic training should aim to achieve: the high number of respondents in the survey relying on self-teaching and non-academic help sources likely indicates a large demand for training or assistance in general. This is supported by the participants asking for teaching with a focus on methodological and practical applications (15.5% and 13.3%, respectively) (Fig. 6A). They also identify further gaps in current training, especially regarding applications in conservation-related policy making (15.5%) (Fig. 6A), thus indicating the wish to address the lower applicability of RS and GIS for policy making described above (Fig. 6B). Besides data access and acquisition, interdisciplinarity, linking technology, methodology and context, state-of-the-art methods, stakeholder involvement, adequate supervision and the creation of support networks play important roles.

State-of-the-art training

Career perspectives are shaped by the methods graduates have been introduced to during their studies. State-of-the-art training is elemental and should not be defined by the specific research at a university but by what students need to succeed. Especially interdisciplinary study programmes cannot teach all aspects of RS simultaneously with their programme content. Nevertheless, RS training should include up-to-date methods such as applications of unmanned aerial vehicles (UAVs) due to their lower costs in comparison to high-resolution imagery (Koh and Wich 2012) and programming languages such as R (already well established in ecology) or Python. While Graphical User Interfaces (GUI) provide a smoother start (Amatulli et al. 2014), programming skills have proven very important for processing big datasets and making research collaborative and replicable through sharing of code. Furthermore, students benefit by gaining flexibility in writing new codes and finding problem solutions, and gain access to cluster or cloud computing facilities. Teaching cloud computing has also been identified as a gap in the survey (9.4%), and should include both functionalities and limits, especially regarding freely accessible cloud-based services.

Data availability and access

Remote sensing provides very diverse products, which has pros and cons when searching for suitable Earth observation data. While new spatial datasets are becoming available every day, tracking what is out there and where to find useful data for one's project is one of the most important skills, and is regarded as a gap by 10.5% of survey respondents. Remotely sensed data from a multitude of sensors with different spatial and temporal resolutions are freely accessible, as well as environmental (e.g. climatic or hydrological) data. The IUCN Red List of Threatened Species and the Global Biodiversity Information Facility (GBIF) provide data on species occurrence. Increasingly, also processed products are available, ranging from ecoregion maps (Olson et al. 2001) or biodiversity hotspot maps (Myers et al. 2000) to Hansen et al.'s (2013) global deforestation data. An overview over what exists, how already processed data can (or cannot) be used, and where their chances and limits are, is a crucial aspect of training. This includes learning to judge which data are applicable for the ecological problem in question, and to assess the uncertainties and limitations of datasets, such as the accuracy of global datasets at regional scale. At the same time, not all available data come free of charge, an issue known to affect conservation (Turner et al. 2015), limiting teaching opportunities and thus students' experiences. Therefore, we strongly echo the call for more freely available data to lay the foundations for successful conservation early on.

Interdisciplinarity

Next to the basis in RS, an understanding of ecology and conservation approaches is necessary. The link between the method, its application and the ultimate goal is sometimes lacking, and even institutionalized. Traditionally, professorships are often awarded for either RS or conservation, and therefore courses offered by these lecturers mostly focus on one or the other. Joint courses, offered by several departments, might require more preparation efforts. However, if well prepared, they pay off through project work that allows students to engage in meaningful topical work, and can help decrease institutional barriers. This was also recognized in the survey, with respondents lacking interdisciplinary lectures and courses (9.4%), knowledge of the use of RS and GIS across fields (9.4%) and the ecological background (6.6%),

Furthermore, the variety of ways in which spatial data can inform conservation should be reflected in the lecture hall. While it is clear that not every method can be taught in detail, a diverse presentation of the many different applications of GIS and RS within a larger conservation context should be striven for.

Linking research and practice

A part of this context should also concern the world outside of academia. Conservation, and especially conservation research, has been recognized to have occurred in a bubble, with developed methods lacking implementation in the field (Knight et al. 2008). We therefore call for courses bringing together different fields of research as well as practitioners, and addressing technical as well as social and economic aspects. Discussing all steps of a real-world conservation project and how spatial conservation data and methods feed into it exemplifies the link of research and practice. This also needs to include a critical discussion of methods and concepts to avoid people thinking in pixels and data only. Real-world projects such as during internship placements reveal the importance of and approaches for cooperating with local institutions or stakeholders, and create benefits through stakeholder knowledge (Brown 2002; Turnhout et al. 2012). Cooperation with local partners can thus make scientific research applicable and create an effect outside of science (Thackway et al. 2013). Lastly, such approaches foster networking skills, a generally regarded essential asset (Schulz 2008; Parsons et al. 2014).

Furthermore, our survey indicated a low applicability of RS in policy contexts and respondents expressed their wish to learn more about the use of RS for conservation-related policy making (15.5%), a highly interdisciplinary aspect. This should comprise also the communication to non-experts, and how to translate data into information. Discussion of impacts and the link between conservation and relevant stakeholders, and how to engage with them, should at least be touched upon or resources be recommended.

Supervision

Lecturers and tutors can provide crucial support in equipping students with the skills discussed. Learning RS and using it to answer complex research questions can be challenging in our experience, and students may feel lost if they do not receive the supervision needed in independent projects. Thus, it is important to break down the complexity and tasks into feasible steps. Supervision should further provide help for self-help, and according to our survey more critical thinking (8.8%).

Also, supervisors need to be aware of gender bias as remote sensing training is a part of academia where social disparities are reflected and well documented (Madera et al. 2009; Moss-Racusin et al. 2012; Roth et al. 2012; Knobloch-Westerwick et al. 2013). It is important that female students are adequately supported to pursue their interests and develop confidence just as much as their male counterparts.

Advanced and continuous learning

Lastly, while this article focuses on graduate and postgraduate education, learning does not stop there. First, not everything can be included in university courses, especially in interdisciplinary programmes. Second, the complexity and wide range of applications of RS make it necessary to keep pace with newly emerging technologies and methods. Training continues after graduation, when young professionals apply and broaden their knowledge in their job. Therefore, more advanced courses that cover specific interests should be offered, not only at the academic level but also for the other kinds of users.

Conclusion

The perspectives provided here cannot be exhaustive given our own shared education and the bias of survey respondents both towards academia and the Global North. Based on our diversity of backgrounds and working experiences, we consider resource availability and capacity crucial issues in many countries. These basic needs are a prior requirement to the ideas discussed in this article and our concluding messages.

What do we want RS education for ecology and conservation to be like? In our perspective, the single biggest answer to this is goal oriented. Developing courses should be focused on the questions: where should students be able to work and what skills do they need to get there? A state-of-the-art, future-oriented training integrating methodology, technology and context forms the basis for this in our eyes. It is crucial to break down the complexity of spatial analyses, so that students develop confidence and strategies to deal with challenges. Training should include field courses or the opportunity for scientific summer schools to provide more realistic experiences. Interdisciplinary and international environments will foster crucial skills and can help to create new perspectives and ideas. A basis of broad general skills and opportunities to specialize according to individual interests would enable students to develop their own skill sets and profiles. Admittedly, we describe a perfect world, and as Eduardo Galeano put it, *utopia can never be reached – its purpose, however, is to make us walk forward* (Galeano 2001).

Acknowledgements

We wish to thank Martin Wegmann for encouragement and critical and helpful comments, and Daniel Kachelriess, Katrin Ziewers and Andrea Hess, as well as two anonymous reviewers, for valuable input and suggestions. Further we would like to thank the respondents of our survey and everyone helping us to distribute the survey. The international study programme Global Change Ecology (M.Sc.) at the University of Bayreuth, where we received our training, strongly shaped our perspective. This publication was funded by the German Research Foundation (DFG) and the University of Bayreuth in the funding programme Open Access Publishing.

References

Amatulli, G., S. Casalegno, R. D'Annunzio, R. Haapanen, P. Kempeneers, E. Lindquist, et al. 2014. *Teaching spatio-temporal analysis and efficient data processing in open source environment.* Pp. 13–26 in A. Jolma, P. Sarkola and L. Lehto, eds. *Proceedings of the 3rd Open Source Geospatial Research & Education Symposium.* OGRS2014, Espoo, Finland. Department of Civil and Environmental Engineering, School of Engineering, Helsinki, Finland.

Anderson, K., and K. J. Gaston. 2013. Lightweight unmanned aerial vehicles will revolutionize spatial ecology. *Front. Ecol. Environ.* **11**, 138–146. doi:10.1890/120150.

Brown, K. 2002. Innovations for conservation and development. *Geogr. J.* **168**, 6–17. doi:10.1111/1475-4959.00034.

Cabello, J., N. Fernández, D. Alcaraz-Segura, C. Oyonarte, G. Piñeiro, A. Altesor, et al. 2012. The ecosystem functioning dimension in conservation: insights from remote sensing. *Biodivers. Conserv.* **21**, 3287–3305. doi:10.1007/s10531-012-0370-7.

Ceccato, P., N. Gobron, S. Flasse, B. Pinty, and S. Tarantola. 2002. Designing a spectral index to estimate vegetation water content from remote sensing data: part 2. Validation and applications. *Remote Sens. Environ.* **82**, 198–207. doi:10.1016/S0034-4257(02)00036-6.

Congalton, R. G., and K. Green. 2008. Assessing the accuracy of remotely sensed data: principles and practices, 2nd edn. CRC Press, Boca Raton, FL.

De Araujo Barbosa, C. C., P. M. Atkinson, and J. A. Dearing. 2015. Remote sensing of ecosystem services: a systematic review. *Ecol. Ind.* **52**, 430–443. doi:10.1016/j.ecolind.2015.01.007.

Fuller, R. M., G. B. Groom, S. Mugisha, P. Ipulet, D. Pomeroy, A. Katende, et al. 1998. The integration of field survey and remote sensing for biodiversity assessment: a case study in the tropical forests and wetlands of Sango Bay, Uganda. *Biol. Conser.* **86**, 379–391. doi:10.1016/S0006-3207(98)00005-6.

Galeano, E. 2001. *Las palabras andantes*, Fifth edn. Catálogos Editora S.R.L, Buenos Aires, Argentina 234.

Hansen, M. C., P. V. Potapov, R. Moore, M. Hancher, S. A. Turubanova, A. Tyukavina, et al. 2013. High-Resolution global maps of 21st-century forest cover change. *Science* **342**, 850–853. doi:10.1126/science.1244693.

Heckman, J., and T. Kautz. 2012. Hard evidence on soft skills. *Labour Econ.* **19**, 451–464. doi:10.3386/w18121.

Jha, M., and V. Chowdary. 2007. Challenges of using remote sensing and GIS in developing nations. *Hydrogeol. J.* **15**, 197–200. doi:10.1007/s10040-006-0117-1.

Kays, R., M. C. Crofoot, W. Jetz, and M. Wikelski. 2015. Terrestrial animal tracking as an eye on life and planet. *Science* **348**, doi:10.1126/science.aaa2478.

Knight, A. T., R. M. Cowling, M. Rouget, A. Balmford, A. T. Lombard, and B. M. Campbell. 2008. Knowing but not doing: selecting priority conservation areas and the research–implementation gap. *Conserv. Biol.* **22**, 610–617. doi:10.1111/j.1523-1739.2008.00914.x.

Knobloch-Westerwick, S., C. J. Glynn, and M. Huge. 2013. The Matilda effect in science communication an experiment on gender bias in publication quality perceptions and collaboration interest. *Sci. Commun.* **35**, 603–625. doi:10.1177/1075547012472684.

Kogut, B. M., and A. Metiu. 2001. Open-source software development and distributed innovation. *Oxford Rev. Econ. Policy* **17**, 248. doi:10.1093/oxrep/17.2.248.

Koh, L. P., and S. A. Wich. 2012. Dawn of drone ecology: low-cost autonomous aerial vehicles for conservation. *Trop. Conserv. Sci.* **5**, 121–132.

Kuenzer, C., A. Bluemel, S. Gebhardt, T. V. Quoc, and S. Dech. 2011. Remote sensing of mangrove ecosystems: a review. *Remote Sens.* **3**, 878–928. doi:10.3390/rs3050878.

Madera, J. M., M. R. Hebl, and R. C. Martin. 2009. Gender and letters of recommendation for academia: agentic and communal differences. *J. Appl. Psychol.* **94**, 1591–1599. doi:10.1037/a0016539.

McDermid, G. J., R. J. Hall, G. A. Sanchez-Azofeifa, S. E. Franklin, G. B. Stenhouse, T. Kobliuk, et al. 2009. Remote sensing and forest inventory for wildlife habitat assessment. *For. Ecol. Manage.* **257**, 2262–2269. doi:10.1016/j.foreco.2009.03.005.

Moss-Racusin, C. A., J. F. Dovidio, V. L. Brescoll, M. J. Graham, and J. Handelsman. 2012. Science faculty's subtle gender biases favour male students. *Proc. Natl. Acad. Sci. U.S. A.* **109**, 16474–16479. doi:10.1073/pnas.1211286109.

Müller, J., and R. Brandl. 2009. Assessing biodiversity by remote sensing in mountainous terrain: the potential of LiDAR to predict forest beetle assemblages. *J. Appl. Ecol.* **46**, 897–905. doi:10.1111/j.1365-2664.2009.01677.x.

Myers, N., R. A. Mittermeier, C. G. Mittermeier, G. A. Da Fonseca, and J. Kent. 2000. Biodiversity hotspots for conservation priorities. *Nature* **403**, 853–858. doi:10.1038/35002501.

Nagendra, H., R. Lucas, J. Honrado, R. Jongman, C. Tarantino, M. Adamo, et al. 2013. Remote sensing for conservation monitoring: assessing protected areas, habitat extent, habitat condition, species diversity, and threats. *Ecol. Ind.* **33**, 45–59. doi:10.1016/j.ecolind.2012.09.014.

O'Connor, B., C. Secades, J. Penner, R. Sonnenschein, A. Skidmore, N. D. Burgess, et al. 2015. Earth observation as a tool for tracking progress towards the Aichi biodiversity targets. *Remote Sensing Ecol. Conserv.* **1**, 19–28. doi:10.1002/rse2.4.

Olson, D. M., E. Dinerstein, E. D. Wikramanayake, N. D. Burgess, G. V. N. Powell, E. C. Underwood, et al. 2001. Terrestrial ecoregions of the world: a new map of life on Earth. *Bioscience* **51**, 933–938. doi:10.1641/0006-3568(2001)051[0933:teotwa]2.0.co;2.

Parsons, E., D. Shiffman, E. Darling, N. Spillman, and A. Wright. 2014. How twitter literacy can benefit conservation scientists. *Conserv. Biol.* **28**, 299–301. doi:10.1111/cobi.12226.

Pettorelli, N., S. Ryan, T. Mueller, N. Bunnefeld, B. Jedrzejewska, M. Lima, et al. 2011. The Normalized difference vegetation index (NDVI): unforeseen successes in animal ecology. *Clim. Res.* **46**, 15–27. doi:10.3354/cr00936.

Pettorelli, N., K. Safi, and W. Turner. 2014. Satellite remote sensing, biodiversity research and conservation of the future. *Philos. Trans. R. Soc. Lond. B: Biol. Sci.* **369**, 20130190. doi:10.1098/rstb.2013.0190.

Rodrigues, A. S., H. R. Akcakaya, S. J. Andelman, M. I. Bakarr, L. Boitani, T. M. Brooks, et al. 2004. Global gap analysis: priority regions for expanding the global protected-area network. *Bioscience* **54**, 1092–1100. doi:10.1641/0006-3568(2004) 054[1092:ggaprf]2.0.co;2.

Roth, P. L., K. L. Purvis, and P. Bobko. 2012. A meta-analysis of gender group differences for measures of job performance in field studies. *J. Manage.* **38**, 719–739. doi:10.1177/0149206310374774.

Schulz, B., 2008. The importance of soft skills: education beyond academic knowledge. *J. Lang. Commun.* **2**, 146–154. doi:10.1016/0006-3207(93)90452-7

Scott, J. M., F. Davis, B. Csuti, R. Noss, B. Butterfield, C. Groves, et al. 1993. Gap analysis: a geographic approach to protection of biological diversity. *Wildl. Monogr.* **123**, 3–41, doi:10.1016/0006-3207(93)90452-7.

Skidmore, A. K., J. Franklin, T. P. Dawson, and P. Pilesjö. 2011. Geospatial tools address emerging issues in spatial ecology: a review and commentary on the Special Issue. *Int. J. Geogr. Inf. Sci.*. **25**, 337–365. doi:10.1080/13658816.2011.554296

Skidmore, A. K., N. Pettorelli, N. C. Coops, G. N. Geller, M. Hansen, R. Lucas, et al. 2015. Environmental science: agree on biodiversity metrics to track from space. *Nature* **523**, 403–405. doi:10.1038/523403a.

Steiniger, S., and G. J. Hay. 2009. Free and open source geographic information tools for landscape ecology. *Ecol. Inform.* **4**, 183–195. doi:10.1016/j.ecoinf.2009.07.004

Steiniger, S., and A. J. S. Hunter. 2013. The 2012 free and open source GIS software map - A guide to facilitate research, development, and adoption. *Comput. Environ. Urban Syst.* **39**, 136–150. doi:10.1016/j.compenvurbsys.2012.10.003

Thackway, R., L. Lymburner, and J. Guerschman. 2013. Dynamic land cover information: bridging the gap between remote sensing and natural resource management. *Ecol. Soc.* **18**, 2. doi:10.5751/es-05229-180102.

Turner, W., C. Rondinini, N. Pettorelli, B. Mora, A. K. Leidner, Z. Szantoi, et al. 2015. Free and open-access satellite data are key to biodiversity conservation. *Biol. Conserv.* **182**, 173–176. doi:10.1016/j.biocon.2014.11.048.

Turnhout, E., B. Bloomfield, M. Hulme, J. Vogel, and B. Wynne. 2012. Conservation policy: listen to the voices of experience. *Nature* **488**, 454–455. doi:10.1038/488454a.

A review of camera trapping for conservation behaviour research

Anthony Caravaggi[1] (iD), Peter B. Banks[2], A Cole Burton[3], Caroline M. V. Finlay[4], Peter M. Haswell[5] (iD), Matt W. Hayward[5,6], Marcus J. Rowcliffe[7] & Mike D. Wood[8]

[1]School of Biological Sciences, Queen's University Belfast, Belfast BT9 7BL, United Kingdom
[2]School of Life and Environmental Sciences, The University of Sydney, Sydney, New South Wales 2006
[3]Department of Forest Resources Management, University of British Columbia, 2215-2424 Main Mall, Vancouver, British Columbia, Canada V6T 1Z4
[4]Ulster Wildlife, McClelland House, 10 Heron Road, Belfast BT3 9LE, United Kingdom
[5]School of Biological Sciences, Bangor University, Bangor, Gwynedd LL57 2UW, United Kingdom
[6]Centre for African Wildlife Ecology, Nelson Mandela University, Port Elizabeth 6031 South Africa; and Centre for African Wildlife Ecology, University of Pretoria, Pretoria 0001,South Africa
[7]ZSL Institute of Zoology, Regent's Park, London NW 4RY,United Kingdom
[8]School of Environment & Life Sciences, University of Salford, Manchester M5 4WT, United Kingdom

Keywords
Anthropogenic impacts, behavioural indicators, ethology, management, monitoring, remote sensing

Correspondence
Anthony Caravaggi, School of Biological Sciences, Queen's University Belfast, Belfast BT9 7BL, United Kingdom.
E-mail: acaravaggi01@qub.ac.uk

Funding Information
The contribution of M.D. Wood was supported by the TREE project (www.ceh.ac.uk/TREE) funded by the Natural Environment Research Council (NERC), Environment Agency and Radioactive Waste Management Limited. The contributions of C.M.V. Finlay were supported by the Red Squirrels United project (http://www.redsquirrelsunited.org.uk/) funded by EU Life and Heritage Lottery Fund.

Editor: Nathalie Pettorelli
Associate Editor: Rahel Sollmann

Abstract

An understanding of animal behaviour is important if conservation initiatives are to be effective. However, quantifying the behaviour of wild animals presents significant challenges. Remote-sensing camera traps are becoming increasingly popular survey instruments that have been used to non-invasively study a variety of animal behaviours, yielding key insights into behavioural repertoires. They are well suited to ethological studies and provide considerable opportunities for generating conservation-relevant behavioural data if novel and robust methodological and analytical solutions can be developed. This paper reviews the current state of camera-trap-based ethological studies, describes new and emerging directions in camera-based conservation behaviour, and highlights a number of limitations and considerations of particular relevance for camera-based studies. Three promising areas of study are discussed: (1) documenting anthropogenic impacts on behaviour; (2) incorporating behavioural responses into management planning and (3) using behavioural indicators such as giving up densities and daily activity patterns. We emphasize the importance of reporting methodological details, utilizing emerging camera trap metadata standards and central data repositories for facilitating reproducibility, comparison and synthesis across studies. Behavioural studies using camera traps are in their infancy; the full potential of the technology is as yet unrealized. Researchers are encouraged to embrace conservation-driven hypotheses in order to meet future challenges and improve the efficacy of conservation and management processes.

Introduction

Animal behaviour is an important component of conservation biology (Berger-Tal et al. 2011) and, hence, is of considerable interest to researchers and wildlife managers (Caro and Durant 1995). For example, behavioural studies can increase our understanding of species' habitat requirements (Pienkowski 1979), reproductive behaviour (Cant 2000) and dispersal or migration (Doerr et al. 2011), and elucidate impacts of habitat fragmentation

(Merckx and Van Dyck 2007) or climate change (Moller 2004). Animal behaviour can also be a useful monitoring tool, with individual- and group-level responses used to evaluate the impacts of management (Morehouse et al. 2016). It is important, therefore, to incorporate behaviour into conservation planning; its omission limits efficacy of conservation actions and could lead to failure (Berger-Tal et al. 2011). The confluence of conservation biology and ethology has come to be known as 'conservation behaviour', wherein conservation problems are addressed by the application of behavioural research (Blumstein and Fernández-Juricic 2004; Berger-Tal et al. 2011).

Quantifying the behaviour of wild animals presents significant challenges. Direct observation of animals can allow the evaluation of individual responses to environmental stimuli. Such studies may be weakened, however, by the influence of the human observer on focal animals (Nowak et al. 2014) and limited by small sample size and logistical constraints (Bridges and Noss 2011). Furthermore, only a limited number of species and habitats are amenable to direct, field-based observations (e.g. larger species and those that can be habituated; and in open and accessible habitats). Many of these have already been the focus of direct behavioural research (Schaller 1967; Kruuk 1972; Caro 1994) or may be atypical of more common habitats and can lead to inconsistent results (Laurenson 1994 vs. Mills and Mills 2014). In cases where focal animal(s) cannot easily be directly observed, the vast majority of field-based behavioural studies have used radio (VHF) or satellite (GPS) telemetry, activity sensors and/or biologgers (e.g. Lewis et al. 2002; Grignolio et al. 2004; Shamoun-Baranes et al. 2012; Bouten et al. 2013). The advantages and disadvantages of these methods, which are currently the gold standards for obtaining spatiotemporal behavioural data, are summarized in Table 1, highlighting that while these devices can provide powerful insights, they also have significant logistical and inferential limitations. Consequently, the suite of species that have had their behaviour quantified is biased and limited. New methods of obtaining behavioural data are, therefore, urgently required.

Camera traps (i.e. cameras that are remotely activated via an active or passive sensor; hereafter referred to as CTs) offer a reliable, minimally invasive, visual means of surveying wildlife that substantially reduces survey effort. CTs are increasingly popular in ecological studies (Burton et al. 2015; Rovero and Zimmermann 2016) and provide a wealth of information that is often of considerable conservation value (e.g. Ng et al. 2004; Di Bitetti et al. 2006; Caravaggi et al. 2016). Continued technological improvements and decreasing equipment costs (Tobler et al. 2008a), combined with their demonstrated versatility (Rovero et al. 2013), mean that CTs will only continue to grow in popularity. CT data take the form of a still image or video of an

Table 1. Potential advantages and disadvantages of three conventional methods commonly used to collect animal behavioural data.

	Method			
	VHF	GPS	ACC	CT
Advantages				
Allows independent data verification			✔	✔
Collection of biometric data during deployment	✔	✔	✔	
Combined analysis of movement and trait-based data	✔[1,2]	✔	✔	✔
Detailed data[2,3,4,10]		✔	✔	✔
Habitat associations	✔	✔		✔
Identification of specific behaviours			✔[10]	✔
Landscape scale	✔	✔		✔
Low cost			✔	✔[10]
Low survey effort		✔[10]	✔[10]	✔[10]
Multi-taxa surveys				✔
Range analyses	✔	✔		✔
Disadvantages				
Bias from handling focal animal(s)[5,6]	✔	✔	✔	
Disturbance effects				✔[10]
Expensive	✔	✔		✔[10]
Limited sample size	✔	✔	✔	
Negative impacts on focal animal(s) during backpack/collar deployment[7]	✔	✔	✔	
Requires ground-truthing to avoid inferential error[4,5,8]			✔	
Simplistic data[10]	✔	✔	✔[9]	✔
Stationary				✔
Technological failure	✔	✔	✔	✔
Triangulation/location error[5]	✔	✔		

These are not necessarily contextual constants. For example, GPS accuracy is affected by vegetation density. Similarly, activity sensors may return detailed or simplistic data, depending on the device used. VHF, Radio telemetry tags; GPS, Global Positioning System tags; ACC, activity sensors; CT, camera traps (still images and video footage, equally).
[1]Grignolio et al. (2004).
[2]Lewis et al. (2002).
[3]Bouten et al. (2013).
[4]Shamoun-Baranes et al. (2012).
[5]Bridges and Noss (2011).
[6]Wilson et al. (1986).
[7]Barron et al. (2010).
[8]Ware et al. (2015).
[9]Coulombe et al. (2006).
[10]Device, environment and/or species dependent.

individual or a group of individuals, of one or more species, which have been detected within the camera and location-specific zone of detection. These images can be linked with additional information, including the date, time and location at which the image was recorded. CT surveys have been effectively used to quantify species diversity (Tobler et al. 2008b), relative abundance (Carbone et al. 2001; Villette et al. 2017), and population parameters (Karanth et al. 2006; Rowcliffe et al. 2008); demonstrate site occupancy of rare or cryptic species (Linkie et al. 2007), and describe species replacement processes (Caravaggi et al. 2016). CTs have also been used in behavioural studies (Maffei et al. 2005; Bridges and Noss 2011). In a recent review of 266 CT studies, Burton et al. (2015) characterized one-third as addressing behavioural questions (e.g. activity patterns, diet; Table 2).

In this paper, we review some of the recent literature on animal behaviour as elucidated by camera trapping studies. We then describe a number of common issues encountered by researchers undertaking such surveys and, finally, suggest future avenues of research that may be of considerable benefit to conservation initiatives. This review serves as a point of reference for researchers and practitioners undertaking conservation-oriented CT surveys of animal behaviour.

Current Applications of Camera Traps to Animal Behaviour

CTs are well suited to ethological studies, providing increasing opportunities to undertake extensive and detailed sampling of wild animal behavioural repertoires (see Fig. 1 and Table 2 for examples). The nature of the technology confers a number of important benefits. For example, CTs facilitate detailed studies of behaviours in species that were previously considered too small or elusive to be reliably observed in the field. CTs have been used to understand burrowing behaviour in <40 g northern hopping mice (*Notomys aquilo*; Diete et al. 2014) and olfactory communication in native and introduced <120 g rats (*Rattus* sp.; Heavener et al. 2014). Furthermore, CTs remove the need for a human observer in situ, thereby reducing the potential for bias as a result of the observer's influence on behaviour. The use of CTs may also lead to further reduction in observer bias as, while a human observer is required to review collected images and assign individual and/or species identities and behaviours, cameras allow independent verification and recurrent analysis of observations. This is in contrast to conventional field methods for documenting behaviour, where it is rarely possible for another scientist to independently verify observational data.

Table 2. Examples of behavioural observations of wildlife via camera trapping. Species are ordered chronologically following the date of corresponding references.

Behaviour	Species	References
Active period	Spotted-tailed quoll (*Dasyurus maculatus*)	Claridge et al. 2004
	Guizhou snub-nosed monkey (*Rhinopithecus brelichi*)	Tan et al. 2013
	Agouti (*Dasyprocta punctata*) and ocelot (*Leopardus pardalis*)	Suselbeek et al. 2014
Antipredator responses	Bush rat (*Rattus fuscipes*)	Carthey and Banks 2016
Bathing/wallowing	Giant anteater (*Myrmecophaga tridactyla*)	Emmons et al. 2004
Crossing roads	Bare-nosed wombats (*Vombatus ursinus*)	Crook et al. 2013
Daily activity	Clouded leopard (*Neofelis nebulosa*), golden cat (*Catopuma temminckii*), and 4 other felids	Azlan and Sharma 2006 Delgado-V. et al. 2011
	Tayra (*Eira barbara*)	Leuchtenberger et al. 2014
	Giant otter (*Pteronura brasiliensis*)	Rowcliffe et al. 2014
	12 terrestrial mammal species	
Denning	American black bear (*Ursus americanus*)	Bridges et al. 2004
Foraging	Yakushima macaque (Macaca fuscata yakui)	Otani 2001
	Tayra (*Eira barbara*)	Delgado-V. et al. 2011
Migration	Bald eagle (*Haliaeetus leucocephalus*), black vulture (*Coragyps atratus*) and 5 other birds of prey	Jachowski et al. 2015
Nest predation	Predators exploiting quail (*Coturnix coturnix*) eggs	Picman and Schriml 1994
Phenological changes	Elk (Cervus elaphus)	Brodie et al. 2012
Positional behaviour	Bare-tailed woolly opossum (*Caluromys philander*)	Dalloz et al. 2012
Resource partitioning	Cape fox (*Vulpes chama*), caracal (*Caracal caracal*), honey badger (*Mellivora capensis*) and 9 other carnivores	Edwards et al. 2015
Response to human-animal conflict	Tiger (*Panthera tigris*) and associated prey species	Johnson et al. 2006
Scent marking	Tayra (*Eira barbara*)	Delgado-V. et al. 2011
	Eurasian lynx (*Lynx lynx*)	Vogt et al. 2014
Social behaviour	Blonde capuchin (*Sapajus flavius*)	Bezerra et al. 2014
	Giant otter (*Pteronura brasiliensis*)	Leuchtenberger et al. 2014
Temporal avoidance	Jaguar (*Panthera onca*) and puma (*Puma concolor*)	Romero-Muñoz et al. 2010
Travel speed	12 terrestrial mammal species	Rowcliffe et al. 2016
Waterhole use	15 species of ungulates, 5 birds, 3 mega-herbivores, 2 primates and 5 carnivores	Hayward and Hayward 2012

Many types of animal behaviours have been studied with CTs (Table 2), including foraging (Otani 2001), daily activity patterns (Tan et al. 2013), scent marking (Delgado-V et al. 2011), movement (Ford et al. 2009), livestock depredation (Bauer et al. 2005), and use of a variety of habitat features including dens/burrows (Clapham et al. 2014), urban habitats (Marks and Duncan 2009), corridors (LaPoint et al. 2013) and waterholes (Hayward and Hayward 2012). CT studies have often yielded key behavioural insights that may otherwise have remained unknown, many of which could be important to conservation processes. For example, studies investigating the efficacy of highway crossings in Banff National Park, Canada, described the effectiveness of under- and over-passes, an expensive and controversial means of impact mitigation (Clevenger and Waltho 2000; Ford et al. 2009), which is now being duplicated in other parts of the world. Picman and Schriml (1994) observed the predators of quail (*Coturnix coturnix*) nests in a variety of habitats, elucidating temporal variation and relative importance of each predatory species. The application of this method to the study of threatened avifauna has clear conservation benefits via the identification of direct impacts on egg success and the development of appropriate mitigation and monitoring techniques. Similarly, cameras provide more accurate post-hibernation den-emergence estimates for American black bears (*Ursus americanus*) than conventional methods, that is, den visits and radio telemetry (Bridges et al. 2004). Long-term monitoring of emergence relative to climate may yield important insights into the effects of climate change on black bears and other hibernating species (sensu Bridges and Noss 2011).

Figure 1. Examples of animal behaviour captured by camera traps: (A) Scent marking by an American black bear (*Ursus americanus*); (B) intraspecific competition in moose (*Alces alces*); (C) interspecific interactions between a European hare (*Lepus europaeus*; anti-predator response), a common buzzard (*Buteo buteo*; avoidance and attempted predation) and a hooded crow (*Corvus cornix*; anti-predator behaviour) captured on video (available at 10.6084/m9.figshare.4508369); (D) predation of a European rabbit (*Oryctolagus cuniculus*) by a red fox (*Vulpes vulpes*); (E) investigation of a squirrel feeding station by a pine marten (*Martes martes*); (F) nut caching by a grey squirrel (*Sciurus carolinensis*). Images provided by A.C. Burton (a, b), A. Caravaggi (c, d) and C.M.V. Finlay (e, f).

The majority of ethological CT studies conducted thus far have been primarily curiosity-driven, rather than being motivated by applied conservation-focussed hypotheses. This is not to say that a large number of these studies do not have conservation value. On the contrary, the conservation relevance of the data is often explicitly discussed. It is apparent, however, that there is an increasing need for conservation-driven studies. CTs are among the most promising and flexible tools available and we are only beginning to explore their potential.

Emerging Directions in Camera-Based Conservation Behaviour

The growth in popularity and application of CT surveys and novel solutions to non-behavioural questions of animal ecology (e.g. Rowcliffe et al. 2008; Martin et al. 2015; Bowler et al. 2016) suggests that creative methodological and analytical solutions will be increasingly used to investigate animal behaviours. If these novel studies are to be developed, it is important that researchers strive for true experimental designs focussed on conservation behaviour. A particular strength of CT surveys is the potential for multiple studies to be carried out concurrently (e.g. estimation of focal species population density and the species richness of the surveyed area). Thus, behaviour can be recorded alongside other important parameters, thereby facilitating insight into processes such as density-dependent behaviours and responses to climate change. New approaches are also being developed to move beyond correlational approaches and incorporate CTs into manipulative experiments, such as measuring animal behavioural responses to introduced stimuli (e.g. predator calls; Suraci et al. 2016).

Berger-Tal et al. (2011) described three ways in which behavioural research can be of conservation benefit: (1) identifying the impact of anthropogenic environmental changes on behaviour; (2) considering behavioural aspects of conservation initiatives ('behaviour-based management'); and (3) identifying behavioural indicators which are suggestive of changes in populations or the environment. We use this framework as a basis for our recommendations, below.

Anthropogenic impacts

An important area of conservation research lies in understanding the influence of anthropogenic stressors on animal behaviours and predicting the resulting population-level responses in order to inform management practices. Stressors such as habitat fragmentation, disturbance, the creation of ecological traps and the introduction of non-native species can have significant effects on behaviour (Robertson and Hutto 2006) and, hence, fitness (Berger-Tal et al. 2011). For example, animals may

exhibit increasing wariness in areas of greater disturbance (Stewart et al. 2016) and may change their daily activity patterns in close proximity to human populations (Carter et al. 2012). While anthropogenic impacts are generally negative, some species show benefits such as increased occupancy in fragmented landscapes (Fleschutz et al. 2016), or using human activity to evade apex predators (Muhly et al. 2011; Steyaert et al. 2016). Impacts on one species may also have spillover effects on the wider ecological community (Wright et al. 2010; Clinchy et al. 2016).

Habitat fragmentation, the division of large, connected habitats into small, isolated fragments separated by dissimilar habitats, is a major conservation issue (Haddad et al. 2015). Fragmentation has a wide range of potential impacts on species and ecosystems (e.g. via edge effects, patch size, shape and complexity and distance from other patches; Fahrig 2003), and these impacts may be mediated through effects on animal behaviour. CTs provide new opportunities for documenting behavioural responses to fragmentation. For example, the activity patterns of nine-banded armadillos (*Dasypus novemcinctus*) varied in association with forest patch size, among other factors, while patch time since isolation was predictive of agouti (*Dasyprocta leporina*) activity (Norris et al. 2010).

The disruption of dispersal behaviour can lead to the endangerment and potential extinction of isolated populations by various mechanisms, including changes to genetic diversity and structure (Keyghobadi 2007), stochastic threats (Fischer and Lindenmayer 2007) and long-term displacement effects (Ewers and Didham 2005). Using CTs to document dispersal behaviour can improve understanding of responses to movement disruption (Blumstein and Fernández-Juricic 2004) and inform design and implementation of mitigation measures that encourage dispersal. Individual-level analysis of dispersal is potentially possible for animals with individually identifiable markings or tags, although designing such a study may be challenging as dispersal routes and, hence, appropriate locations for CT deployment may not be known a priori. Inferences about dispersal, however, can also be drawn without individual identification. For example, cameras are well suited to quantifying use of presumed dispersal routes or movement corridors, including mitigations designed to promote connectivity (e.g. highway crossings; Clevenger and Waltho 2005; Ford et al. 2009). CTs can also be used to identify colonization of new habitat patches (including range expansions or species invasions) and parameterize landscape connectivity models (Brodie et al. 2015).

No studies have integrated environmental sensors into CT studies investigating anthropogenic impacts on behaviour, and we believe this is a promising area for future development. Local temperature, precipitation and humidity can readily be recorded, and phenocams can be used to document vegetation and environmental changes (Brown et al. 2016). Collecting such information alongside CT-based behavioural data will allow us to increase our understanding of how animals respond to changing conditions at both large (population) and small (localities within home ranges) spatial scales. This is particularly important given the rapid changes that are predicted to occur under climate change.

Behaviour-based management

Berger-Tal et al. (2011) suggested that behaviour-sensitive management and behavioural modification are two key pathways through which ethology can inform active management for conservation. The former considers animal behaviour in the design of reserves and corridors, planning species reintroductions and translocations, and epidemiology with the goal of stabilizing or increasing threatened populations or controlling pest or invasive species. Behavioural modification focuses on changing or preserving key behaviours within a focal population. CT surveys have the potential to inform both of these areas.

Considering social dynamics is one important area in which CT surveys can inform behaviour-sensitive management. Social species, that is, those that interact and/or live together, often exhibit complex inter-group relationships and social structure (Rowell 1966; Creel et al. 1997; Archie et al. 2006; Wolf et al. 2007; Wey et al. 2008), that are susceptible to rapid change via the social displacement or death of one or more individuals. This can have severe consequences for the species and/or their environment (e.g. Nyakaana et al. 2001). Social Network Analysis (SNA) facilitates the study of relationships between nodes (i.e. individuals), within networks (i.e. social groups; Sueur et al. 2011). The methodology is increasingly used to study animal behaviour (Lusseau et al. 2006; Whitehead 2008; Voelkl and Kasper 2009; Jacoby and Freeman 2016). Examples of SNA demonstrating a direct benefit to conservation, however, are few. SNA studies are limited in that they require the reliable identification of individuals and, hence, are only applicable with CTs where animals exhibit individual characteristics or markings, or where marks (e.g. tags) can be attached. However, placing cameras in areas frequented by social groups such as feeding or resting sites, and with a sufficient number of units, could yield a considerable amount of important data for behaviour-sensitive management. Such site-specific studies have some limitations and incur biases that require evaluation. For example, individuals may not be equally detectable, or full groups may not be observed. Furthermore, it would be difficult to account for behaviours and social

interactions which occur while away from the focal site. However, SNA analyses do not require constant observation of all group members to be effective (see Jacoby and Freeman 2016). Assessing potential bias with calibration by direct observation or other methods and placing observations in appropriate contexts is, therefore, important.

SNA has the potential to increase our understanding of disease or pathogen transmission and individual or group vulnerability (Krause et al. 2007), an issue of particular relevance to the conservation of species which are susceptible to outbreaks (e.g. Hamede et al. 2009). SNA studies have demonstrated that the removal of certain individuals (e.g. via hunting) can have a considerable effect on the stability of the social network (e.g. Flack et al. 2006), thus demonstrating their potential utility in elucidating the impacts of the bushmeat trade on inter- and intra-group dynamics in primates, for example. Furthermore, SNA has implications for reintroduction programmes, where the (re)construction of cohesive social structures in a captive setting would be necessary for the return of the focal species to the wild (Abell et al. 2013). Studies of the relationships between individuals, therefore, can help us to understand how social behaviour is influenced by a variety of factors and, hence, provide an additional means by which practitioners can build an evidence base to address conservation questions.

CTs can also be applied to studies of behavioural modification. For example, Davies et al. (2016) used CTs to investigate responses of African herbivores to changes in predation risk resulting from recently reintroduced lions. Cameras are also well suited to monitoring animal responses to conflict mitigation measures and have been used to demonstrate the efficacy of bees as a deterrant of crop-raiding elephants (Ngama et al. 2016).

Behavioural indicators

The ways in which animals adapt their foraging behaviour in human-impacted environments have important implications for their abilities to adapt and persist under increasing pressures. Behavioural indicators can be used to assess the state of animals and the environments they inhabit, highlighting important conservation issues such as population decline or habitat degradation, or being used to monitor the efficacy of management (Berger-Tal et al. 2011). Behaviour effectively acts as an early-warning system, indicating changes to processes before they are evident through, for example, population decline.

The giving up density (GUD; that is, the amount of food left behind from a known starting quantity; Brown 1988) is one such behavioural indicator that has been used to study predation risk (Orrock 2004; Severud et al. 2011), energetic costs (Nolet et al. 2006), forager state and forage quality (Hayward et al. 2015), plant toxins (Emerson and Brown 2015), competition (Brown et al. 1997) and predator–prey dynamics (Andruskiw et al. 2008). It is also central to describing the 'landscape of fear' (i.e. relative levels of predation risk within an area of use) of an animal and its habitat preferences, which are direct behavioural indicators with significant conservation implications (Kotler et al. 2016). CTs offer a relatively reliable way of using the GUD technique to ask more in-depth questions of conservation relevance. For example, CTs have been used to calculate GUDs for multiple species (Lerman et al. 2012), examine (Mella et al. 2015) and differentiate individual versus group foraging habits (Carthey and Banks 2015). These observations can then be used to inform the development of hypotheses relating to the broader effects of local food and predator abundance, predation pressure and inter- and intra-specific competition. With advancements in CT technology and creative experimental design, a wealth of conservation-focussed GUD applications are now possible.

A key strength of CTs lies in collecting data on multiple species, either as bycatch in a focal study, or as part of a specific multi-taxa investigation. Accordingly, there has been an increasing focus on assessing species interactions and niche partitioning via comparisons of co-occurrence and activity patterns (de Almeida Jacomo et al. 2004; Kukielka et al. 2013; Farris et al. 2014; Wang et al. 2015; Bu et al. 2016; Cusack et al. 2016; Sweitzer and Furnas 2016). Animal activity patterns are shaped by a number of factors, including foraging efficiency (Lode 1995), predator/prey activity (Middleton et al. 2013), photoperiodism (McElhinny et al. 1997) and competition (Rychlik 2005). Conservation-focussed studies using these methodologies, however, are scarce. Changes in the way species interact and use the landscape may be indicative of responses to changing environmental pressures and, hence, can direct development of early conservation strategies. For example, brown bears (*Ursus arctos*; Ordiz et al. 2013) altered their movement patterns, and wolverines (*Gulo gulo*; Stewart et al. 2016) behaved differently when faced with human disturbance, potentially impacting their ecosystem roles and, hence, associated species and habitats. Disturbance of the activity patterns of one or more species in a dynamic interaction, particularly ecological competitors or predators and prey, can, therefore, be interpreted as indicative of environmental changes and, hence, suggest additional lines of enquiry and highlight areas of conservation concern.

Scaling-up

Cameras can be used to monitor large-scale biodiversity conservation processes (O'Brien et al. 2010; Ahumada

et al. 2013) and investigate animal behaviour on a landscape scale. Scaling-up CT networks would provide stronger, larger-scale inferences on spatiotemporal variation in behaviours (Steenweg et al. 2016). Studies conducted on a broader scale have inherent limitations, however, that are not necessarily considerations for more localized investigations. The trade-off between the scale of investigation and camera array density has spatiotemporal implications which must be considered when designing a study, formulating hypotheses and deriving inferences from resultant data. Broad-scale studies are also ostensibly limited by the number of researchers available to place and check cameras and process data. The recruitment of volunteers (i.e. citizen scientists), however, offers a means of expanding the scope of research (Cohn 2008), greatly expanding spatial coverage and delivering a wealth of temporally comparable data (McShea et al. 2016). Emerging large-scale camera monitoring initiatives, such as Snapshot Serengeti (www.snapshotserengeti.org; Swanson et al. 2015) and Wildcam Gorongosa (www.wildcamgorongosa.org), demonstrate the benefits of this approach. CT projects utilizing citizen science have the potential to deliver a substantial amount of behavioural data (McShea et al. 2016) and inform conservation processes. However, few large-scale studies utilizing citizen science involve behavioural analyses. CT video data can produce vast amounts of video footage, but the extraction of key behavioural data from video footage is time consuming, imposing a major obstacle. Crowdsourcing video interpretations can overcome this limitation, however, and the use of robust ethograms, simple training regimes and blinding of observers to treatments can assuage concerns about the reliability of citizen science interpretations (e.g. Carthey 2013).

Synthesizing across projects offers another means of conducting broader analyses (Steenweg et al. 2016). We recommend that researchers embrace emerging CT metadata standards and associated opportunities to use common data repositories such as Wildlife Insights (www.wildlifeinsights.org; Forrester et al. 2016), thus increasing the potential for the synthesis of inferences across large scales. The value of current data repositories is reduced, however, by their reliance on static images and omission of video. Expenses notwithstanding, it is in the interests of conservation behaviour researchers to establish a digital repository for video data.

Relevant Limitations and Considerations

Despite the great promise of new insights in conservation behaviour from CTs, it is important to consider potential limitations. CTs are passive instruments; thus, while it is possible to identify animals according to species, age class

(Clapham et al. 2014), sex (Bezerra et al. 2014) or, indeed, identify individuals (Karanth et al. 2006; Zheng et al. 2016), the collection of biometric, genetic and other data of interest requires the application of supplementary or alternative methodologies. Furthermore, CTs are frequently considered to be non-intrusive, causing little to no disturbance. However, while the sound produced by recording units is largely inaudible to humans, it is frequently detected by wildlife (Meek et al. 2014a). Similarly, CTs which utilize visible light (as opposed to infra-red) increase the chances of the camera being detected by animals, potentially disrupting their natural behaviour (Meek et al. 2016a).

Camera failure, although rare, can result in the loss of large quantities of data. Similarly, camera theft is becoming increasingly common (Meek et al. 2016b). It is, therefore, necessary to balance the frequency of visits to maintain CTs with risk of data loss. To accommodate this, it is advisable to build some redundancy into the study design, such as the use of cameras that allow the transmission of images via Global Packet Radio Service (GRPS) and/or Wi-Fi and can, therefore, facilitate remote data collection and inform the timing of maintenance visits.

Researchers utilizing CTs have the option of recording data in the form of still images or video footage. In many cases, still images are adequate; it is possible to derive important behavioural data from them and, indeed, the format offers some obvious benefits. For example, still images require considerably less memory than video footage (Glen et al. 2013) and, hence, may be more suitable for studies which require CTs to be deployed, without intervention, for a prolonged period of time. However, similar capture success rates can be achieved with either format (Glen et al. 2013) and the majority of operational limitations apply equally to both. For example, some cameras have a slow trigger time meaning that initial behaviours, which might be the most important in terms of measuring detection of a stimulus (rather than the response), can be missed. Furthermore, many cameras offer only a limited number of high-speed 'burst' (i.e. sequentially captured) images or length of video (e.g. 60 sec), requiring the camera to be retriggered to continue the capture of the behaviours and, hence, creating gaps in the observation. However, video footage opens up new opportunities, for example, observing interactions at focal sites, or measuring the duration of behavioural bouts. While both formats can be effectively used in most of the applications described herein (SNA being the one exception, with video being preferred), videos are undoubtedly more informative and an important future direction for CT-based behavioural research.

Sampling the behaviours of small species can be particularly challenging, with CTs typically designed for deer-sized game species (Weerakoon et al. 2014), a problem that will require novel solutions. For example, flash-illuminated images are frequently obscured by overexposure when close enough to small mammals to observe behaviour clearly, whereas at the correctly exposed distance, animals can be too far away to reliably identify species or discern behaviours. Furthermore, understanding the reliability of camera surveys for addressing multi-species objectives remains an important area of methodological research (see Burton et al. 2015). Multi-taxa studies also require careful planning to ensure that CTs are appropriately located and adequately spaced to maximize the chances of capturing a diverse species assemblage while meeting analytical assumptions such as independence of sampling sites. The choice and placement of cameras should, therefore, be dictated by the objectives of the study, the ecology of the study species, the statistical sampling framework and associated considerations.

An oft-repeated concern relates to study repeatability; specific details of study design (e.g. how survey sites were chosen, use of lures) and camera protocols (e.g. camera model, deployment details) are often lacking (Meek et al. 2014b; Burton et al. 2015). A number of factors influence the detection of individuals (see Burton et al. 2015), and sampling details may have important implications for analytical assumptions such as effective sampling area and site independence (Harmsen et al. 2010; Mccoy et al. 2011; du Preez et al. 2014; Newey et al. 2015). Comprehensive methodological descriptions and utilization of emerging CT metadata standards (Forrester et al. 2016) are important for facilitating reproduction, comparison and synthesis across studies.

Finally, as with any survey method, observations from CTs are incomplete and may contain biases that affect inferences. As noted above, species and individuals may vary in their detectability by CTs according to attributes such as body size, movement speed, curiosity and wariness. Behaviours observed by CTs may also not always be representative of behaviours more generally. Thus it is incumbent upon researchers to remain vigilant for potential biases and test CT-based inferences through comparison and calibration with more established ethological methods.

Conclusions

CTs are rapidly increasing in popularity, and their application to conservation behaviour is growing. Recent efforts to coordinate camera studies across large scales through methodological standardization and/or better reporting of methodologies and metadata will facilitate broader ethological inferences on species' behavioural responses to environmental change. The development and application of new techniques and analytical methods explicitly focussed on anthropogenic impacts, behaviour-based management and behavioural indicators would undoubtedly benefit conservation programmes. CTs are not a panacea, but they confer many benefits to researchers and the diversity of possible applications is gradually being realized. We hope that this paper will act as a catalyst, advancing the adoption of CT technology within conservation behaviour. It is important, therefore, that potentially profitable avenues of investigation are identified and pursued if we are to maximize the generation of valuable data and, hence, improve the conservation outlook for the ever-increasing number of threatened or endangered species.

Acknowledgments

The authors thank the journal for their encouragement and support of this review and, specifically, the editor and reviewers for their valuable comments which strengthened the final manuscript. Authors are ordered alphabetically by surname, except for A. Caravaggi who conceived, organized and compiled the manuscript. The contribution of M.D. Wood was supported by the TREE project (www.ce h.ac.uk/TREE) funded by the Natural Environment Research Council (NERC), Environment Agency and Radioactive Waste Management Limited. The contributions of C.M.V. Finlay were supported by the Red Squirrels United project (http://www.redsquirrelsunited.org.uk/) funded by EU Life and Heritage Lottery Fund.

References

Abell, J., M. W. B. Kirzinger, Y. Gordon, J. Kirk, R. Kokeŝ, K. Lynas, B. Mandinyenya, et al. 2013. A social network analysis of social cohesion in a constructed pride: implications for ex situ reintroduction of the African lion (*Panthera leo*). *PLoS ONE* **8**, e82541.

Ahumada, J. A., J. Hurtado, and D. Lizcano. 2013. Monitoring the status and trends of tropical forest terrestrial vertebrate communities from camera trap data: a tool for conservation. *PLoS ONE* **8**, e73707.

de Almeida Jacomo, A. T., L. Silveira, and J. A. F. Diniz-Filho. 2004. Niche separation between the maned wolf (*Chrysocyon brachyurus*), the crab-eating fox (*Dusicyon thous*) and the hoary fox (*Dusicyon vetulus*) in central Brazil. *J. Zool.* **262**, 99–106.

Andruskiw, M., J. M. Fryxell, I. D. Thompson, and J. A. Baker. 2008. Habitat-mediated variation in predation risk by the American marten. *Ecology* **89**, 2273–2280.

Archie, E. A., T. A. Morrison, C. A. H. Foley, C. J. Moss, and S. C. Alberts. 2006. Dominance rank relationships among wild female African elephants, *Loxodonta africana*. *Anim. Behav.* **71**, 117–127.

Azlan, J. M., and D. S. K. Sharma. 2006. The diversity and activity patterns of wild felids in a secondary forest in Peninsular Malaysia. *Oryx* **40**, 36.

Barron, D. G., J. D. Brawn, and P. J. Weatherhead. 2010. Meta-analysis of transmitter effects on avian behaviour and ecology: Meta-analysis of avian transmitter effects. *Methods Ecol. Evol.* **1**, 180–187.

Bauer, J. W., K. A. Logan, L. L. Sweanor, and W. M. Boyce. 2005. Scavenging behaviour in puma. *The Southwestern Naturalist* **50**, 466–471.

Berger-Tal, O., T. Polak, A. Oron, Y. Lubin, B. P. Kotler, and D. Saltz. 2011. Integrating animal behavior and conservation biology: a conceptual framework. *Behav. Ecol.* **22**, 236–239.

Bezerra, B. M., M. Bastos, A. Souto, M. P. Keasey, P. Eason, N. Schiel, et al. 2014. Camera trap observations of nonhabituated critically endangered wild blonde capuchins, Sapajus flavius (formerly Cebus flavius). *Int. J. Primatol.* **35**, 895–907.

Blumstein, D. T., and E. Fernández-Juricic. 2004. The emergence of conservation behavior. *Conserv. Biol.* **18**, 1175–1177.

Bouten, W., E. W. Baaij, J. Shamoun-Baranes, and K. C. J. Camphuysen. 2013. A flexible GPS tracking system for studying bird behaviour at multiple scales. *J. Ornithol.* **154**, 571–580.

Bowler, M. T., Tobler M. W., Endress B. A., Gilmore M. P., and Anderson M. J., 2016. Estimating mammalian species richness and occupancy in tropical forest canopies with arboreal camera traps. *Remote Sens. Ecology and Conserv.* [online]. Available at: http://doi.wiley.com/10.1002/rse2.35 [accessed 7 December. 2016].

Bridges, A. S., and A. J. Noss. 2011. Behaviour and activity patterns. Pp. 57–69 in: A. F. O'Connell, J. D. Nichols, K. U. Karanth. eds., *Camera traps in animal ecology: methods and analyses.* Springer, New York.

Bridges, A. S., M. R. Vaughan, and S. Klenzendorf. 2004. Seasonal variation in American black bear Ursus americanus activity patterns: quantification via remote photography. *Wildlife Biol.* **10**, 277–284.

Brodie, J., E. Post, F. Watson, and J. Berger. 2012. Climate change intensification of herbivore impacts on tree recruitment. *Proceed. Royal Soc. B: Biol. Sci.* **279**, 1366–1370.

Brodie, J. F., A. J. Giordano, B. Dickson, M. Hebblewhite, H. Bernard, J. Mohd-Azlan, et al. 2015. Evaluating multispecies landscape connectivity in a threatened tropical mammal community: multispecies habitat corridors. *Conserv. Biol.* **29**, 122–132.

Brown, J. S. 1988. Patch use as an indicator of habitat preference, predation risk, and competition. *Behav. Ecol. Sociobiol.* **22**, 37–47.

Brown, J. S., B. P. Kotler, and W. A. Mitchell. 1997. Competition between birds and mammals: a comparison of giving-up densities between crested larks and gerbils. *Evol. Ecol.* **11**, 757–771.

Brown, T. B., K. R. Hultine, H. Steltzer, E. G. Denny, M. W. Denslow, J. Granados, et al. 2016. Using phenocams to monitor our changing Earth: toward a global phenocam network. *Front. Ecol. Environ.* **14**, 84–93.

Bu, H., F. Wang, W. J. McShea, Z. Lu, D. Wang, and S. Li. 2016. Spatial Co-Occurrence and Activity Patterns of Mesocarnivores in the Temperate Forests of Southwest China. *PLoS ONE* **11**, e0164271.

Burton, A. C., E. Neilson, D. Moreira, A. Ladle, R. Steenweg, J. T. Fisher, et al. 2015. Wildlife camera trapping: a review and recommendations for linking surveys to ecological processes. *J. Appl. Ecol.* **52**, 675–685.

Cant, M. A. 2000. Social control of reproduction in banded mongooses. *Anim. Behav.* **59**, 147–158.

Caravaggi, A., M. Zaccaroni, F. Riga, S. C. Schai-Braun, J. T. A. Dick, W. I. Montgomery, et al. 2016. An invasive-native mammalian species replacement process captured by camera trap survey random encounter models. *Remote Sens. Ecology Conserv.* **2**, 45–58.

Carbone, C., S. Christie, K. Conforti, T. Coulson, N. Franklin, J. R. Ginsberg, et al. 2001. The use of photographic rates to estimate densities of tigers and other cryptic mammals. *Anim. Conserv.* **4**, 75–79.

Caro, T. M. 1994. *Cheetahs of the Serengeti Plains: group living in an asocial species.* University of Chicago Press, Chicago.

Caro, T., and S. M. Durant. 1995. The importance of behavioral ecology for conservation biology: examples from Serengeti carnivores. Pp. 451–472 in A. R. E. Sinclair and P. Arcese, eds. *Serengeti II: Dynamics, Management, and Conservation of an Ecosystem.* University of Chicago, Chicago.

Carter, N. H., B. K. Shrestha, J. B. Karki, N. M. B. Pradhan, and J. Liu. 2012. Coexistence between wildlife and humans at fine spatial scales. *Proc. Natl Acad. Sci.* **109**, 15360–15365.

Carthey, A. J. R., 2013. *Naivete, novelty and native status: mismatched ecological interactions in the Australian environment.* University of Sydney, Australia.

Carthey, A. J. R., and P. B. Banks. 2015. Foraging in groups affects giving-up densities: solo foragers quit sooner. *Oecologia* **178**, 707–713.

Carthey, A. J. R., and P. B. Banks. 2016. Naiveté is not forever: responses of a vulnerable native rodent to its long term alien predators. *Oikos* **125**, 918–926.

Clapham, M., O. T. Nevin, A. D. Ramsey, and F. Rosell. 2014. Scent-marking investment and motor patterns are affected by the age and sex of wild brown bears. *Anim. Behav.* **94**, 107–116.

Claridge, A. W., G. Mifsud, J. Dawson, and M. J. Saxon. 2004. Use of infrared digital cameras to investigate the behaviour of cryptic species. *Wildl. Res.* **31**, 645–650.

Clevenger, A. P., and N. Waltho. 2000. Factors influencing the effectiveness of wildlife underpasses in Banff National Park, Alberta, Canada. *Conserv. Biol.* **14**, 47–56.

Clevenger, A. P., and N. Waltho. 2005. Performance indices to identify attributes of highway crossing structures facilitating movement of large mammals. *Biol. Cons.* **121**, 453–464.

Clinchy, M., L. Y. Zanette, D. Roberts, J. P. Suraci, C. D. Buesching, C. Newman, et al. 2016. Fear of the human 'super predator' far exceeds the fear of large carnivores in a model mesocarnivore. *Behav. Ecol.* **27**, 1826–1832.

Cohn, J. P. 2008. Citizen science: can volunteers do real research? *Bioscience* **58**, 192.

Coulombe, M., A. Massé, and S. Côté. 2006. Quantification and accuracy of activity data measured with VHF and GPS telemetry. *Wildl. Soc. Bull.* **34**, 81–92.

Creel, S. 1997. Handling of African wild dogs and chronic stress: reply to East et al. *Conserv. Biol.* **11**, 1454–1456.

Crook, N., S. C. Cairns, and K. Vernes. 2013. Bare-nosed wombats (*Vombatus ursinus*) use drainage culverts to cross roads. *Aust. Mammal.* **35**, 23–29.

Cusack, J. J., et al. 2016. Revealing kleptoparasitic and predatory tendencies in an African mammal community using camera traps: a comparison of spatiotemporal approaches. *Oikos.* Available at: http://doi.wiley.com/10.1111/oik.03403 (accessed: 18 December 2016).

Dalloz, M. F., D. Loretto, B. Papi, P. Cobra, and M. V. Vieira. 2012. Positional behaviour and tail use by the bare-tailed woolly opossum *Caluromys philander* (Didelphimorphia, Didelphidae). *Mamm. Biol. - Zeitschrift für Säugetierkunde* **77**, 307–313.

Davies, A. B., C. J. Tambling, G. I. H. Kerley, and G. P. Asner. 2016. Limited spatial response to direct predation risk by African herbivores following predator reintroduction. *Ecol. Evol.* **6**, 5728–5748.

Delgado-V., C. A., Arias Alzate A., Botero S., and Sánchez-Londoño J. D. 2011. Behaviour of the *Tayra Eira barbara* near Medellín, Colombia: preliminary data from a video-capturing survey. *Small Carnivore Conserv.* **44**:19–21.

Di Bitetti, M. S., A. Paviolo, and C. De Angelo. 2006. Density, habitat use and activity patterns of ocelots (*Leopardus pardalis*) in the Atlantic Forest of Misiones, Argentina. *J. Zool.* **270**, 153–163.

Diete, R. L., P. D. Meek, C. R. Dickman, and L. K.-P. Leung. 2014. Burrowing behaviour of the northern hopping-mouse (*Notomys aquilo*): field observations. *Aust. Mammal.* **36**, 242–246.

Doerr, V. A. J., Barrett T., and Doerr E. D. 2011. Connectivity, dispersal behaviour and conservation under climate change: a response to Hodgson et al.: Connectivity and dispersal behaviour. *J. Appl. Ecol.*, **48**:143–147.

Edwards, S., A. C. Gange, and I. Wiesel. 2015. Spatiotemporal resource partitioning of water sources by African carnivores on Namibian commercial farmlands: Carnivore resource partitioning at water sources. *J. Zool.* **297**, 22–31.

Emerson, S. E., and J. S. Brown. 2015. The influence of food chemistry on food-safety tradeoffs in samango monkeys. *J. Mammal.* **96**, 237–244.

Emmons, L. H., R. P. Flores, S. A. Alpirre, and M. J. Swarner. 2004. Bathing behavior of giant anteaters (*Myrmecophaga tridactyla*). *Edentata* **6**, 41–43.

Ewers, R. M., and R. K. Didham. 2005. Confounding factors in the detection of species responses to habitat fragmentation. *Biol. Rev.* **81**, 117–142.

Fahrig, L. 2003. Effects of habitat fragmentation on biodiversity. *Annu. Rev. Ecol. Evol. Syst.* **34**, 487–515.

Farris, Z. J., S. M. Karpanty, F. Ratelolahy, and M. J. Kelly. 2014. Predator–primate distribution, activity, and co-occurrence in relation to habitat and human activity across fragmented and contiguous forests in northeastern Madagascar. *Int. J. Primatol.* **35**, 859–880.

Fischer, J., and D. B. Lindenmayer. 2007. Landscape modification and habitat fragmentation: a synthesis. *Glob. Ecol. Biogeogr.* **16**, 265–280.

Flack, J. C., M. Girvan, F. B. M. de Waal, and D. C. Krakauer. 2006. Policing stabilizes construction of social niches in primates. *Nature* **439**, 426–429.

Fleschutz, M. M., N. Gálvez, G. Pe'er, DaviesZ. G., K. Henle, and E. Schüttler. 2016. Response of a small felid of conservation concern to habitat fragmentation. *Biodivers. Conserv.* **25**, 1447–1463.

Ford, A. T., A. P. Clevenger, and A. Bennett. 2009. Comparison of methods of monitoring wildlife crossing-structures on highways. *J. Wildl. Manage.* **73**, 1213–1222.

Forrester, T., T. O'Brien, E. Fegraus, P. Jansen, J. Palmer, R. Kays, et al. 2016. An open standard for camera trap data. *Biodiversity Data Journal* **4**, e10197.

Glen, A. S., S. Cockburn, M. Nichols, J. Ekanayake, and B. Warburton. 2013. Optimising camera traps for monitoring small mammals. *PLoS ONE* **8**, e67940.

Grignolio, S., I. Rossi, B. Bassano, F. Parrini, and M. Apollonio. 2004. Seasonal variations of spatial behaviour in female Alpine ibex (*Capra ibex ibex*) in relation to climatic conditions and age. *Ethology Ecol. Evol.* **16**, 255–264.

Haddad, N. M., L. A. Brudvig, J. Clobert, K. F. Davies, A. Gonzalez, R. D. Holt, et al. 2015. Habitat fragmentation and its lasting impact on Earth's ecosystems. *Sci. Adv.* **1**, e1500052.

Hamede, R. K., J. Bashford, H. McCallum, and M. Jones. 2009. Contact networks in a wild Tasmanian devil (*Sarcophilus harrisii*) population: using social network analysis to reveal seasonal variability in social behaviour and its implications for transmission of devil facial tumour disease. *Ecol. Lett.* **12**, 1147–1157.

Harmsen, B. J., R. J. Foster, S. Silver, L. Ostro, and C. P. Doncaster. 2010. Differential use of trails by forest mammals and the implications for camera-trap studies: a case study from Belize: trail use by Neotropical forest mammals. *Biotropica* **42**, 126–133.

Hayward, M. W., and M. D. Hayward. 2012. Waterhole use by African fauna. *South African J. Wildlife Res.* **42**, 117–127.

Hayward, M. W., S. Ortmann, and R. Kowalczyk. 2015. Risk perception by endangered European bison Bison bonasus is context (condition) dependent. *Landscape Ecol.* **30**, 2079–2093.

Heavener, S. J., A. J. R. Carthey, and P. B. Banks. 2014. Competitive naïveté between a highly successful invader and a functionally similar native species. *Oecologia* **175**, 73–84.

Jachowski, D. S., T. Katzner, J. L. Rodrigue, and W. M. Ford. 2015. Monitoring landscape-level distribution and migration phenology of raptors using a volunteer camera-trap network. *Wildl. Soc. Bull.* **39**, 553–563.

Jacoby, D. M. P., and R. Freeman. 2016. Emerging Network-Based Tools in Movement Ecology. *Trends Ecol. Evol.* **31**, 301–314.

Johnson, A., C. Vongkhamheng, M. Hedemark, and T. Saithongdam. 2006. Effects of human-carnivore conflict on tiger (*Panthera tigris*) and prey populations in Lao PDR. *Anim. Conserv.* **9**, 421–430.

Karanth, K. U., J. D. Nichols, N. S. Kumar, and J. E. Hines. 2006. Assessing tiger population dynamics using photographic capture-recapture sampling. *Ecology* **87**, 2925–2937.

Keyghobadi, N. 2007. The genetic implications of habitat fragmentation for animals. *Can. J. Zool.* **85**, 1049–1064.

Kotler, B. P., D. W. Morris, and J. S. Brown. 2016. Direct behavioral indicators as a conservation and management tool. *Conservation Behavior* **21**, 307.

Krause, J., D. P. Croft, and R. James. 2007. Social network theory in the behavioural sciences: potential applications. *Behav. Ecol. Sociobiol.* **62**, 15–27.

Kruuk, H. 1972. *The Spotted Hyaena*. University of Chicago Press, Chicago.

Kukielka, E., J. A. Barasona, C. E. Cowie, J. A. Drewe, C. Gortazar, I. Cotarelo, et al. 2013. Spatial and temporal interactions between livestock and wildlife in South Central Spain assessed by camera traps. *Prev. Vet. Med.* **112**, 213–221.

LaPoint, S., P. Gallery, M. Wikelski, and R. Kays. 2013. Animal behavior, cost-based corridor models, and real corridors. *Landscape Ecol.* **28**, 1615–1630.

Laurenson, M. K. 1994. High juvenile mortality in cheetahs (*Acinonyx jubatus*) and its consequences for maternal care. *J. Zool.* **234**, 387–408.

Lerman, S. B., P. S. Warren, H. Gan, and E. Shochat. 2012. Linking foraging decisions to residential yard bird composition. *PLoS ONE* **7**, e43497.

Leuchtenberger, C., C. A. Zucco, C. Ribas, W. Magnusson, and G. Mourão. 2014. Activity patterns of giant otters recorded by telemetry and camera traps. *Ethology Ecol. & Evol.* **26**, 19–28.

Lewis, S., S. Benvenuti, L. Dall-Antonia, R. Griffiths, L. Money, T. N. Sherratt, et al. 2002. Sex-specific foraging behaviour in a monomorphic seabird. *Proceed Royal Soc. London B: Biol. Sci.* **269**, 1687–1693.

Linkie, M., Y. Dinata, A. Nugroho, and I. A. Haidir. 2007. Estimating occupancy of a data deficient mammalian species living in tropical rainforests: Sun bears in the Kerinci Seblat region, Sumatra.. *Biol. Cons.* **137**, 20–27.

Lode, T. 1995. Activity pattern of polecats Mustela putorius L. in relation to food habits and prey activity. *Ethology* **100**, 295–308.

Lusseau, D., B. Wilson, P. S. Hammond, K. Grellier, J. W. Durban, K. M. Parsons, et al. 2006. Quantifying the influence of sociality on population structure in bottlenose dolphins. *J. Anim. Ecol.* **75**, 14–24.

Maffei, L., A. J. Noss, E. Cuéllar, and D. I. Rumiz. 2005. Ocelot (*Felis pardalis*) population densities, activity, and ranging behaviour in the dry forests of eastern Bolivia: data from camera trapping. *J. Trop. Ecol.* **21**, 349–353.

Marks, B. K., and R. S. Duncan. 2009. Use of forest edges by free-ranging cats and dogs in an urban forest fragment. *Southeast. Nat.* **8**, 427–436.

Martin, E. H., N. Cavada, V. G. Ndibalema, and F. Rovero. 2015. Modelling fine-scale habitat associations of medium-to-large forest mammals in the Udzungwa Mountains of Tanzania using camera trapping. *Trop. Zool.* **28**, 137–151.

Mccoy, J. C., S. S. Ditchkoff, and T. D. Steury. 2011. Bias associated with baited camera sites for assessing population characteristics of deer. *J. Wildl. Manag.* **75**, 472–477.

McElhinny, T. L., L. Smale, and K. E. Holekamp. 1997. Patterns of body temperature, activity, and reproductive behavior in a tropical murid rodent. *Arvicanthis niloticus.. Physiol. Behav.* **62**, 91–96.

McShea, W. J., T. Forrester, R. Costello, Z. He, and R. Kays. 2016. Volunteer-run cameras as distributed sensors for macrosystem mammal research. *Landscape Ecol.* **31**, 55–66.

Meek, P. D., G. Ballard, A. Claridge, R. Kays, K. Moseby, T. O'Brien, et al. 2014a. Recommended guiding principles for reporting on camera trapping research. *Biodivers. Conserv.* **23**, 2321–2343.

Meek, P. D., G.-A. Ballard, P. J. S. Fleming, M. Schaefer, W. Williams, and G. Falzon. 2014b. Camera traps can be heard and seen by animals. *PLoS ONE* **9**, e110832.

Meek, P. D., G. Ballard, P. Fleming, and G. Falzon. 2016a. Are we getting the full picture? Animal responses to camera traps and implications for predator studies. *Ecol. Evol.* **6**, 3216–3225.

Meek, P. D., G. A. Ballard, and G. Falzon. 2016b. The higher you go the less you will know: placing camera traps high to avoid theft will affect detection. *Remote Sens. Ecology Conserv.* **2**, 204–211.

Mella, V. S. A., A. J. W. Ward, P. B. Banks, and C. McArthur. 2015. Personality affects the foraging response of a mammalian herbivore to the dual costs of food and fear. *Oecologia* **177**, 293–303.

Merckx, T., and H. Van Dyck. 2007. Habitat fragmentation affects habitat-finding ability of the speckled wood butterfly *Pararge aegeria* L. *Anim. Behav.* **74**, 1029–1037.

Middleton, A. D., M. J. Kauffman, D. E. McWhirter, M. D. Jimenez, R. C. Cook, J. G. Cook, et al. 2013. Linking anti-predator behaviour to prey demography reveals limited risk effects of an actively hunting large carnivore. *Ecol. Lett.* **16**, 1023–1030.

Mills, M. G. L., and M. E. J. Mills. 2014. Cheetah cub survival revisited: a re-evaluation of the role of predation, especially by lions, and implications for conservation. *J. Zool.* **292**, 136–141.

Moller, A. P. 2004. Protandry, sexual selection and climate change. *Glob. Change Biol.* **10**, 2028–2035.

Morehouse, A. T., T. A. Graves, N. Mikle, and M. S. Boyce. 2016. Nature vs. nurture: evidence for social learning of conflict behaviour in grizzly bears. *PLoS ONE* **11**, e0165425.

Muhly, T. B., C. Semeniuk, A. Massolo, L. Hickman, and M. Musiani. 2011. Human activity helps prey win the predator-prey space race. *PLoS ONE* **6**, e17050.

Newey, S., P. Davidson, S. Nazir, G. Fairhurst, F. Verdicchio, R. J. Irvine, et al. 2015. Limitations of recreational camera traps for wildlife management and conservation research: A practitioner's perspective. *Ambio* **44**, 624–635.

Ng, S. J., J. W. Dole, R. M. Sauvajot, S. P. Riley, and T. J. Valone. 2004. Use of highway under crossings by wildlife in southern California. *Biol. Cons.* **115**, 499–507.

Ngama, S., L. Korte, J. Bindelle, C. Vermeulen, and J. R. Poulsen. 2016. How bees deter elephants: beehive trials with forest elephants (*Loxodonta africana cyclotis*) in Gabon. *PLoS ONE* **11**, e0155690.

Nolet, B. A., V. N. Fuld, and M. E. C. van Rijswijk. 2006. Foraging costs and accessibility as determinants of giving-up densities in a swan-pondweed system. *Oikos* **112**, 353–362.

Norris, D., F. Michalski, and C. A. Peres. 2010. Habitat patch size modulates terrestrial mammal activity patterns in Amazonian forest fragments. *J. Mammal.* **91**, 551–560.

Nowak, K., A. le Roux, S. A. Richards, C. P. J. Scheijen, and R. A. Hill. 2014. Human observers impact habituated samango monkeys' perceived landscape of fear. *Behav. Ecol.* **25**, 1199–1204.

Nyakaana, S., E. L. Abe, P. Arctander, and H. R. Siegismund. 2001. DNA evidence for elephant social behaviour breakdown in Queen Elizabeth National Park, Uganda. *Anim. Conserv.* **4**, 231–237.

O'Brien, T. G., J. E. M. Baillie, L. Krueger, and M. Cuke. 2010. The Wildlife Picture Index: monitoring top trophic levels. *Anim. Conserv.* **13**, 335–343.

Ordiz, A., O.-G. Støen, S. Saebø, V. Sahlén, B. E. Pedersen, J. Kindberg, et al. 2013. Lasting behavioural responses of brown bears to experimental encounters with humans. *J. Appl. Ecol.* **50**, 306–314.

Orrock, J. L. 2004. Rodent foraging is affected by indirect, but not by direct, cues of predation risk. *Behav. Ecol.* **15**, 433–437.

Otani, T. 2001. Measuring fig foraging frequency of the Yakushima macaque by using automatic cameras. *Ecol. Res.* **16**, 49–54.

Picman, J., and L. M. Schriml. 1994. A camera study of temporal patterns of nest predation in different habitats. *The Wilson Bulletin* 456–465.

Pienkowski, M. W. 1979. Differences in habitat requirements and distribution patterns of plovers and sandpipers as investigated by studies of feeding behaviour. *Verhandlungen der Ornithologische Gesellschaft in Bayern* **23**, 105–124.

du Preez, B. D., A. J. Loveridge, and D. W. MacDonald. 2014. To bait or not to bait: a comparison of camera-trapping methods for estimating leopard *Panthera pardus* density. *Biol. Cons.* **176**, 153–161.

Robertson, B. A., and R. L. Hutto. 2006. A framework for understanding ecological traps and an evaluation of existing evidence. *Ecology* **87**, 1075–1085.

Romero-Muñoz, A., L. Maffei, E. Cuéllar, and A. J. Noss. 2010. Temporal separation between jaguar and puma in the dry forests of southern Bolivia. *J. Trop. Ecol.* **26**, 303–311.

Rovero, F., and F. Zimmermann. 2016. *Camera trapping for wildlife research.* Pelagic Publishing Ltd., UK.

Rovero, F., F. Zimmermann, D. Berzi, and P. Meek. 2013. 'Which camera trap type and how many do I need?' A review of camera features and study designs for a range of wildlife research applications. *Hystrix, Italian J. Mammal* **24**, 148–156.

Rowcliffe, J. M., J. Field, S. T. Turvey, and C. Carbone. 2008. Estimating animal density using camera traps without the need for individual recognition. *J. Appl. Ecol.* **45**, 1228–1236.

Rowcliffe, J. M., R. Kays, B. Kranstauber, C. Carbone, and P. A. Jansen. 2014. Quantifying levels of animal activity using camera trap data. *Methods Ecol. Evol.* **5**, 1170–1179.

Rowcliffe, J. M., P. A. Jansen, R. Kays, B. Kranstauber, and C. Carbone. 2016. Wildlife speed cameras: measuring animal travel speed and day range using camera traps. *Remote Sens. Ecology Conserv.* **2**, 84–94.

Rowell, T. E. 1966. Hierarchy in the organization of a captive baboon group. *Anim. Behav.* **14**, 430–443.

Rychlik, L. 2005. Overlap of temporal niches among four sympatric species of shrews. *Acta Theriologica* **50**, 175–188.

Schaller, G. B. 1967. *The deer and the tiger: a study of wildlife in India.* University of Chicago Press, Chicago.

Severud, W. J., J. L. Belant, J. G. Bruggink, and S. K. Windels. 2011. Predator cues reduce American beaver use of foraging trails. *Human-Wildlife Interact.* **5**, 296–305.

Shamoun-Baranes, J., R. Bom, E. E. van Loon, B. J. Ens, K. Oosterbeek, and W. Bouten. 2012. From sensor data to animal behaviour: an oystercatcher example. *PLoS ONE* **7**, e37997.

Steenweg, R., M. Hebblewhite, R. Kays, J. Ahumada, J. T. Fisher, and C. Burton. 2016. Scaling up camera traps: monitoring the planet's biodiversity with networks of remote sensors. *Front. Ecol. Environ.* **15**, 26–34.

Stewart, F. E. C., N. A. Heim, A. P. Clevenger, J. Paczkowski, J. P. Volpe, and J. T. Fisher. 2016. Wolverine behavior varies spatially with anthropogenic footprint: implications for conservation and inferences about declines. *Ecol. Evol.* **6**, 1493–1503.

Steyaert, S. M. J. G., M. Leclerc, F. Pelletier, J. Kindberg, S. Brunberg, J. E. Swenson, et al. 2016. Human shields mediate sexual conflict in a top predator. *Proceed. Royal Soc. B: Biol. Sci.* **283**, 20160906.

Sueur, C., A. Jacobs, F. Amblard, O. Petit, and A. J. King. 2011. How can social network analysis improve the study of primate behavior? *Am. J. Primatol.* **73**, 703–719.

Suraci, J. P., Clinchy M., Mugerwa B., Delsey M., Macdonald D. W., Smith J. A., et al., 2016. A new automated behavioural response system to integrate playback experiments into camera trap studies. *Methods Ecol. Evol.* [online], Available at: http://doi.wiley.com/10.1111/2041-210x.12711 (accessed: 22 December 2016).

Suselbeek, L., W.-J. Emsens, B. T. Hirsch, R. Kays, J. M. Rowcliffe, V. Zamora-Gutierrez, et al. 2014. Food acquisition and predator avoidance in a Neotropical rodent. *Anim. Behav.* **88**, 41–48.

Swanson, A. M., M. Kosmala, C. Lintott, R. Simpson, A. Smith, and C. Packer. 2015. Snapshot Serengeti, high-frequency annotated camera trap images of 40 mammalian species in an African savanna. *Scientific Data* **2**, 150026.

Sweitzer, R. A., and B. J. Furnas. 2016. Data from camera surveys identifying co-occurrence and occupancy linkages between fishers (*Pekania pennanti*), rodent prey, mesocarnivores, and larger predators in mixed-conifer forests. *Data in Brief* **6**, 783–792.

Tan, C. L., Y. Yang, and K. Niu. 2013. Into the night: camera traps reveal nocturnal activity in a presumptive diurnal primate, *Rhinopithecus brelichi.. Primates* **54**, 1–6.

Tobler, M. W., S. E. Carrillo-Percastegui, R. Leite Pitman, R. Mares, and G. Powell. 2008a. An evaluation of camera traps for inventorying large- and medium-sized terrestrial rainforest mammals. *Anim. Conserv.* **11**, 169–178.

Tobler, M. W., S. E. Carrillo-Percastegui, R. Leite Pitman, R. Mares, and G. Powell. 2008b. Further notes on the analysis of mammal inventory data collected with camera traps. *Anim. Conserv.* **11**, 187–189.

Villette, P., Krebs C. J., and Jung T. S. 2017. Evaluating camera traps as an alternative to live trapping for estimating the density of snowshoe hares (*Lepus americanus*) and red squirrels (*Tamiasciurus hudsonicus*). *Eur. J. Wild. Res.*, **6**:3. Available at: http://link.springer.com/10.1007/s10344-016-1064-3. (accessed 18 December 2016).

Voelkl, B., and C. Kasper. 2009. Social structure of primate interaction networks facilitates the emergence of cooperation. *Biol. Let.* **5**, 462–464.

Vogt, K., F. Zimmermann, M. Kölliker, and U. Breitenmoser. 2014. Scent-marking behaviour and social dynamics in a wild population of Eurasian lynx *Lynx lynx. Behav. Proc.* **106**, 98–106.

Wang, Y., M. L. Allen, and C. C. Wilmers. 2015. Mesopredator spatial and temporal responses to large predators and human development in the Santa Cruz Mountains of California. *Biol. Cons.* **190**, 23–33.

Ware, J. V., K. D. Rode, A. M. Pagano, J. Bromaghin, C. T. Robbins, J. Erlenbach, et al. 2015. Validation of mercury tip-switch and accelerometer activity sensors for identifying resting and active behavior in bears. *Ursus* **26**, 86–96.

Weerakoon, M. K., L. Ruffino, G. P. Cleary, S. Heavener, J. P. Bytheway, et al. 2014. Can camera traps be used to estimate small mammal population size. Pp. 307–316 in P. Fleming, P. Meek, P. B. Banks, A. Claridge, J. Sanderson and D. Swann, eds. *Camera trapping: wildlife management and research*. CSIRO Publishing, Collingwood, Australia.

Wey, T., D. T. Blumstein, W. Shen, and F. Jordán. 2008. Social network analysis of animal behaviour: a promising tool for the study of sociality. *Anim. Behav.* **75**, 333–344.

Whitehead, H. 2008. *Analyzing animal societies: quantitative methods for vertebrate social analysis*. University of Chicago Press, USA.

Wilson, R. P., W. S. Grant, and D. C. Duffy. 1986. Recording devices on free-ranging marine animals: does measurement affect foraging performance? *Ecology* **67**, 1091–1093.

Wolf, J. B. W., D. Mawdsley, F. Trillmich, and R. James. 2007. Social structure in a colonial mammal: unravelling hidden structural layers and their foundations by network analysis. *Anim. Behav.* **74**, 1293–1302.

Wright, J. T., J. E. Byers, L. P. Koukoumaftsis, P. J. Ralph, and P. E. Gribben. 2010. Native species behaviour mitigates the impact of habitat-forming invasive seaweed. *Oecologia* **163**, 527–534.

Zheng, X., M. A. Owen, Y. Nie, Y. Hu, R. R. Swaisgood, L. Yan, et al. 2016. Individual identification of wild giant pandas from camera trap photos - a systematic and hierarchical approach. *J. Zool.* **300**, 247–256.

Upland vegetation mapping using Random Forests with optical and radar satellite data

Brian Barrett[1], Christoph Raab[2], Fiona Cawkwell[3] & Stuart Green[4]

[1]School of Geographical and Earth Sciences, University of Glasgow, Scotland, United Kingdom
[2]Centre of Biodiversity and Sustainable Land Use, University of Göttingen, Göttingen, Germany
[3]School of Geography and Archaeology, University College Cork (UCC), Cork, Ireland
[4]Teagasc, Irish Agriculture and Food Development Authority, Ashtown Dublin 15, Dublin, Ireland

Keywords
Radar, random forests, remote sensing, satellite data, uplands, vegetation mapping

Correspondence
Brian Barrett, School of Geographical and Earth Sciences, University of Glasgow, Scotland, UK.
E-mail: brian.barrett@glasgow.ac.uk

Funding Information
This research was funded by the Environmental Protection Agency (Ireland) as part of the Science, Technology, Research and Innovation for the Environment (STRIVE) Programme 2007–2013.

Editor: Harini Nagendra
Associate Editor: Ned Horning

Abstract

Uplands represent unique landscapes that provide a range of vital benefits to society, but are under increasing pressure from the management needs of a diverse number of stakeholders (e.g. farmers, conservationists, foresters, government agencies and recreational users). Mapping the spatial distribution of upland vegetation could benefit management and conservation programmes and allow for the impacts of environmental change (natural and anthropogenic) in these areas to be reliably estimated. The aim of this study was to evaluate the use of medium spatial resolution optical and radar satellite data, together with ancillary soil and topographic data, for identifying and mapping upland vegetation using the Random Forests (RF) algorithm. Intensive field survey data collected at three study sites in Ireland as part of the National Parks and Wildlife Service (NPWS) funded survey of upland habitats was used in the calibration and validation of different RF models. Eight different datasets were analysed for each site to compare the change in classification accuracy depending on the input variables. The overall accuracy values varied from 59.8% to 94.3% across the three study locations and the inclusion of ancillary datasets containing information on the soil and elevation further improved the classification accuracies (between 5 and 27%, depending on the input classification dataset). The classification results were consistent across the three different study areas, confirming the applicability of the approach under different environmental contexts.

Introduction

Regular monitoring of vegetation in upland areas is important for biodiversity conservation, land management, carbon storage and within a European context, European Union (EU) policy compliance. Approximately 19% of the area of the Republic of Ireland supports upland habitats and these have not been adequately described or their distribution adequately mapped (Perrin et al. 2009). These upland areas contain the nation's largest expanse of semi-natural habitats and provide many benefits to society – water supply, climate regulation, maintenance of biodiversity, and provision of recreational activities to name but a few. Notwithstanding this, the uplands are under increasing pressure from a myriad of issues; grazing management, scrub encroachment, diminished supports, ageing farming population and abandonment of land that will lead to major landscape changes into the future (MacDonald et al. 2000; Reed et al. 2009). These stresses have serious consequences upon the composition, extent and conservation status of important vegetation habitats in these areas (Mehner et al. 2004). The inaccessibility and scale of the uplands, along with constraints in time and finance, make monitoring changes in vegetation covering large spatial areas difficult using traditional field-based surveys (Lees and Ritman 1991; Buchanan et al. 2005; Rhodes et al. 2015). The use of Earth Observation (EO) data can help overcome this problem (Kerr and Ostrovsky 2003; Gillespie et al. 2008; Vanden Borre et al. 2011) and help comply with reporting obligations under the Birds Directive (Council Directive 79/409/EEC 1979) and the Habitats Directive (Council Directive 92/43/EEC 1992). The number of EO satellites orbiting the Earth is increasing, and concurrent with algorithmic advances in information extraction capabilities (Lausch et al. 2015), EO datasets offer a real possibility to provide reliable, high-quality and spatially explicit maps of habitat distribution and monitor habitat fragmentation at intervals determined by management needs (Nagendra et al. 2013; Barrett et al. 2014; Pettorelli et al. 2014a).

In order to obtain sufficient information at the ecotope level, hyperspectral data can be the preferred choice in some studies and has been successfully demonstrated under various conditions (Lawrence et al. 2006; Chan and Paelinckx 2008; Chan et al. 2012; Delalieux et al. 2012; Lucas et al. 2015). However, hyperspectral data is not always or commonly available, and in its absence, multispectral data have also shown their use for habitat mapping (e.g. Feilhauer et al. [2014] and Corbane et al. [2015]). The difficulty of acquiring cloud-free observations in temperate areas prone to persistent cloud coverage, especially during spring and summer periods often limits the capability of classifying habitats with optical data as it is not possible to capture the seasonal variability in the spectral response of the vegetation (Lucas et al. 2011). Additionally, upland areas are usually more prone to cloud cover due to the effects of orographic lift. Consequently, Synthetic Aperture Radar (SAR) data are increasingly being investigated for landscape monitoring as they are largely unaffected by atmospheric conditions (Baghdadi et al. 2009; Waske and Braun 2009; Evans and Costa 2013; Barrett et al. 2014). SAR data are sensitive to vegetation structure and moisture content and in combination with optical data, may help to further improve discrimination of habitats that are structurally different but spectrally similar.

The choice and availability of suitable EO data will ultimately determine the amount of information that can be extracted to map and monitor habitats to varying degrees of resolution. In general, most studies are concerned with using the highest spatial resolution data possible, which often introduces further challenges in terms of the maximum coverage that is attainable and financial constraints in acquiring the data. In many cases, extremely detailed imagery may not be needed for a widespread conservation status assessment and the use of medium spatial resolution (>10 and ≤20 m) data may be sufficient to capture the broad extent and spatial patterns of habitats and meet the local needs of stakeholders along with national requirements in terms of reporting under the EU Directives (Lucas et al. 2007; Varela et al. 2008; Nagendra et al. 2013). Within this context, the objective of this study is to evaluate the use of medium spatial resolution optical and radar satellite data, together with ancillary soil and topographic data for identifying and mapping upland vegetation to complement field studies and help contribute to national policy in the area of upland management and the future sustainable development of the uplands. The definition used in this study for upland habitats is taken from Perrin et al. (2009) and is the same as that used by the NPWS of Ireland, whereby uplands are defined as unenclosed areas of land over 150 m in elevation, and contiguous areas of related habitats that descend below this value. Consequently, the study also includes areas below the 150 m cut-off to include the broad band of transitional vegetation and land management that exists between lowland and upland habitats.

Materials and Methods

Study sites

Suitable study areas were selected from a list of candidate sites for an upland monitoring network, proposed by Perrin et al. (2009) that is designed to meet part of Ireland's obligations under Articles 6, 11 and 17 of the Habitats Directive (92/43/EEC). Figure 1 displays the three areas selected for this study; Mount Brandon, the Galtee Mountains, and the Comeragh Mountains.

Mount Brandon

Mount Brandon is located on the Dingle Peninsula in west Kerry, in south western Ireland. It is a mountainous area that includes the second highest peak in Ireland – Mount Brandon at 952 m. It is a designated candidate Special Area of Conservation (cSAC 000375, Lat: 52.22, Long: −10.07) and the area has an oceanic climate with a mean temperature range of between 7 and 13°C and a mean annual rainfall of 1560 mm (calculated from the 1981–2010 averages of the nearest synoptic weather station at Valentia), although the upland summits often receive over 3000 mm per annum. The area of the studied upland site is 162 km^2 (16,212 ha) while the area of the entire region in Figure 1(A) is 1030 km^2.

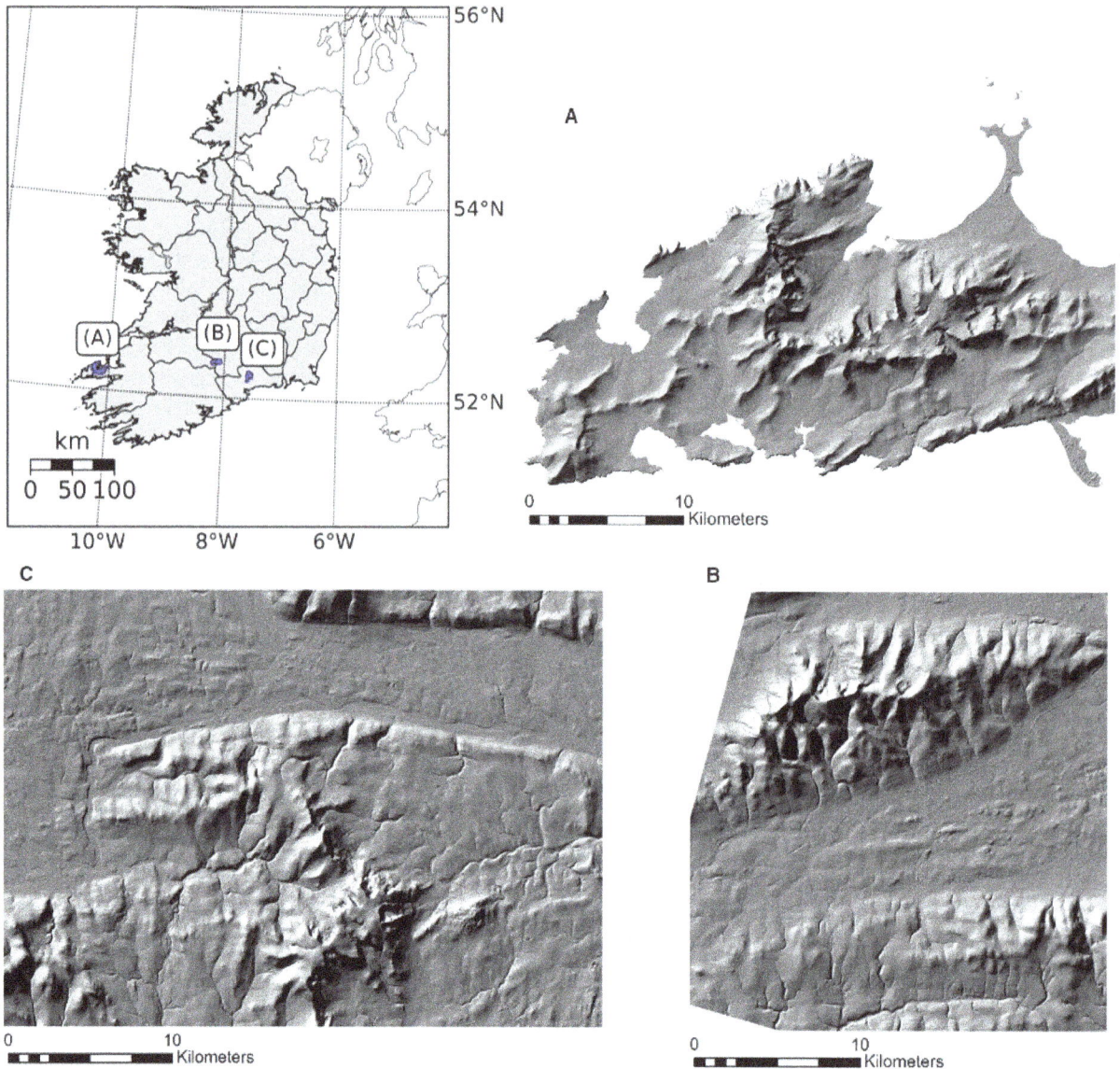

Figure 1. Location of the three upland study sites in Ireland. (A) Mount Brandon, (B) Galtee Mountains, and (C) Comeragh Mountains showing topography in shaded relief.

Galtee Mountains

The Galtee Mountains span across three counties: Cork, Tipperary and Limerick, and are the highest inland mountain range in Ireland (Galtymore at 920 m). It is a designated candidate Special Area of Conservation (cSAC 000646, Lat: 52.36, Long: −8.14), where the mean recorded temperature range is 5–13°C with a mean annual precipitation of 820 mm for the lowlands, rising to 1900 mm for upland regions (meteorological data recorded at the closest synoptic weather station at Moorepark in Fermoy, Cork). The area of the studied upland

site is 83 km^2 (8279 ha) while the area of the entire region in Figure 1(B) is 619 km^2.

Comeragh Mountains

The Comeragh Mountains are located in county Waterford and are a designated Special Area of Conservation (SAC 001952, Lat: 52.23, Long: −7.56). The central area of the mountains features a boggy plateau and reaches a maximum elevation of 792 m. Moorepark is also the closest synoptic weather station to the Comeragh Mountains and so the historical meteorological data is the same as for the Galtee Mountains. The area of the studied upland site is 103 km^2 (10,329 ha) while the area of the entire region in Figure 1(C) is 943 km^2.

Satellite data

The Advanced Visible and Near Infrared Radiometer type 2 (AVNIR-2) instrument onboard the Advanced Land Observation Satellite (ALOS) satellite was a multispectral sensor that acquired data in four visible and near infrared wavebands corresponding to the blue (0.42–0.50 μm), green (0.52–0.60 μm), red (0.61–0.69 μm) and near infra-red (0.76–0.89 μm) spectral channels. The ALOS satellite was launched by the Japanese Aerospace Exploration Agency (JAXA) on 19th January 2006 and operated until 12th May 2011. The satellite also had a Phased Array-type L-band Synthetic Aperture Radar (PALSAR) instrument onboard that operated at L-band (wavelength (λ) = 23.6 cm). PALSAR level 1.1 data (single-look-complex [SLC]) obtained from the European Space Agency (ESA) were used in this study. Two different modes were selected: fine beam single (FBS) and fine beam dual (FBD) polarization. The characteristics of the satellite data used in this study are displayed in Table 1. AVNIR-2 and PALSAR data were selected for this study based on their spatial resolution, availability and closeness in acquisition to the field measurements.

Image pre-processing

Avnir-2

All data were received as level 1B2 products: two acquisitions from 2009 and one acquisition from 2010 were analysed for this study. The spatial resolution of the four AVNIR-2 bands is 10 m and these were resampled to 15 m using a bilinear resampling to match the resolution of the PALSAR data. Each of the AVNIR-2 scenes were geo-rectified using ground control points (GCPs) collected from the Ordnance Survey of Ireland (OSi) orthophotography and yielded a root-mean-square (rms) error of less than 0.56 pixel. Atmospheric correction was performed using the MODTRAN® correction model as implemented in ATCOR-3® (Richter and Schlapfer 2011). A C-factor topographic correction (Teillet et al. 1982) was applied to the data using a sun illumination terrain model derived from a NextMap® 5 m Digital Elevation Model (DEM) (Intermap Technologies, 2008) covering the scene. The topographic correction was implemented in GRASS (GRASS Development Team, 2012). Cloud cover was present in each of the scenes and a mask (manually digitized on screen) was applied in order to exclude cloud and cloud shadow affected areas. Shadows cast by topography were identified using the shadow file output from ATCOR-3 and subsequently masked.

Vegetation indices

Vegetation indices (VIs) have been used extensively for monitoring, analysing, and mapping vegetation dynamics and are often used to remove the variability caused by bare soil, illumination angles and atmospheric conditions when estimating vegetation parameters (Sarker and Nichol 2011). A selection of commonly used vegetation indices were generated using the atmospherically corrected AVNIR-2 data in order to assess their additional information contribution to the classification process (see Table 2). In addition to the VIs, simple reflectance ratios

Table 1. Satellite data used for each of the study sites. Azimuth corresponds to the solar azimuth and elevation corresponds the sun elevation angle, both in degrees. D corresponds to acquisitions from a descending orbit and A corresponds to acquisitions from an ascending orbit.

Site	Sensor	Date	Track	Frame	Pass	Azimuth	Elevation
Mount Brandon	AVNIR-2	2009-09-14	358	2540/2550	D	166.53/ 166.19	40.09/ 40.54
	PALSAR FBD	2010-05-14	7	1030/1040	A	—	—
		2010-06-29	7	1030/1040	A		
	PALSAR FBS	2010-03-29	7	1030/1040	A	—	—
Galtee Mountains	AVNIR-2	2010-10-11	354	2540	D	169.91	30.02
	PALSAR FBD	2010-06-07	3	1040	A	—	—
	PALSAR FBS	2010-03-07	3	1040	A	—	—
		2011-03-10	3	1040	A		
Comeragh Mountains	AVNIR-2	2010-10-11	354	2540	D	169.91	30.02
	PALSAR FBD	2010-05-21	2	1030/1040	A		
		2010-07-06	2	1030/1040	A	—	—
	PALSAR FBS	2011-02-21	2	1030/1040	A	—	—

AVNIR-2, advanced visible and near infrared radiometer type 2; FBD, fine beam dual; FBS, fine beam single; PALSAR, phased array-type L-band synthetic aperture radar.

Table 2. Vegetation Indices selected for this study.

Vegetation Index	Reference
Renormalized difference vegetation index (RDVI)	(Roujean and Breon 1995)
Difference vegetation index (DVI)	(Tucker 1979)
Modified nonlinear index (MNLI)	(Yang et al. 2008)
Normalized difference vegetation index (NDVI)	(Rouse et al. 1974)
Soil adjusted vegetation index (SAVI)	(Huete 1988)
Optimized soil adjusted vegetation index (OSAVI)	(Rondeaux et al. 1996)
Transformed vegetation index (TVI)	(Deering and Rouse 1975)
Corrected transformed vegetation index (CTVI)	(Perry and Lautenschlager 1984)
Thiam's transformed vegetation index (TTVI)	(Thiam 1997)

were calculated for all four bands (blue/green; blue/red; blue/NIR; green/red; green/NIR; and red/NIR) and included as input in the classifications. Although many of the VIs are highly correlated, the use of multiple VIs could offer a more complete characterization of the upland vegetation classes.

Texture measures

Eight texture measures (mean, homogeneity, contrast, variance, dissimilarity, entropy, correlation, and second moment) based on the grey-level co-occurrence matrix (GLCM) (Haralick et al. 1973) of the near infra-red band and radar backscatter were created using a 3 × 3 kernel size. These measures were included as they often provide unique information concerning the spatial pattern and variation in surface features and have been shown to improve classification accuracy (Lu and Weng 2007; Paneque-Gálvez et al. 2013).

Palsar

The FBS and FBD data were multi-looked by a factor of 2 in range and 4 in azimuth, and 1 in range and 4 in azimuth, respectively, to generate 15 × 15 m pixels. The FBS and FBD scenes for each study area were co-registered and speckle filtered using a multi-temporal de Grandi filter (De Grandi et al. 1997), and subsequently radiometrically and geometrically calibrated and converted to dB using a range-doppler approach and a Next-Map 5 m spatial resolution DEM. The radar backscatter returned to the sensor is affected by the topography of the surface where certain terrain-induced distortions are present in areas with increased topographic relief. These areas were subsequently masked for each of the study

sites. For Mt Brandon, 15.08 km² of the upland area out of a total of 162 km² was masked (due to the presence of cloud and/or shadow and terrain-induced distortions), corresponding to 9.3% of the total area. For the larger scene (see Fig. 4), 62.29 km² was masked out of a total of 581.18 km² (land area only) which corresponds to 10.7% of the total land surface area of the scene. 2.74 km² of the Comeraghs upland area of 103 km² and 13.38 km² of the total area of 943 km² was masked, corresponding to 2.7% and 1.4% respectively. 2.27 km² of the Galtees was masked, corresponding to 2.7% and 0.4% of the upland (83 km²) and total area (619 km²) respectively.

Ancillary variables

Two different groups of ancillary variables were chosen for inclusion in the classifications: (1) Topographic – elevation and slope, and (2) Soils. Soil and subsoil information was derived from the Teagasc-EPA Soils and Subsoils dataset (Fealy et al. 2009) and have a nominal working scale of 1:50,000 and elevation and slope data were obtained from a NextMap 5 m DEM. These parameters can influence the spatial distribution of upland vegetation species by affecting the amount of solar radiation and rainfall intercepted by the surface (Bennie et al. 2008, 2010), along with soil nutrient availability and moisture-holding capacity (Franklin 1995).

Classification schema and reference data

The broad-scale habitat classification scheme of Fossitt (2000) has been widely adopted by government authorities and the ecological community for habitat mapping in Ireland. The classification schema adopted for the NPWS-funded National Survey of Upland Habitats (NSUH) is principally based on Fossitt (2000) and has been used in this study (see Table 3). A total of 15 classes (level 2) were identified and a stratified random sampling approach adopted for the selection of sample points. The three study sites have different class distributions and the proportion of each class varies relative to each site. Some classes are not present at some sites (e.g. lowland blanket bog) and some classes have a lower occurrence at other sites (e.g. exposed rock and montane heath at the Galtees). The sample set reflects these differences and as much as possible, the class proportions of the sample data are representative of actual class proportions in the study area landscape. User interpretation of NSUH field survey data for Mount Brandon (collected between May – Aug 2011), Galtee Mountains (Aug –Sept 2011), Comeragh Mountains (Mar – May 2010), Forest Inventory and Planning System (FIPS) and Microsoft® Bing Imagery aided the distinction between the different classes.

Table 3. Classification Schema and number of training samples. Class descriptions are adopted from Fossitt (2000).

Level 0		Level 1		Level 2		BR	GT	CM	Description
G	Grassland	GA	Improved	GA1	Improved	340	507	407	Grassland on well drained soils, usually consists of highly managed pastures
		GS	Semi-improved	GS3	Dry humid grassland	266	186	237	Semi-improved grassland over acid soils
				GS4	Wet grassland	101	106	114	Semi-improved grassland on poorly drained soils
H	Heath	HH	Heath	HH1	Dry siliceous heath	213	258	158	Usually occurs on free-draining acid soils where the vegetation is open and dwarf shrubs are present
				HH3	Wet heath	236	171	122	Usually found on lower slopes of upland areas on peaty soils
				HH4	Montane heath	116	57	111	Substantial cover of dwarf shrubs occurring at high elevation and/or very exposed locations
				HD1	Dense bracken	111	162	135	Areas of open vegetation dominated by Bracken
P	Peatland	PBR	Raised Bog	PB4	Cutover bog	/	/	/	Mostly located in the lowlands of central and mid-west Ireland where there are accumulations of deep peat (3–12 m)
		PBB	Blanket Bog	PB2	Upland blanket bog	271	383	467	Usually occurs on flat or gently sloping ground (above 150 m elevation) on variable peat depths (>0.5 m depth)
				PB3	Lowland blanket bog	129	/	/	Usually confined to wetter regions along the western seaboard. Occurs on flat or gently sloping ground below 150 m elevation
W	Woodland					381	311	674	Areas dominated by trees and woody vegetation
E	Exposed Rock	ER	Exposed Rock	ER1/ER3	Exposed siliceous rock/scree and loose rock	163	55	109	Areas of natural and artificial exposure of bedrock and loose rock (excluding sea cliffs)
		DG	Disturbed Ground	ED1/ED2	Exposed sand, gravel or till.	/	67	102	Areas of exposed sand, gravel or till
B	Built land					139	252	338	All developed land, including transportation infrastructure and human settlements
C	Coastland					134	/	/	Includes sea cliffs and sand dunes
M	Water body					305	45	242	Bodies of permanent fresh and/or salt water
					Total	2905	2560	3216	

Classification

The Random Forests (RF) machine learning classifier (Breiman 2001) was used to relate the vegetation types to the satellite and ancillary data. RF was chosen as the preferred classification method as it has consistently demonstrated its skill for vegetation mapping using various types of data (Cutler et al. 2007; Chapman et al. 2010; Bradter et al. 2011; Rodriguez-Galiano et al. 2012; Barrett et al. 2014; Feilhauer et al. 2014) and can handle high-dimensional datasets and not suffer from overfitting (Belgiu and Drăguţ 2016). RF builds an ensemble of individual decision-like trees from which a final prediction is made using a majority voting scheme. The individual trees are trained using a bootstrap sample of the training data (2/3 of samples) with the remaining 1/3 of samples used to test the classification and estimate the out-of-bag (OOB) error. In this study, RF models consisted of 200 trees. Separate models were generated to analyse the performance of the different data types separately and collectively. Eight different datasets were analysed to compare the change in classification accuracy depending on the selected input variables. These models concentrated on the use of optical only, radar only and various combinations of optical-derived and radar-derived variables along with certain ancillary variables. The influence of the different input variables was calculated and the variable importances (based on the Gini importance) in the initial models were used to improve model fit and model parsimony. RF was implemented in Python 2.7.8 using the sci-kit learn library (Pedregosa et al. 2011).

Accuracy assessment

The results of all classifications were assessed using a standard confusion matrix to calculate the overall accuracy and the user's and producer's accuracies (Congalton 1991). An additional independent validation was also carried out for comparison to the RF OOB accuracies. A total of 876, 839, and 881 samples were randomly selected throughout the Mt Brandon, Galtee Mts, and Comeragh Mts study areas to create the independent accuracy assessment dataset. The statistical significance of the differences between the classification datasets was evaluated, using the Mc Nemar's test (Foody 2004), using the following formula:

$$z = \frac{f_{12} - f_{21}}{\sqrt{f_{12} + f_{21}}}, \qquad (1)$$

where f_{12} indicates the number of samples correctly classified in the first classification, but incorrectly in the second classification, and f_{21} represents the number of samples correctly classified in the second classification, but incorrectly classified in the first classification. The Mc Nemar's test has been commonly used in previous studies to evaluate the variability between classifications (Duro et al. 2012; Belgiu et al. 2014). In this study, the significance level α is set at 0.05 (z critical value = 1.96).

Results

The accuracy assessments (overall accuracy, user's and producer's accuracy) for the different classification datasets are shown in Table 4. The overall accuracy (OA) values among the datasets vary from 59.8% to 94.3% across the three study locations. The highest overall accuracies (93.2–94.3%) were obtained for the combined optical, radar and ancillary data (viii), across all three study areas. Most datasets achieved high accuracies (>~85%) with the exception of the radar and texture measures dataset (≤68%). The RF classifier displayed a relatively consistent overall performance across the three study sites, however certain differences between classes were observed.

Figure 2 displays the producer's (PA) and user's (UA) accuracies for each of the eight classification datasets and provides more insight into the classification errors that are unique to specific classes. The producer's and user's accuracy represent the omission and commission errors respectively. The radar dataset (ii) displays the highest variation between PA and UA for many of the vegetation classes indicating that the radar and texture data tend to overestimate and when used alone, cannot reliably separate these classes. The lowest values for most datasets are confined to the heath classes, where differences between the study areas become more readily apparent. When both optical and radar datasets are combined with the ancillary datasets (viii), these differences between the study areas are less obvious.

It can be seen that the increases in the accuracies achieved in some of the datasets by the addition of certain variables are not large. RF produces a measure of the variable importance by analyzing the deterioration of the predictive ability of the model when each predictor variable is replaced in turn by random noise (Vincenzi et al. 2011). In general, the texture measures and radar data have low importance scores. The class-specific contributions of different variables to the models are shown in Figure 3. Due to their negligible influence, the texture measures (optical and radar) have been omitted. In all three study areas, all models strongly relied on distinct spectral bands and band ratios. The influence of the ancillary data is variable between classes and study sites. The RF models were applied across the entire study areas to obtain vegetation cover for the whole regions (see Fig. 4), while the upland subsets in these study areas are shown in Figure 5. These maps were created using the (vii) dataset, without the inclusion of the soils and elevation ancillary data. A 3 × 3 pixel majority filter was applied to the thematic outputs to improve the homogeneity of the final product. As can be seen from Figure 4, the dominant vegetation cover in all areas is grasslands, and this is relatable to most areas in Ireland. There is very little forest cover on the Dingle Peninsula, while both the Galtee and Comeragh study areas have considerably larger forest areas, especially along the lower slopes of the upland areas. These areas usually represent lands that are marginal for agriculture and since the 1950s, large extents have been afforested, supported through various government and EU incentive programmes. Concentrating on the upland subsets in Figure 5, the true value of upland areas in terms of habitat diversity is apparent. Mount Brandon (Fig. 5A) has extensive areas of wet heath, semi-improved (dry-humid acid) grasslands, blanket bog and dry siliceous heath. Large areas of montane heath are observed, especially along the western edge of the area making it quite distinctive when compared to the Galtee and Comeragh Mountains. From Figure 5(B), the dominant classes for the Galtee Mountains are dry-humid acid grassland along the north-west of the area, dry siliceous heath and blanket bog. Wet heath occurs less frequently, compared to the Mount Brandon area, though there are increased areas of wet grassland. Similar to the Galtees, the dominant classes in the Comeragh Mountains area (Fig. 5C) are blanket bog, dry siliceous heath and dry-humid acid grassland. Small areas of wet heath are scattered throughout the area and areas of dense bracken are prevalent along the eastern edges of the upland area.

The results of the Mc Nemar's test between classification (vii) and the others are displayed in Table 5 for all study sites. McNemar's test is non parametric and based on the classifier's confusion matrices with the null

Table 4. Level 2 classification results.

	(i) Optical and texture			(ii) Radar and texture			(iii) Optical and texture and VIs			(iv) Optical and ancillary data			(v) Radar and ancillary data			(vi) Optical and Radar			(vii) Optical and Radar (inc texture and VIs)			(viii) Optical and Radar and ancillary data		
	BR	GT	CM	BR	GT	CM	BR	GT	CM	BR	GT	CM	BR	GT	CM	BR	GT	CM	BR	GT	CM	BR	GT	CM
Overall Accuracy (%)	85.3	86.1	85.6	60.0	59.8	68.0	84.6	85.5	84.9	93.0	92.4	93.1	87.4	83.2	88.7	85.6	85.8	86.2	88.4	86.4	88.5	94.3	93.2	93.8
Improved grassland (GA1)																								
PA	0.98	0.97	0.96	0.63	0.65	0.70	0.97	0.96	0.95	0.97	0.99	0.98	0.95	0.94	0.95	0.98	0.97	0.96	0.97	0.97	0.97	0.99	0.99	0.97
UA	0.97	0.95	0.98	0.83	0.82	0.87	0.96	0.95	0.98	0.98	1.00	1.00	0.96	0.98	0.99	0.98	0.96	0.99	0.97	0.97	0.99	0.98	1.00	1.00
Dry humid grassland (GS3)																								
PA	0.94	0.78	0.87	0.40	0.49	0.45	0.92	0.78	0.87	0.95	0.91	0.91	0.80	0.81	0.72	0.93	0.77	0.88	0.92	0.80	0.86	0.95	0.94	0.92
UA	0.98	0.76	0.93	0.51	0.22	0.36	0.97	0.75	0.91	0.98	0.90	0.94	0.80	0.68	0.83	0.98	0.76	0.91	0.97	0.77	0.92	0.98	0.91	0.93
Wet grassland (GS4)																								
PA	0.94	0.75	0.63	0.32	0.56	0.74	0.93	0.75	0.59	0.96	0.96	0.83	0.88	0.84	0.80	0.94	0.77	0.63	0.95	0.76	0.71	0.97	0.97	0.87
UA	0.92	0.77	0.54	0.12	0.13	0.12	0.91	0.75	0.49	0.91	0.88	0.78	0.65	0.61	0.71	0.94	0.77	0.58	0.94	0.75	0.54	0.94	0.87	0.81
Dry siliceous heath (HH1)																								
PA	0.78	0.77	0.71	0.44	0.46	0.49	0.76	0.77	0.69	0.87	0.86	0.81	0.79	0.73	0.71	0.83	0.76	0.74	0.84	0.76	0.75	0.90	0.85	0.84
UA	0.75	0.82	0.63	0.54	0.54	0.22	0.78	0.83	0.61	0.91	0.88	0.68	0.85	0.74	0.65	0.79	0.86	0.65	0.83	0.84	0.68	0.92	0.89	0.72
Wet heath (HH3)																								
PA	0.58	0.65	0.57	0.32	0.35	0.43	0.57	0.60	0.60	0.83	0.77	0.76	0.70	0.62	0.77	0.56	0.56	0.56	0.63	0.60	0.71	0.86	0.81	0.81
UA	0.65	0.51	0.32	0.36	0.15	0.05	0.63	0.49	0.36	0.73	0.71	0.71	0.70	0.66	0.42	0.62	0.40	0.27	0.72	0.43	0.32	0.79	0.78	0.70
Montane heath (HH4)																								
PA	0.78	0.69	0.60	0.57	0.27	0.29	0.76	0.68	0.58	0.86	0.81	0.81	0.92	0.92	0.81	0.81	0.72	0.49	0.83	0.75	0.70	0.87	0.84	0.88
UA	0.78	0.72	0.36	0.35	0.05	0.11	0.75	0.74	0.34	0.83	0.77	0.84	0.73	0.58	0.55	0.83	0.68	0.39	0.82	0.70	0.52	0.90	0.75	0.77
Dense bracken (HD1)																								
PA	0.80	0.87	0.71	0.63	0.48	0.33	0.81	0.84	0.70	0.96	0.90	0.88	0.92	0.78	0.81	0.84	0.86	0.69	0.88	0.84	0.79	0.87	0.91	0.88
UA	0.77	0.89	0.55	0.26	0.17	0.27	0.75	0.88	0.59	0.93	0.94	0.87	0.81	0.81	0.84	0.78	0.86	0.64	0.74	0.90	0.76	0.90	0.95	0.94
Upland blanket bog (PB2)																								
PA	0.60	0.77	0.78	0.38	0.42	0.47	0.58	0.79	0.79	0.84	0.87	0.87	0.74	0.75	0.82	0.59	0.76	0.77	0.70	0.76	0.76	0.83	0.88	0.92
UA	0.60	0.88	0.93	0.54	0.71	0.89	0.59	0.86	0.92	0.93	0.93	0.92	0.92	0.88	0.94	0.61	0.87	0.92	0.73	0.86	0.95	0.95	0.93	0.96
Lowland blanket bog (PB3)																								
PA	0.66	/	/	0.32	/	/	0.67	/	/	0.84	/	/	0.81	/	/	0.65	/	/	0.77	/	/	0.93	/	/
UA	0.55	/	/	0.05	/	/	0.50	/	/	0.75	/	/	0.61	/	/	0.53	/	/	0.56	/	/	0.78	/	/
Woodland (W)																								
PA	0.99	0.97	0.99	0.93	0.86	0.95	1.00	0.97	0.98	1.00	0.98	0.99	0.96	0.86	0.99	1.00	0.99	0.99	1.00	0.99	0.99	1.00	0.99	0.99
UA	0.99	0.99	0.98	0.98	0.98	0.95	0.99	0.98	0.97	0.99	0.99	0.98	0.99	1.00	0.97	1.00	0.99	0.98	1.00	0.99	0.99	1.00	1.00	0.99
Exposed Rock (ER1)																								
PA	0.84	0.83	0.70	0.61	0.64	0.62	0.84	0.81	0.65	0.90	0.93	0.75	0.86	0.85	0.78	0.84	0.80	0.69	0.89	0.90	0.71	0.94	0.91	0.79
UA	0.82	0.69	0.68	0.26	0.25	0.32	0.84	0.64	0.64	0.90	0.76	0.83	0.87	0.60	0.78	0.80	0.67	0.72	0.88	0.67	0.77	0.91	0.76	0.85
Disturbed ground (ED1)																								
PA	/	0.87	0.55	/	0.65	0.39	/	0.87	0.55	/	0.93	0.93	/	1.00	0.89	/	0.93	0.71	/	0.96	0.75	/	0.96	0.85
UA	/	0.78	0.97	/	0.16	0.13	/	0.72	0.97	/	0.81	0.89	/	0.22	0.91	/	0.76	0.77	/	0.78	0.87	/	0.79	0.96
Builtland (B)																								
PA	0.89	1.00	0.99	0.69	0.80	0.87	0.91	1.00	0.99	0.97	1.00	0.99	0.94	0.95	0.95	0.91	0.98	0.99	0.94	0.99	0.99	1.00	1.00	1.00
UA	0.93	0.98	1.00	0.71	0.89	0.92	0.92	0.98	1.00	0.99	0.99	1.00	0.90	0.92	0.98	0.92	0.97	1.00	0.95	0.99	1.00	0.99	1.00	1.00

(Continued)

Table 4. Continued.

	(i) Optical and texture			(ii) Radar and texture			(iii) Optical and texture and VIs			(iv) Optical and ancillary data			(v) Radar and ancillary data			(vi) Optical and Radar			(vii) Optical and Radar (inc texture and VIs)			(viii) Optical and Radar and ancillary data		
	BR	GT	CM	BR	GT	CM	BR	GT	CM	BR	GT	CM	BR	GT	CM	BR	GT	CM	BR	GT	CM	BR	GT	CM
Coastland (C)																								
PA	0.95	/	/	0.79	/	/	0.94	/	/	1.00	/	/	1.00	/	/	0.96	/	/	0.98	/	/	0.99	/	/
UA	0.93	/	/	0.66	/	/	0.92	/	/	0.97	/	/	0.99	/	/	0.93	/	/	0.96	/	/	1.00	/	/
Water body (M)																								
PA	1.00	1.00	0.99	0.94	0.98	0.90	1.00	1.00	0.98	1.00	1.00	0.99	1.00	1.00	1.00	1.00	1.00	0.98	1.00	1.00	0.99	1.00	1.00	1.00
UA	1.00	0.87	0.89	0.95	0.91	0.94	1.00	0.91	0.88	1.00	0.98	0.88	1.00	1.00	0.99	1.00	0.98	0.99	1.00	0.96	1.00	0.98	0.98	1.00

PA, producer accuracy; UA, user accuracy for the different datasets at each of the three study sites. BR, Mount Brandon; GT, Galtee Mountains; CM, Comeragh Mountains.

hypothesis of no significant differences between classifications (e.g. (i) = (vii)). For all three sites, the difference between (vii) and (ii) and (vii) and (v) were significantly different ($P < 0.001$). The difference between (vii) and (iii) was significantly different ($P < 0.001$) at both the Mt Brandon and Comeragh Mts sites. Mt Brandon displayed significant differences ($P < 0.05$) between all classifications except (vii) and (iv) while the Comeragh Mts also displayed significant differences ($P < 0.001$) between (vii) and (iii) and (vii) and (vi).

Discussion

The results from this study demonstrate the advantage of integrating EO satellite data from multiple sensors to improve vegetation mapping in upland regions. Even though it may not be surprising that the multispectral data outperforms the radar data, there is merit in incorporating both data types in the classifier models. One of the first published studies to investigate radar differences between upland and lowland vegetation was by Krohn et al. (1983) using L-band SEASAT data. Since then, few published studies on the use of radar for mapping uplands can be found in the literature. The results from this study reveal that a short time series of L-band radar data cannot exclusively separate all the distinct vegetation classes used in this analysis. The results show that combined optical and radar data obtain the highest classification accuracies, in agreement with previous studies (e.g. Bagan et al. (2012)). The inclusion of ancillary datasets containing information on the soil and elevation further improves the classification accuracies (between 5 and 27%, depending on the input classification dataset) and is similar to that found in previous studies for both optical (Sesnie et al. 2008) and radar data (Barrett et al. 2014). When several vegetation classes are grouped into broader habitat types, classification accuracies also show an improvement. There is little difference between level 0 and level 1 accuracies and in most cases, the lower level classifications show only a marginal improvement upon level 2 accuracies (see Fig. S1 and Tables S1, S2). To determine the stability of the level 2 classification results, 25 iterations of the RF classifications were run for the optical and radar dataset (vii) where the maximum variation observed in OA for Mt Brandon was 1.01%, Galtee Mts was 0.71%, and Comeragh Mts was 0.69%.

Relative importance of explanatory variables

It can be seen from Figure 3 that the radar data has low importance scores for most of the vegetation classes, with the lowest scores obtained for the GS3 and PB2

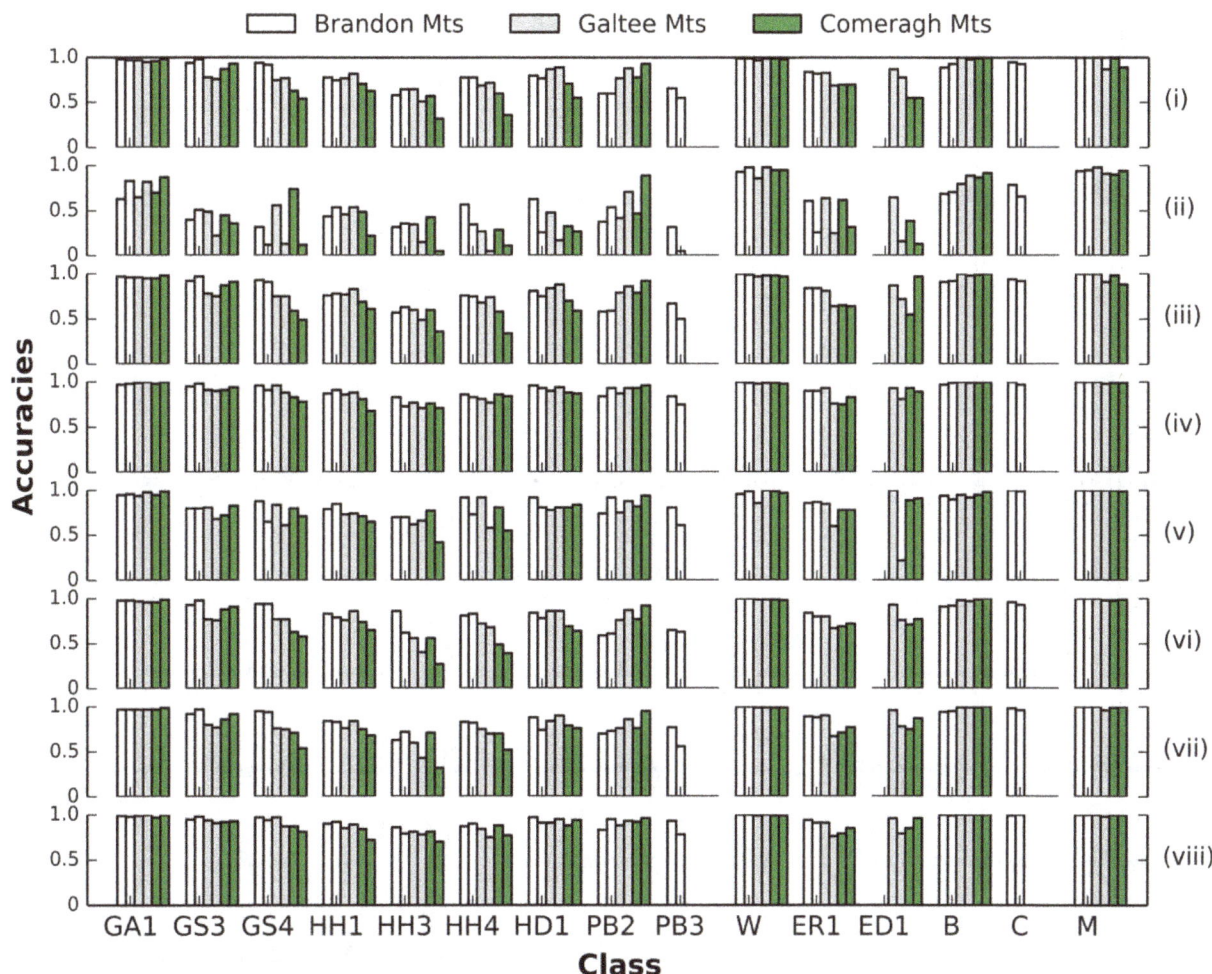

Figure 2. Producers and User's accuracies, represented as the first and second column at each of the three study sites is displayed for the eight different classification datasets (i–viii) and correspond with those as presented in Table 4.

classes. This is likely due to the long wavelength of the radar signal ($\lambda = 23.6$ cm) which penetrates through the vegetation canopy and returns mostly information about the underlying soil properties. Shorter wavelength (e.g. C- or X-band) backscatter is influenced more by the vegetation canopy and may provide more information on the plant geometry that could facilitate the distinction of different upland vegetation classes. Within the optical domain, the NIR signal is particularly useful for discriminating between grassland types (GA1, GS3 and GS4) while the green and red spectral bands perform well for distinguishing between the heath classes (HH1, HH3 and HH4) and blanket bog (PB2). The spectral band ratios (blue/red and green/red) performed especially well in separating dense bracken (HD1), and in general performed better than the vegetation indices. The greater importance of these band ratios is likely due to the higher reflectance of bracken compared to other vegetation in autumn, especially in the red wavelengths

due to the higher amount of underlying dead litter. Similar findings were observed by Holland and Aplin (2013) for winter acquisitions.

Factors such as the bare soil, moisture conditions, solar zenith angle and the atmosphere can impact on the effective use of VIs for distinguishing vegetation types (Jackson and Huete 1991). Soil-adjusted indices such as SAVI and OSAVI minimise the soil background influence but do not outperform other VIs, indicating a likely negligible influence of bare soil on the classifications. In fact, the nine VIs investigated in this study perform similarly across the different study areas. The exception is for the improved grassland (GA1) class where the DVI and renormalized difference vegetation index (RDVI) revealed the highest discriminatory power for the Mount Brandon and Comeragh Mountains. In both of these areas, the NIR channel also had a higher influence than other spectral bands or indices. This is likely due to the strong absorption of electromagnetic

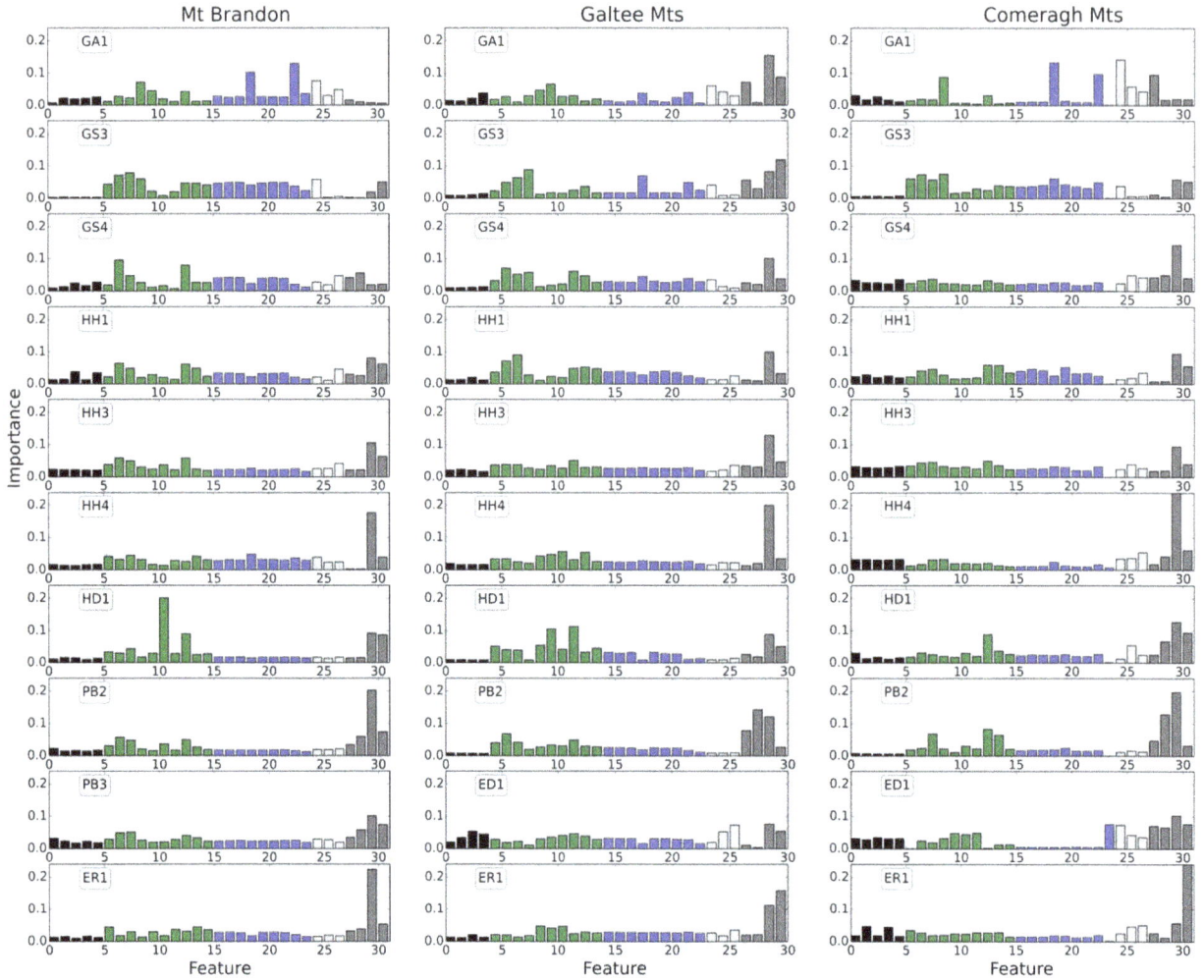

Figure 3. Variable importance scores of the different classes for the three study areas. Apart from the mean, all texture measures were excluded as their importance was negligible. Radar backscatter data (black) represent the first four (Galtee Mountains) and five (Mount Brandon and Comeragh Mountains) variables followed by the four spectral bands (b1, b2, b3, b4) and spectral band ratios (b1b2, b1b3, b1b4, b2b3, b2b4, b3b4) in green. The vegetation indices (NDVI, SAVI, OSAVI, DVI, CTVI, TVI, TTVI, RDVI, and MNLI) are in blue with the band 4 mean, HH polarization mean, and HV polarization in light grey. The final four variables are the soil, subsoil, elevation, and slope (dark grey). NDVI, normalized difference vegetation index; OSAVI, optimized soil adjusted vegetation index; RDVI, renormalized difference vegetation index; SAVI, soil adjusted vegetation index.

radiation in the red wavelengths (0.61–0.69 μm) by chlorophyll in pastures and it's high reflectance in the NIR region. RDVI is similar to NDVI but tends to be more sensitive to changes in vegetation coverage under low leaf area index conditions.

Elevation is one of the most important factors determining the broad-scale distribution of upland vegetation as it influences precipitation and temperature. Thus, elevation controls the ecological and physiological adaptations of various plant species (Lomolino 2001) and the significance of this variable and to a lesser extent, the slope can be seen across most of the classes. The high explanatory power of these variables is not surprising, as

upland grasslands and heaths tend to occur on sloping ground, and montane heaths generally occur at high elevations. Similarly, blanket bogs usually occur on level ground or gentle slopes. Furthermore, they generally occur on deep peaty soils and the results indicate that the soil and subsoil variables had a high importance in this class also. The particular importance of soil characteristics for vegetation mapping has been demonstrated in previous studies by Rogan et al. (2003), Barrett et al. (2014), and Gartzia et al. (2014).

Studies within different scientific disciplines (e.g. bioinformatics, statistics, ecology) suggest RF variable importance measures can display a bias towards highly

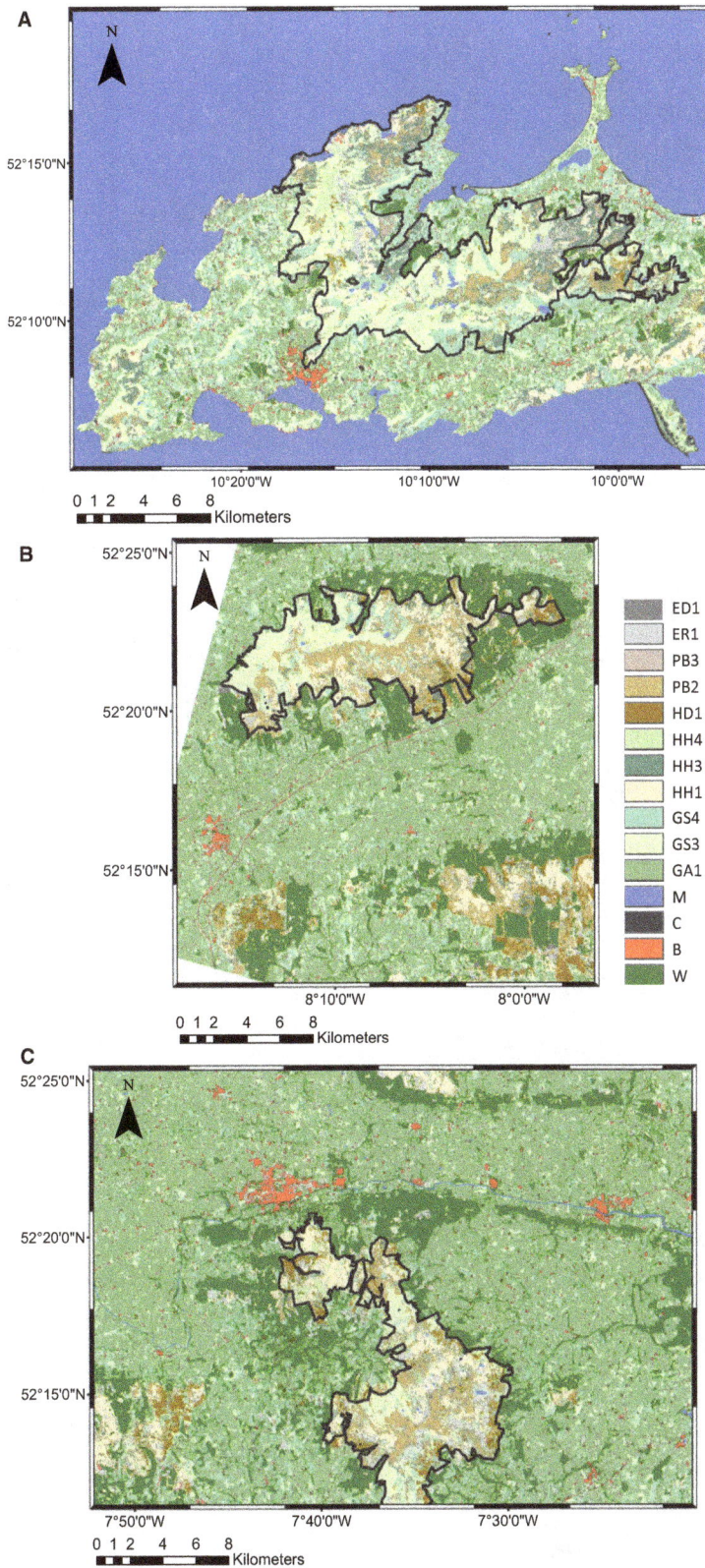

Figure 4. Maps derived from the optical and radar datasets (vii) for (A) Mount Brandon, (B) Galtee Mountains, and (C) Comeragh Mountain study areas. The delineated regions correspond to the upland areas of interest within each area.

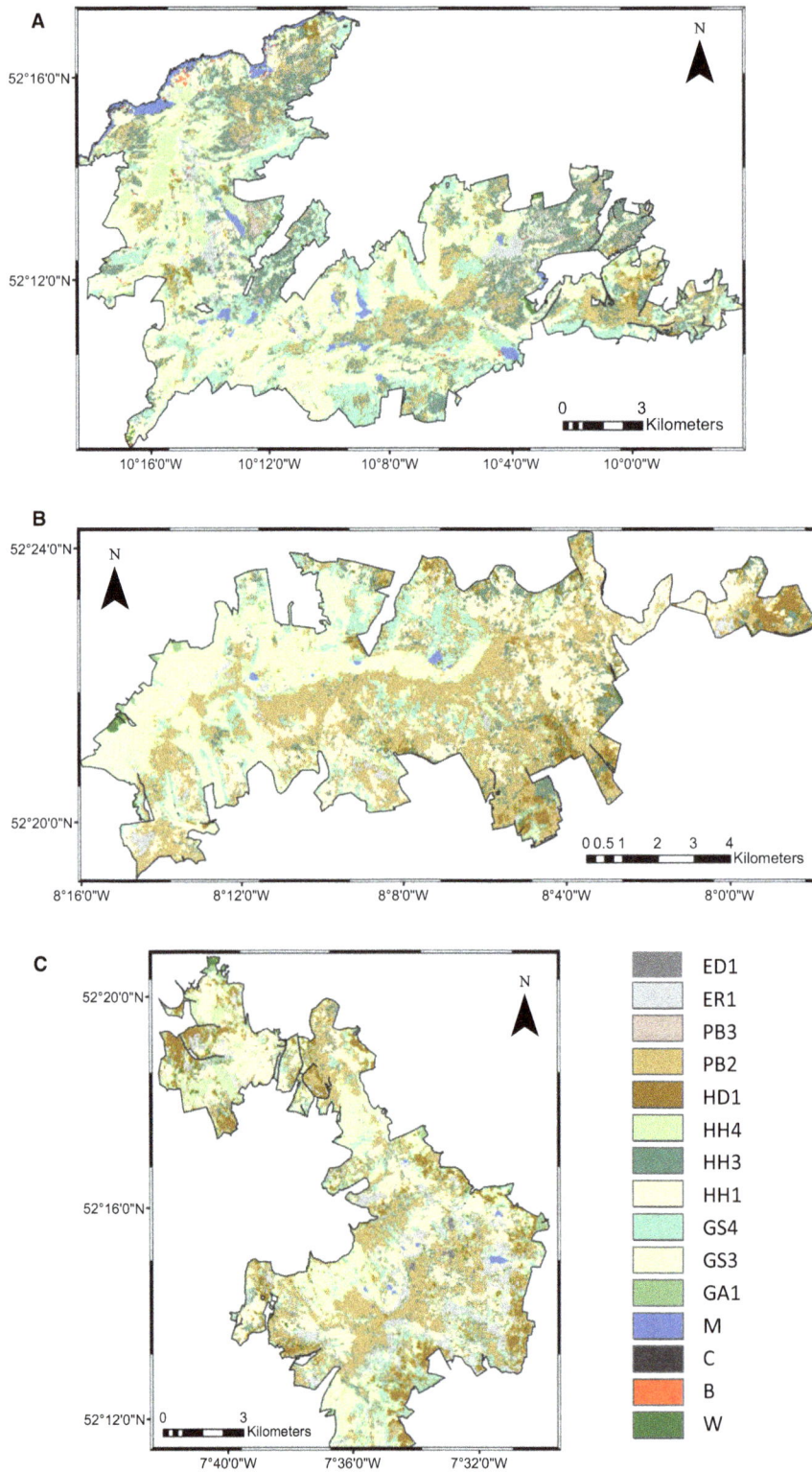

Figure 5. Maps of the upland areas of (A) Mount Brandon, (B) Galtee Mountains, and (C) Comeragh Mountain study areas. These areas correspond to the delineated regions in Figure 3.

Table 5. Summary of the classification comparisons for the three study areas.

Class 1	Class 2	Mt Brandon		Galtees Mts		Comeragh Mts	
		|z|	P value	|z|	P value	|z|	P value
(i)	(vii)	3.035	0.002	1.331	0.183	2.373	0.176
(ii)	(vii)	14.284	<0.001	13.844	<0.001	12.736	<0.001
(iii)	(vii)	3.428	<0.001	1.825	0.068	3.582	<0.001
(iv)	(vii)	0.447	0.655	1.281	0.200	0.681	0.496
(v)	(vii)	6.167	<0.001	6.972	<0.001	3.618	<0.001
(vi)	(vii)	2.331	0.020	1.543	0.122	2.592	<0.001
(viii)	(vii)	4.587	<0.001	1.643	0.100	1.709	0.087

correlated variables (Strobl et al. 2008; Genuer et al. 2010; Ellis et al. 2012). This bias can be lessened by increasing the subsample size of input variables at each node but at the expense of increasing the generalization error and decreasing the overall accuracy (Breiman 2001). Although not considered here, approaches such as the conditional permutation method (Strobl et al. 2008) could be explored as an alternative importance measure in future studies.

Predicted output map uncertainty

The retrieval of habitat information in upland areas using EO data is challenging due to the variable topography and the difficulty of obtaining cloud-free acquisitions in these regions. Furthermore, habitat delineation is more difficult to achieve as the landscape is more heterogeneous (in terms of composition and structure) and consists of a number of interlinked habitats at different scales (spatial, temporal and spectral) (Varela et al. 2008). In this study, misclassification has occurred within and between subclasses of the main vegetation classes of interest (grassland, heaths, and blanket bog). An important feature of the RF algorithm is the ability to compute class probabilities in order to quantify the level of uncertainty in the predicted output maps. The probability of correct classification for each class was calculated to make this uncertainty explicitly available, whereby the relative proportion of each vegetation class per pixel is provided in Figure 6. The predicted probabilities of the main vegetation classes are shown, where the darkest areas represent the pixels with the lowest uncertainty of the assigned class. The classes with the highest overall probabilities in each of the study sites are dry humid acid grasslands (GS3), blanket bogs (PB2), and dry siliceous heath (HH1). Ireland is the most important European country for blanket bog habitats and contains almost 8% of the worldwide blanket bog resource, thus these areas are of prime conservation value. Furthermore, these expanses represent a significant active natural carbon sink (Tomlinson 2005; Bullock et al. 2012).

Comparison with additional independent validation dataset

Evaluation of classification accuracy, using the OOB accuracies reported in the RF algorithm have generally been shown to be a reliable measure of classification accuracy (Lawrence et al. 2006; Devaney et al. 2015). Belgiu and Drăguţ (2016) suggest that this claim requires further validation using a variety of datasets and application areas. In this study, an additional independent validation was performed and the results are presented in Table S3. In all cases, the accuracies obtained for the independent validation were, on average 5.1 ± 2.5% lower than the achieved OOB accuracies for all three study areas. The radar and ancillary dataset (v) had the largest differences, ranging between 8.4 and 11.6%, while the optical and radar (including texture measures and VIs) (vii) had the lowest, ranging between 2.5 and 3.3%. Although many studies have demonstrated the ability of RF to perform well on high dimensional data, Millard and Richardson (2015) found that RF can underestimate the error and recommend reducing the dimensionality of high dimensional datasets to significantly reduce the difference between OOB and independent assessment accuracies.

EO data acquisition timing and spatial resolution

The similarity of accuracies between the study areas may be attributable in part to the similar acquisition periods of the optical and radar data for each of the study areas. The AVNIR-2 scenes were acquired in September (Mount Brandon) and in October (Galtee and Comeragh Mountains) while the radar acquisitions were acquired between February and March and May and July for the FBS and FBD mode data respectively. The different modes of PALSAR data were only available for certain times of year, as part of JAXA's systematic observation strategy, whereby FBS mode acquisitions were available between January and April, and FBD mode acquisitions were available between May and October.

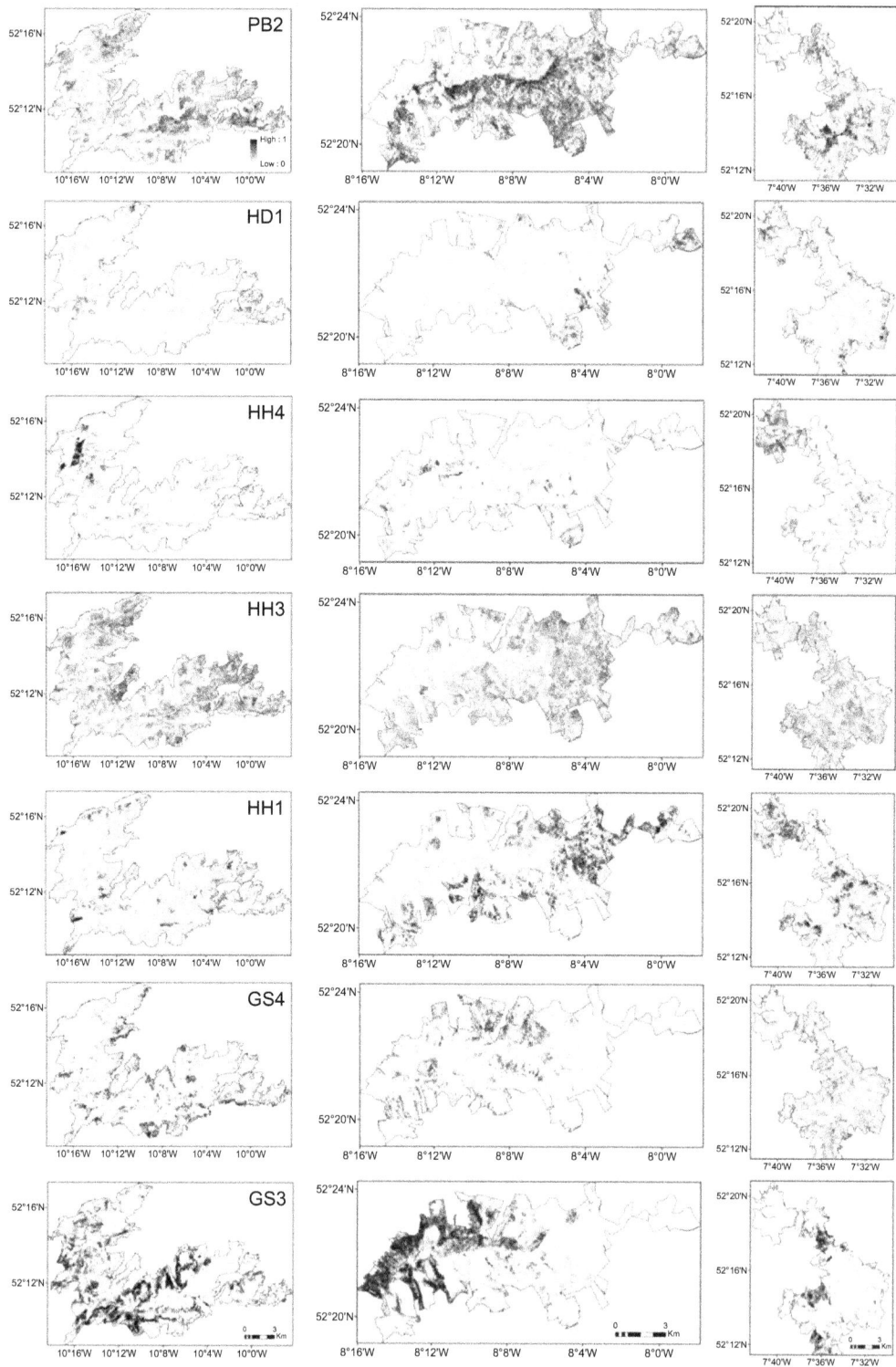

Figure 6. Prediction probabilities for the main classes of interest for the upland areas of Mount Brandon (left), Galtee Mountains (middle) and Comeragh Mountains (right). Darker areas represent higher probabilities while the lighter areas indicate low probabilities. Class designations correspond to those in Table 3.

Vegetation has unique spectral signatures which evolve with the plant life cycle during the year. Characteristics such as pigmentation, water content and physiological structure affect the reflectance, absorption, and transmittance of plant leaves, stems and flowers. In this regard, the time of year of image acquisition will have a strong bearing on the classification accuracy and the ability to distinguish different types of vegetation. Nonetheless, it is difficult to identify an optimal temporal window for operational monitoring of all upland vegetation types (Cole et al. 2014), although acquisitions around September are considered optimum as most upland vegetation types are fully developed (Mills et al. 2006). The spectral similarity between different vegetation types during the summer often limits the ability of acquisitions during these months to reliably distinguish between vegetation types. Ideally, a dense time series of data would allow this to be investigated further as the use of multitemporal data can account for the seasonal variation in vegetation and provide more accurate classifications (Gillanders et al. 2008). This could also open up the possibilities of monitoring grazing management (under- and over-grazing) more effectively and identify burning.

In addition to multitemporal data, a higher discrimination between classes where misclassifications were high could be achieved with data from several spectral bands. For example, Feilhauer et al. (2014) successfully demonstrated the use of simulated multispectral data at 6 m, 10 m, 20 m and 60 m spatial resolution in providing detailed information on the distribution of habitat types. Similarly, Holland and Aplin (2013) found 4 m spatial resolution IKONOS imagery not to be comprehensively superior to Landsat (30 m spatial resolution) for mapping bracken at an uplands site in the UK. Similar findings were observed by Rocchini (2007) and Nagendra et al. (2010). All of these studies found spectral information to be much more important than spatial resolution. With the successful launch of medium spatial resolution sensors such as Sentinel-2 on 23rd June 2015 and future launch of the environmental mapping and analysis program (EnMAP) hyperspectral satellite (providing global coverage at 30 m spatial resolution in 232 spectral channels) in 2018, a valuable and inexpensive source of information to derive spatially complete vegetation information for upland areas in a consistent and regular manner can be provided. Moreover, the perceived inadequacy of medium spatial resolution data may be overcome by incorporating information on the class probabilities as a measure of quantifying the level of uncertainty in the predicted output maps.

Conclusion

In upland areas, meteorological, hydrological and ecological conditions often change substantially over relatively short distances and thus contain a high diversity of habitats and species. Improving our knowledge on upland environments will give valuable insights into holistic environmental processes, aiding the development of sustainable land management strategies for managing the effects of climate change, dormancy and promote conservation of terrestrial and aquatic biodiversity (Nogués-Bravo et al. 2007; Ramchunder et al. 2009; Hodd et al. 2014). EO provides the only means of measuring the characteristics of habitats across broad areas and detecting environmental changes that occur as a result of human or natural processes in these areas on a frequent basis (Kerr and Ostrovsky 2003; Turner et al. 2003; Duro et al. 2007; Nagendra et al. 2014). With the current availability of satellite EO data at low or no cost and an increased number of satellites in orbit or planned, there has never been a better time to incorporate EO data into operational vegetation mapping and monitoring programmes. EO data will never likely provide the fine-scale information that only field measurements can provide but can offer a powerful complimentary information source (Spanhove et al. 2012; Feilhauer et al. 2014; Pettorelli et al. 2014b; O'Connor et al. 2015). From this study, it can be concluded that medium spatial resolution (~15 m) satellite data acquired from optical and microwave sensors offers a basis for supporting mapping and monitoring of upland vegetation. The mapping approach has been demonstrated over large areas in three distinctive upland regions, indicating the consistency and the transferability of the method.

Acknowledgements

This study was funded by the Environmental Protection Agency (EPA) under the Science, Technology, Research and Innovation for the Environment (STRIVE) programme 2007–2013. The authors thank the National Parks and Wildlife Service (NPWS) for providing the field survey data from the National Survey of Upland Habitats (NSUH) programme and also the European Space Agency (ESA) for providing the satellite data through Cat-1 proposal ID 28407. The authors are grateful to John Finn in Teagasc for his assistance throughout this study and would also like to thank the referees and the associate editor for their valuable comments which have helped improve the original manuscript.

Conflicts of Interest

The authors declare no conflicts of interest.

References

Bagan, H., T. Kinoshita, and Y. Yamagata. 2012. Combination of AVNIR-2, PALSAR, and polarimetric parameters for land cover classification. *IEEE Trans. Geosci. Remote Sens.* **50**, 1318–1328.

Baghdadi, N., N. Boyer, P. Todoroff, M. El Hajj, and A. Bégué. 2009. Potential of SAR sensors TerraSAR-X, ASAR/ENVISAT and PALSAR/ALOS for monitoring sugarcane crops on Reunion Island. *Remote Sens. Environ.* **113**, 1724–1738.

Barrett, B., I. Nitze, S. Green, and F. Cawkwell. 2014. Assessment of multi-temporal, multi-sensor radar and ancillary spatial data for grasslands monitoring in Ireland using machine learning approaches. *Remote Sens. Environ.* **152**, 109–124.

Belgiu, M., and L. Drăguţ. 2016. Random forest in remote sensing: a review of applications and future directions. *ISPRS J. Photogramm. Remote Sens.* **114**, 24–31.

Belgiu, M., L. Drăguţ, and J. Strobl. 2014. Quantitative evaluation of variations in rule-based classifications of land cover in urban neighbourhoods using WorldView-2 imagery. *ISPRS J. Photogramm. Remote Sens.* **87**, 205–215.

Bennie, J., B. Huntley, A. Wiltshire, M. O. Hill, and R. Baxter. 2008. Slope, aspect and climate: spatially explicit and implicit models of topographic microclimate in chalk grassland. *Ecol. Model.* **216**, 47–59.

Bennie, J., A. Wiltshire, A. Joyce, D. Clark, A. Lloyd, J. Adamson, et al. 2010. Characterising inter-annual variation in the spatial pattern of thermal microclimate in a UK upland using a combined empirical–physical model. *Agric. For. Meteorol.* **150**, 12–19.

Bradter, U., T. J. Thom, J. D. Altringham, W. E. Kunin, and T. G. Benton. 2011. Prediction of National Vegetation Classification communities in the British uplands using environmental data at multiple spatial scales, aerial images and the classifier random forest. *J. Appl. Ecol.* **48**, 1057–1065.

Breiman, L. 2001. Random forests. *Mach. Learn* **45**, 5–32.

Buchanan, G., J. Pearce-Higgins, M. Grant, D. Robertson, and T. Waterhouse. 2005. Characterization of moorland vegetation and the prediction of bird abundance using remote sensing. *J. Biogeogr.* **32**, 697–707.

Bullock, C. H., M. J. Collier, and F. Convery. 2012. Peatlands, their economic value and priorities for their future management – The example of Ireland. *Land Use Policy* **29**, 921–928.

Chan, J. C.-W., and D. Paelinckx. 2008. Evaluation of random forest and adaboost tree-based ensemble classification and spectral band selection for ecotope mapping using airborne hyperspectral imagery. *Remote Sens. Environ.* **112**, 2999–3011.

Chan, J. C.-W., P. Beckers, T. Spanhove, and J. V. Borre. 2012. An evaluation of ensemble classifiers for mapping Natura 2000 heathland in Belgium using spaceborne angular hyperspectral (CHRIS/Proba) imagery. *Int. J. Appl. Earth Obs. Geoinf.* **18**, 13–22.

Chapman, D. S., A. Bonn, W. E. Kunin, and S. J. Cornell. 2010. Random Forest characterization of upland vegetation

and management burning from aerial imagery. *J. Biogeogr.* **37**, 37–46.

Cole, B., J. McMorrow, and M. Evans. 2014. Spectral monitoring of moorland plant phenology to identify a temporal window for hyperspectral remote sensing of peatland. *ISPRS J. Photogramm. Remote Sens.* **90**, 49–58.

Congalton, R. G. 1991. A review of assessing the accuracy of classifications of remotely sensed data. *Remote Sens. Environ.* **37**, 35–46.

Corbane, C., S. Lang, K. Pipkins, S. Alleaume, M. Deshaves, V. García Millán, et al. 2015. Remote sensing for mapping natural habitats and their conservation status–New opportunities and challenges. *Int. J. Appl. Earth Obs. Geoinf.* **37**, 7–16.

Council Directive 79/409/EEC. 1979. Council Directive 79/409/EEC of 2 April 1979 on the conservation of wild birds.

Council Directive 92/43/EEC. 1992. Council Directive 92/43/EEC of 21 May 1992 on the conservation of natural habitats and of wild fauna and flora

Cutler, D. R., T. Edwards, K. Beard, A. Cutler, K. Hess, J. Gibson, et al. 2007. Random forests for classification in ecology. *Ecology* **88**, 2783–2792.

De Grandi, G., M. Leysen, J. Lee, and D. Schuler. 1997. Radar reflectivity estimation using multiple SAR scenes of the same target: technique and applications, Proceedings of the International Geoscience and Remote Sensing Symposium (IGARSS'97), Singapore, pp. 1047-1050.

Deering, D., and J. Rouse. 1975. Measuring 'forage production' of grazing units from Landsat MSS data, Proceedings of the 10th International Symposium on Remote Sensing of Environment, Ann Arbor, Mich, pp. 1169-1178.

Delalieux, S., B. Somers, B. Haest, T. Spanhove, J. Vanden Borre, and C. A. Mücher, et al. 2012. Heathland conservation status mapping through integration of hyperspectral mixture analysis and decision tree classifiers. *Remote Sens. Environ.* **126**, 222–231.

Devaney, J., B. Barrett, F. Barrett, J. Redmond, and O. John. 2015. Forest cover estimation in Ireland using radar remote sensing: a comparative analysis of forest cover assessment methodologies. *PLoS ONE* **10**, e0133583.

Duro, D. C., N. C. Coops, M. A. Wulder, and T. Han. 2007. Development of a large area biodiversity monitoring system driven by remote sensing. *Prog. Phys. Geogr.* **31**, 235–260.

Duro, D. C., S. E. Franklin, and M. G. Dubé. 2012. A comparison of pixel-based and object-based image analysis with selected machine learning algorithms for the classification of agricultural landscapes using SPOT-5 HRG imagery. *Remote Sens. Environ.* **118**, 259–272.

Ellis, N., S. J. Smith, and C. R. Pitcher. 2012. Gradient forests: calculating importance gradients on physical predictors. *Ecology* **93**, 156–168.

Evans, T. L., and M. Costa. 2013. Landcover classification of the Lower Nhecolândia subregion of the Brazilian Pantanal Wetlands using ALOS/PALSAR, RADARSAT-2 and ENVISAT/ASAR imagery. *Remote Sens. Environ.* **128**, 118–137.

Fealy, R., S. Green, R. Meehan, T. Radford, C. Cronin, and M. Bulfin. 2009. *Teagasc EPA Soil and Subsoils Mapping Project-Final Report.* Teagasc 1 1-126 23123 R1. Teagasc, Dublin.

Feilhauer, H., C. Dahlke, D. Doktor, A. Lausch, S. Schmidtlein, G. Schulz, et al. 2014. Mapping the local variability of Natura 2000 habitats with remote sensing. *Appl. Veg. Sci.* **17**, 765–779.

Foody, G. M. 2004. Thematic map comparison. *Photogramm. Eng. Remote Sensing* **70**, 627–633.

Fossitt, J. A. 2000. *A guide to habitats in Ireland.* The Heritage Council, Dublin.

Franklin, J. 1995. Predictive vegetation mapping: geographic modelling of biospatial patterns in relation to environmental gradients. *Prog. Phys. Geogr.* **19**, 474–499.

Gartzia, M., C. L. Alados, and F. Pérez-Cabello. 2014. Assessment of the effects of biophysical and anthropogenic factors on woody plant encroachment in dense and sparse mountain grasslands based on remote sensing data. *Prog. Phys. Geogr.* **38**, 201–217.

Genuer, R., J.-M. Poggi, and C. Tuleau-Malot. 2010. Variable selection using random forests. *Pattern Recogn. Lett.* **31**, 2225–2236.

Gillanders, S. N., N. C. Coops, M. A. Wulder, S. E. Gergel, and T. Nelson. 2008. Multitemporal remote sensing of landscape dynamics and pattern change: describing natural and anthropogenic trends. *Prog. Phys. Geogr.* **32**, 503–528.

Gillespie, T. W., G. M. Foody, D. Rocchini, A. P. Giorgi, and S. Saatchi. 2008. Measuring and modelling biodiversity from space. *Prog. Phys. Geogr.* **32**, 203–221.

GRASS Development Team. 2012. Geographic Resources Analysis Support System (GRASS) Software, Open Source Geospatial Foundation Project. Avaliable at http://grass.osgeo.org. (accessed 21 October 2016)

Haralick, R. M., K. Shanmugam, and I. H. Dinstein. 1973. Pp. 610–621 *Textural features for image classification.* IEEE Transactions on Systems, Man and Cybernetics.

Hodd, R. L., D. Bourke, and M. S. Skeffington. 2014. Projected range contractions of European protected oceanic montane plant communities: focus on climate change impacts is essential for their future conservation. *PLoS ONE* **9**, e95147.

Holland, J., and P. Aplin. 2013. Super-resolution image analysis as a means of monitoring bracken (Pteridium aquilinum) distributions. *ISPRS J. Photogramm. Remote Sens.* **75**, 48–63.

Huete, A. R. 1988. A soil-adjusted vegetation index (SAVI). *Remote Sens. Environ.* **25**, 295–309.

Intermap Technologies. 2008. Comprehensive 3D Data for More Effective Analysis. Available at: http://www.intermap.com/data/nextmap (accessed 4 April 2016).

Jackson, R. D., and A. R. Huete. 1991. Interpreting vegetation indices. *Prev. Vet. Med.* **11**, 185–200.

Kerr, J. T., and M. Ostrovsky. 2003. From space to species: ecological applications for remote sensing. *Trends Ecol. Evol.* **18**, 299–305.

Krohn, M. D., N. Milton, and D. B. Segal. 1983. Seasat synthetic aperture radar (SAR) response to lowland vegetation types in eastern Maryland and Virginia. *J. Geophy. Res. Oceans (1978–2012)* **88**(C3), 1937–1952.

Lausch, A., A. Schmidt, and L. Tischendorf. 2015. Data mining and linked open data–New perspectives for data analysis in environmental research. *Ecol. Model.* **295**, 5–17.

Lawrence, R. L., S. D. Wood, and R. L. Sheley. 2006. Mapping invasive plants using hyperspectral imagery and Breiman Cutler classifications (RandomForest). *Remote Sens. Environ.* **100**, 356–362.

Lees, B. G., and K. Ritman. 1991. Decision-tree and rule-induction approach to integration of remotely sensed and GIS data in mapping vegetation in disturbed or hilly environments. *Environ. Manage.* **15**, 823–831.

Lomolino, M. 2001. Elevation gradients of species-density: historical and prospective views. *Glob. Ecol. Biogeogr.* **10**, 3–13.

Lu, D., and Q. Weng. 2007. A survey of image classification methods and techniques for improving classification performance. *Int. J. Remote Sens.* **28**, 823–870.

Lucas, R., A. Rowlands, A. Brown, S. Keyworth, and P. Bunting. 2007. Rule-based classification of multi-temporal satellite imagery for habitat and agricultural land cover mapping. *ISPRS J. Photogramm. Remote Sens.* **62**, 165–185.

Lucas, R., K. Medcalf, A. Brown, P. Bunting, J. Breyer, D. Clewley, et al. 2011. Updating the Phase 1 habitat map of Wales, UK, using satellite sensor data. *ISPRS J. Photogramm. Remote Sens.* **66**, 81–102.

Lucas, R., P. Blonda, P. Bunting, G. Jones, J. Inglada, M. Arias et al. 2015. The Earth observation data for habitat monitoring (EODHaM) system. *Int. J. Appl. Earth Obs. Geoinf.* **37**, 17–28.

MacDonald, D., J. Crabtree, G. Wiesinger, T. Dax, N. Stamou, P. Fleury, et al. 2000. Agricultural abandonment in mountain areas of Europe: environmental consequences and policy response. *J. Environ. Manage.* **59**, 47–69.

Mehner, H., M. Cutler, D. Fairbairn, and G. Thompson. 2004. Remote sensing of upland vegetation: the potential of high spatial resolution satellite sensors. *Glob. Ecol. Biogeogr.* **13**, 359–369.

Millard, K., and M. Richardson. 2015. On the importance of training data sample selection in random forest image classification: a case study in peatland ecosystem mapping. *Remote Sens.* **7**, 8489–8515.

Mills, H., M. Cutler, and D. Fairbairn. 2006. Artificial neural networks for mapping regional-scale upland vegetation from high spatial resolution imagery. *Int. J. Remote Sens.* **27**, 2177–2195.

Nagendra, H., D. Rocchini, R. Ghate, B. Sharma, and S. Pareeth. 2010. Assessing plant diversity in a dry tropical forest: comparing the utility of Landsat and IKONOS satellite images. *Remote Sens.* **2**, 478–496.

Nagendra, H., R. Lucas, R. J. Pradinho Honrado, R. Jongman, C. Tarantino, M. Adamo, et al. 2013. Remote sensing for conservation monitoring: assessing protected areas, habitat extent, habitat condition, species diversity, and threats. *Ecol. Ind.* **33**, 45–59.

Nagendra, H., P. Mairota, C. Marangi, R. Lucas, P. Dimopoulos, J. Pradinho Honrado, et al. 2014. Satellite earth observation data to identify anthropogenic pressures in selected protected areas. *Int. J. Appl. Earth Obs. Geoinf.* **37**, 124–132.

Nogués-Bravo, D., M. B. Araújo, M. Errea, and J. Martinez-Rica. 2007. Exposure of global mountain systems to climate warming during the 21st Century. *Glob. Environ. Change* **17**, 420–428.

O'Connor, B., C. Secades, J. Penner, R. Sonnenschein, A. Skidmore, N. D. Burgess, et al. 2015. Earth observation as a tool for tracking progress towards the Aichi Biodiversity Targets. *Remote Sens. Ecol. Conserv.* **1**, 19–28.

Paneque-Gálvez, J., J.-F. Mas, G. Moré, J. Cristóbal, M. Orta-Martínez, A. Luz et al. 2013. Enhanced land use/cover classification of heterogeneous tropical landscapes using support vector machines and textural homogeneity. *Int. J. Appl. Earth Obs. Geoinf.* **23**, 372–383.

Pedregosa, F., G. Varoquanx, A. Gramfort, V. Michel, B. Thirion, O. Grisel, et al. 2011. Scikit-learn: Machine learning in Python. *J. Machine Learn. Res.* **12**, 2825–2830.

Perrin, P. M., B. O'Hanrahan, J. R. Roche, and S. J. Barron. 2009. *Scoping study and pilot survey for a National Survey and Conservation Assessment of Upland Vegetation and Habitats in Ireland.* Unpublished report to National Parks & Wildlife Service, Department of Environment, Heritage and Local Government, Dublin.

Perry, C. R., and L. F. Lautenschlager. 1984. Functional equivalence of spectral vegetation indices. *Remote Sens. Environ.* **14**, 169–182.

Pettorelli, N., W. F. Laurance, T. G. O'Brien, M. Wegmann, H. Nagendra, and W. Turner. 2014a. Satellite remote sensing for applied ecologists: opportunities and challenges. *J. Appl. Ecol.* **51**, 839–848.

Pettorelli, N., K. Safi, and W. Turner. 2014b. Satellite remote sensing, biodiversity research and conservation of the future. *Philos. Trans. R. Soc. B Biol. Sci.* **369**, 20130190.

Ramchunder, S., L. Brown, and J. Holden. 2009. Environmental effects of drainage, drain-blocking and prescribed vegetation burning in UK upland peatlands. *Prog. Phys. Geogr.* **33**, 49–79.

Reed, M., A. Bonn, W. Slee, N. Beharry-Borg, J. Birch, I. Brown, et al. 2009. The future of the uplands. *Land Use Policy* **26**, S204–S216.

Rhodes, C. J., P. Henrys, G. M. Siriwardena, M. J. Whittingham, and L. R. Norton. 2015. The relative value of field survey and remote-sensing for biodiversity assessment. *Methods Ecol. Evol.* **6**, 772–781.

Richter, R., and D. Schlapfer. 2011. Atmospheric/topographic correction for satellite imagery. *DLR-German Aerospace Center: Wessling, Germany* **2011**, 202.

Rocchini, D. 2007. Effects of spatial and spectral resolution in estimating ecosystem α-diversity by satellite imagery. *Remote Sens. Environ.* **111**, 423–434.

Rodriguez-Galiano, V. F., B. Ghimire, J. Rogan, M. Chica-Olmo, and J. P. Rigol-Sanchez. 2012. An assessment of the effectiveness of a random forest classifier for land-cover classification. *ISPRS J. Photogramm. Remote Sens.* **67**, 93–104.

Rogan, J., J. Miller, D. Stow, J. Franklin, L. Levien, and C. Fischer. 2003. Land-cover change monitoring with classification trees using Landsat TM and ancillary data. *Photogramm. Eng. Remote Sensing* **69**, 793–804.

Rondeaux, G., M. Steven, and F. Baret. 1996. Optimization of soil-adjusted vegetation indices. *Remote Sens. Environ.* **55**, 95–107.

Roujean, J.-L., and F.-M. Breon. 1995. Estimating PAR absorbed by vegetation from bidirectional reflectance measurements. *Remote Sens. Environ.* **51**, 375–384.

Rouse, J.W., R. H. Haas, J. A. Schell, D. W. Deering, and J. C. Harlan. 1974. Monitoring the vernal advancement retrogradation of natural vegetation. NASA/GSFC (Type III, Final Report), Greenbelt, MD.

Sarker, L. R., and J. E. Nichol. 2011. Improved forest biomass estimates using ALOS AVNIR-2 texture indices. *Remote Sens. Environ.* **115**, 968–977.

Sesnie, S. E., P. E. Gessler, B. Finegan, and S. Thessler. 2008. Integrating Landsat TM and SRTM-DEM derived variables with decision trees for habitat classification and change detection in complex neotropical environments. *Remote Sens. Environ.* **112**, 2145–2159.

Spanhove, T., J. Vanden Borre, S. Delalieux, B. Haest, and D. Paelinckx. 2012. Can remote sensing estimate fine-scale quality indicators of natural habitats? *Ecol. Ind.* **18**, 403–412.

Strobl, C., A.-L. Boulesteix, T. Kneib, T. Augustin, and A. Zeileis. 2008. Conditional variable importance for random forests. *BMC Bioinformatics* **9**, 1.

Teillet, P. M., B. Guindon, and D. G. Goodenough. 1982. On the slope-aspect correction of multispectral scanner data. *Can. J. Remote Sen.* **8**, 84–106.

Thiam, A. K. 1997. Geographic Information System and Remote Sensing Methods for Assessing and Monitoring Land Degradation in the Shale: The Case of Southern Mauritania, Darks University.

Tomlinson, R. W. 2005. Soil carbon stocks and changes in the Republic of Ireland. *J. Environ. Manage.* **76**, 77–93.

Tucker, C. J. 1979. Red and photographic infrared linear combinations for monitoring vegetation. *Remote Sens. Environ.* **8**, 127–150.

Turner, W., S. Spector, N. Gardiner, M. Fladeland, E. Sterling, and M. Steininger. 2003. Remote sensing for biodiversity science and conservation. *Trends Ecol. Evol.* **18**, 306–314.

Vanden Borre, J., D. Paelinckxa, C. A. Mücherb, L. Kooistrac, B. Haestd, G. De Blusta, et al. 2011. Integrating remote sensing in Natura 2000 habitat monitoring: Prospects on the way forward. *J. Nat. Conserv.* **19**, 116–125.

Varela, R. D., P. R. Rego, S. C. Iglesias, and C. M. Sobrino. 2008. Automatic habitat classification methods based on satellite images: A practical assessment in the NW Iberia coastal mountains. *Environ. Monit. Assess.* **144**, 229–250.

Vincenzi, S., M. Zucchetta, P. Franzoi, M. Pellizzato, F. Pranovi, G. A. De Leo, et al. 2011. Application of a random forest algorithm to predict spatial distribution of the potential yield of Ruditapes philippinarum in the Venice lagoon, Italy. *Ecol. Model.* **222**, 1471–1478.

Waske, B., and M. Braun. 2009. Classifier ensembles for land cover mapping using multitemporal SAR imagery. *ISPRS J. Photogramm. Remote Sens.* **64**, 450–457.

Yang, Z., P. Willis, and R. Mueller. 2008. Impact of band-ratio enhanced AWIFS image to crop classification accuracy. Proceedings of Pecora 17, Available at: http://www. asprs. org/a/publications/proceedings/pecora17/0041. pdf.

PERMISSIONS

LIST OF CONTRIBUTORS

Susana Baena & Justin Moat
1Royal Botanic Gardens, Kew, Richmond, Surrey TW9 3AB, United Kingdom
2School of Geography, University of Nottingham, Nottingham NG7 2RD, United Kingdom

Doreen S. Boyd & Giles M. Foody
School of Geography, University of Nottingham, Nottingham NG7 2RD, United Kingdom

Paul Smith
Botanic Gardens Conservation International, Descanso House, 199 Kew Road, Richmond, Surrey TW9 3BW, United Kingdom

Amber R. Carver, Diana F. Tomback & Michael B. Wunder
Department of Integrative Biology, University of Colorado Denver, Campus Box 171, P.O. Box 173364, Denver, Colorado 80217-3364

Jeremy D. Ross
Sutton Avian Research Center, P.O. Box 2007, Bartlesville, Oklahoma 74005
Oklahoma Biological Survey, University of Oklahoma, 111 E. Chesapeake St, Norman, Oklahoma 73019

David J. Augustine
Rangeland Resources Research Unit, USDA-Agricultural Research Service, 1701 Centre Ave., Fort Collins, Colorado 80526

Susan K. Skagen
U.S. Geological Survey, Fort Collins Science Center, 2150 Centre Ave, Bldg. C, Fort Collins, Colorado 80526

Angela M. Dwyer
Bird Conservancy of the Rockies, 230 Cherry St. Fort Collins, Colorado 80521

T. Trevor Caughlin, Sami W. Rifai, Sarah J. Graves & Stephanie A. Bohlman
University of Florida, School of Forest Resources and Conservation, Gainesville, Florida 32601

Gregory P. Asner
Department of Global Ecology, Carnegie Institution for Science, 260 Panama Street, Stanford, California, 94305

Bethany L. Clark
Environment and Sustainability Institute, University of Exeter, Penryn Campus, TR10 9FE, Penryn, Cornwall, United Kingdom

Mirjana Bevanda
Animal Ecology and Tropical Biology - Ecological Field Station, Institute of Biology, University of W€urzburg, 97074 W€urzburg, Germany

Eneko Aspillaga
Departament d'Ecologia, Universitat de Barcelona, 08029 Barcelona, Spain

Nicolai H. Jørgensen
Department of Animal and Aquacultural Science, Norwegian University of Life Science, 1430 As, Norway

Sandra Frey & John P. Volpe
School of Environmental Studies, University of Victoria, 3800 Finnerty Rd., Victoria, BC V8W 2Y2, Canada

Jason T. Fisher
School of Environmental Studies, University of Victoria, 3800 Finnerty Rd., Victoria, BC V8W 2Y2, Canada
Ecosystem Management Unit, InnoTech Alberta, 3-4476 Markham St., Victoria, BC V8Z 7X8, Canada

A. Cole Burton
Department of Forest Resources Management, University of British Columbia, Vancouver, BC V6T 1Z4, Canada

Kate S. He
Department of Biological Sciences, Murray State University, Murray, Kentucky 42071, USA

Bethany A. Bradley
Department of Environmental Conservation, University of Massachusetts, Amherst, Massachusetts 01003, USA

Anna F. Cord
Department of Computational Landscape Ecology, Helmholtz Centre for Environmental Research – UFZ, Permoserstraße 15, 04318, Leipzig, Germany

Duccio Rocchini
GIS and Remote Sensing Unit, Department of Biodiversity and Molecular Ecology, Research and Innovation Center, Fondazione Edmund Mach, Via E. Mach 1, S. Michele all'Adige, Trento 38010, Italy

Mao-Ning Tuanmu
Department of Ecology and Evolutionary Biology, Yale University, New Haven, Connecticut 06520, USA

Sebastian Schmidtlein
Institute of Geography and Geoecology, Karlsruhe Institute of Technology (KIT), 76131 Karlsruhe Germany

Woody Turner
Earth Science Division, NASA, Washington, District of Columbia, USA

Martin Wegmann
Department of Remote Sensing, University of Wuerzburg, 97074, Wurzburg, Germany
CEOS Biodiversity at German Remote Sensing Data Centre, German Aerospace Centre DLR, 82234 Wessling, Germany

Nathalie Pettorelli
Institute of Zoology, Zoological Society of London, Regent's Park, London, NW1 4RY, UK

Helen Margaret de Klerk
Department of Geography and Environmental Studies, Stellenbosch University, P. Bag X1 Matieland, Stellenbosch 7602, South Africa

Graeme Buchanan
Conservation Science, RSPB Scotland Headquarters, 2 Lochside View, Edinburgh Park, Edinburgh EH12 9DH, United Kingdom

Ilaria Palumbo
European Commission – Joint Research Centre, Institute for Environment and Sustainability, 21027 Ispra (VA), Italy

Robert A. Rose
Center for Geospatial Analysis, College of William & Mary, Williamsburg, Virginia 23185

Rachel M. K. Headley
Cobblestone Science, Spearfish, South Dakota 57783

Janet Nackoney
Department of Geographical Sciences, University of Maryland, College Park, Maryland 20742

Anthony Vodacek
Rochester Institute of Technology, Chester F. Carlson Center for Imaging Science, Rochester, New York 14623

Martin Wegmann
Department of Remote Sensing, Remote Sensing and Biodiversity Research, Institute of Geography, University of Wuerzburg, Oswald Külpe Weg 86, 97074 Wuerzburg, Germany

Stephanie Pau
Department of Geography, Florida State University, Tallahassee, Florida 32306

Laura E. Dee
Institute on the Environment, University of Minnesota, St. Paul, Minnesota 55108

Ian Paynter, Edward Saenz, Daniel Genest, Francesco Peri, Angela Erb, Zhan Li, Kara Wiggin & Crystal Schaaf
School for the Environment, University of Massachusetts Boston, Boston, Massachusetts

Jasmine Muir
Department of Science, Information Technology and Innovation, Queensland, Brisbane, Australia

Pasi Raumonen
Department of Mathematics, Tampere University of Technology, Tampere, Finland

Erica Skye Schaaf
Department of Geography, McGill University, Quebec, Montreal, Canada

Alan Strahler
Department of Earth and Environment, Boston University, Boston, Massachusetts

Duccio Rocchini
Department of Biodiversity and Molecular Ecology, Research and Innovation Centre, Fondazione Edmund Mach, Via E. Mach 1, 38010, S. Michele all'Adige, TN, Italy

Doreen S. Boyd & Giles M. Foody
School of Geography, University of Nottingham, Nottingham, NG7 2RD, United Kingdom

Jean-Baptiste Féret
UMR-TETIS, IRSTEA Montpellier, Maison de la Teledetection, 500 rue JF Breton, 34093, Montpellier Cedex 5, France

Kate S. He
Department of Biological Sciences, Murray State University, Murray, Kentucky, 42071, USA

Angela Lausch
Department of Computational Landscape Ecology, Helmholtz Centre for Environmental Research-UFZ, Permoserstreet 15, D-04318, Leipzig, Germany

Harini Nagendra
Azim Premji University, PES Institute of Technology Campus, Pixel Park, B Block, Electronics City, Hosur Road, Bangalore, 560100, India

Martin Wegmann
Department of Remote Sensing, Remote Sensing and Biodiversity Research Group, University of Wuerzburg, Wuerzburg, Germany

Nathalie Pettorelli
Institute of Zoology, The Zoological Society of London, Regent's Park, London, United Kingdom

Zoltan Szantoi, Andreas Brink, Lucy Bastin, Andrea Lupi, Dario Simonetti, Stephen Peedell & James Davy
European Commission, Joint Research Centre, Institute for Environment and Sustainability, Ispra, Italy

Graeme Buchanan
RSPB Centre for Conservation Science, RSPB, Edinburgh, UK

Philippe Mayaux
Climate Change, Environment, Natural Resources, Water Unit, Directorate General for International Cooperation and Development, European Commission, Brussels, Belgium

Zachary A. Weishampel, Wan-Hwa Cheng & John F. Weishampel
Department of Biology, University of Central Florida, Orlando, Florida, USA

Asja Bernd
EcoDev/ALARM, Yangon, Myanmar
Department of Biogeography, University of Bayreuth, Bayreuth, Germany

Daniela Braun
Remote Sensing Laboratories, Department of Geography, University of Zurich, Zurich, Switzerland

Antonia Ortmann
Food and Agriculture Organization of the United Nations, Rome, Italy

Yrneh Z. Ulloa-Torrealba
Department of Biogeography, University of Bayreuth, Bayreuth, Germany

Christian Wohlfart
Company for Remote Sensing and Environmental Research (SLU), German Aerospace Center (DLR), Oberpfaffenhofen, Germany

Alexandra Bell
University of Cambridge Conservation Research Institute (UCCRI) and Ecosystems and Global Change Group, University of Cambridge, Cambridge, UK

Anthony Caravaggi
School of Biological Sciences, Queen's University Belfast, Belfast BT9 7BL, United Kingdom

Peter B. Banks
School of Life and Environmental Sciences, The University of Sydney, Sydney, New South Wales 2006

A Cole Burton
Department of Forest Resources Management, University of British Columbia, 2215-2424 Main Mall, Vancouver, British Columbia, Canada V6T 1Z4

Caroline M. V. Finlay
Ulster Wildlife, McClelland House, 10 Heron Road, Belfast BT3 9LE, United Kingdom

Peter M. Haswell
School of Biological Sciences, Bangor University, Bangor, Gwynedd LL57 2UW, United Kingdom

Matt W. Hayward
School of Biological Sciences, Bangor University, Bangor, Gwynedd LL57 2UW, United Kingdom

Centre for African Wildlife Ecology, Nelson Mandela University, Port Elizabeth 6031 South Africa; and Centre for African Wildlife Ecology, University of Pretoria, Pretoria 0001,South Africa

Marcus J. Rowcliffe
ZSL Institute of Zoology, Regent's Park, London NW 4RY,United Kingdom

Mike D. Wood
School of Environment & Life Sciences, University of Salford, Manchester M5 4WT, United Kingdom

Brian Barrett
School of Geographical and Earth Sciences, University of Glasgow, Scotland, United Kingdom

Christoph Raab
Centre of Biodiversity and Sustainable Land Use, University of G€ottingen, G€ottingen, Germany

Fiona Cawkwell
School of Geography and Archaeology, University College Cork (UCC), Cork, Ireland

Stuart Green
Teagasc, Irish Agriculture and Food Development Authority, Ashtown Dublin 15, Dublin, Ireland

Index

A

Academic Education, 156

Academic Programs, 71, 78

Accuracy Assessment, 1, 3, 6, 8-9, 11, 185

Activity Patterns, 46-55, 166, 170, 172, 174-177

Alpha-diversity, 121-125, 127-128, 131

Animal Movement, 38-40, 43, 67, 89

Anthropogenic Impacts, 52, 166, 170, 174

Artificial Light, 142-149, 151-154

B

Behavioural Indicators, 166, 170-172, 174

Beta-diversity, 104, 121-131

Bio-logging, 38-39

Biodiversity, 2, 11-12, 27,46, 52-56, 66, 68-70, 72, 78, 89, 97, 99, 104, 127, 129-136, 138, 140-142, 154, 162, 165, 172, 176, 178, 180, 198

Biodiversity Monitoring, 2, 86, 94, 101, 129, 141, 196

C

Camera Trapping, 46, 51, 53, 166, 168, 175, 177-179

Capacity Building, 43, 85, 92

Change Detection, 1, 5-6, 10-12, 25, 36-37, 84, 119, 137

Chelonia Mydas, 143, 154

Compact Biomass Lidar, 105-106, 109-110, 113-118

Competition, 26, 36, 46, 49-51, 53, 62, 64, 98, 169, 172, 174-175

Conservation, 1-3, 12, 21-23, 35, 38-39, 47, 52-53, 56, 62, 69, 71, 89, 91-94, 100-104, 128, 130, 134, 137-138, 140-142, 144, 153-154, 156, 158, 161-178, 182, 193, 198

Conservation Implementation, 71

Continuous Change Detection, 25

D

Deforestation, 1, 6-11, 35-36, 91-92, 100, 134, 140, 162

Dermochelys Coriacea, 142-143, 155

Distance Decay Models, 121

Dmsp, 142, 145, 147-148, 152-154

Duplication, 133

E

Earth Observation Data, 1-3, 67, 92, 162, 197-198

Ecosystem Functions, 94-95, 99-100

Ecosystem Properties, 105, 107

Essential Biodiversity Variables, 45, 85

Ethology, 166-167, 171, 176-177

Extreme Weather Events, 13, 23

F

Forest Landscape Restoration, 25

G

Gis, 15, 44-45, 60, 70, 76-79, 82-89, 93, 104, 156, 158-165, 197

Grassland Birds, 13-14, 21-22, 24

Grazing Management, 180, 195

H

Habitat Fragmentation, 166, 170, 176-177

Hail, 13-16, 18-24

Human Capacity, 71

I

Information System, 45, 76-78, 84, 93, 133-134, 136, 138, 198

Itigi-sumbu Thicket, 1-9, 11-12

L

Land Cover Change, 25-27, 30-31, 34-35, 62, 66, 79, 81-82, 91, 133, 135, 137-140

Landsat, 1-6, 10-12, 25-38, 41, 57-58, 62-63, 65, 68-69, 71-74, 76, 78, 81, 86, 91-93, 99-100, 102, 123, 125-126, 131-136, 138, 140, 195-196, 198

Lidar, 25-27, 29, 31-32, 34-37, 39, 56-70, 72-73, 92, 99, 105-111, 113-120, 127, 152, 154-155, 164

Light Pollution, 142-143, 151, 153-155

M

Misidentification, 59, 122, 131

Monitoring, 1, 11, 14, 16, 21-22, 53, 62, 65, 70-72, 76, 78-79, 81, 86-87, 89, 95, 102, 118, 123, 127, 129, 131, 138, 142, 154, 156, 169, 172, 174, 178, 181, 183, 199

N

Nest Mortality, 13, 15, 20-21

Nest Surveys, 142

Nest Survival, 13, 15-16, 18, 20-23

Nest Vegetation, 13

Nexrad, 13-16, 18, 20-21

Niche Partitioning, 46-47, 49-53, 172

Nutrient Cycling, 94, 99

O

Online Survey, 38-39, 156

Open-source Software, 85, 88-89, 91, 164

Operational Definition, 122

P

Plant Conservation, 1

Priority Assessment, 133

Protected Areas, 2, 12, 76, 125, 131-134, 136-140, 165, 198

Q

Quantitative Structure Models, 108, 110, 112, 120

R

Radar, 13, 15, 21-23, 35, 39, 56-57, 60-64, 66, 68, 72-73, 76, 99, 120, 180-181, 183-191, 193, 196-197

Random Forests, 180, 185, 196-197

Reforestation, 25-26, 33-36

Remote Sensing, 8, 10 14, 26, 29-30, 38, 42, 57, 62-73, 79, 81, 95, 97, 104, 106, 119, 127, 138, 140, 153, 156, 158, 166, 180, 199

Reproducibility, 166

S

Satellite Imagery, 26, 38, 40, 69, 78, 82-83, 104, 121, 131-134, 136, 141, 197-198

Severe Weather, 13, 15

Spatial Ecology, 65, 89, 121, 164

Species Abundance, 36, 66, 94, 103

Species Coexistence, 53-54

Species Composition, 5, 27, 36, 67, 97, 102, 121-122, 126-128, 130

Species Distribution Models, 44, 56, 64, 66-67, 69, 95, 102, 130, 154

Species Interactions, 46, 51-52

Spectral Mixture Analysis, 28, 31, 59, 95

Statistical Population, 122

Survey, 4, 11-13, 38-40, 42, 44, 54, 60, 67, 76, 83, 85-88, 122, 143-144, 156, 158-164, 166-167, 173, 175, 180, 183-184, 195, 197-198

T

Temporal Resolution, 12, 65, 152

Terrestrial Lidar Scanners, 105-106

Topographic Data, 61, 180-181

Tracking, 38-40, 45, 52, 59, 62, 64-65, 67-68, 84, 94-95, 98, 100, 102-103, 134, 138, 140-141, 158, 162, 164-165, 174, 198

Training, 6-7, 10, 29-31, 38-39, 43-44, 60, 71-73, 76-77, 79, 81-83, 85-92, 97, 134, 156, 158-164, 172, 185, 197

Tropical Secondary Forest, 25

Turtles, 142-146, 148-149, 151-155

V

Validation, 7-8, 11-12, 29, 59-60, 65, 70, 105-106, 118, 124, 130, 164, 179-180, 186, 193

Vegetation Mapping, 67, 180, 185, 188, 190, 197

Vulnerability, 85, 171

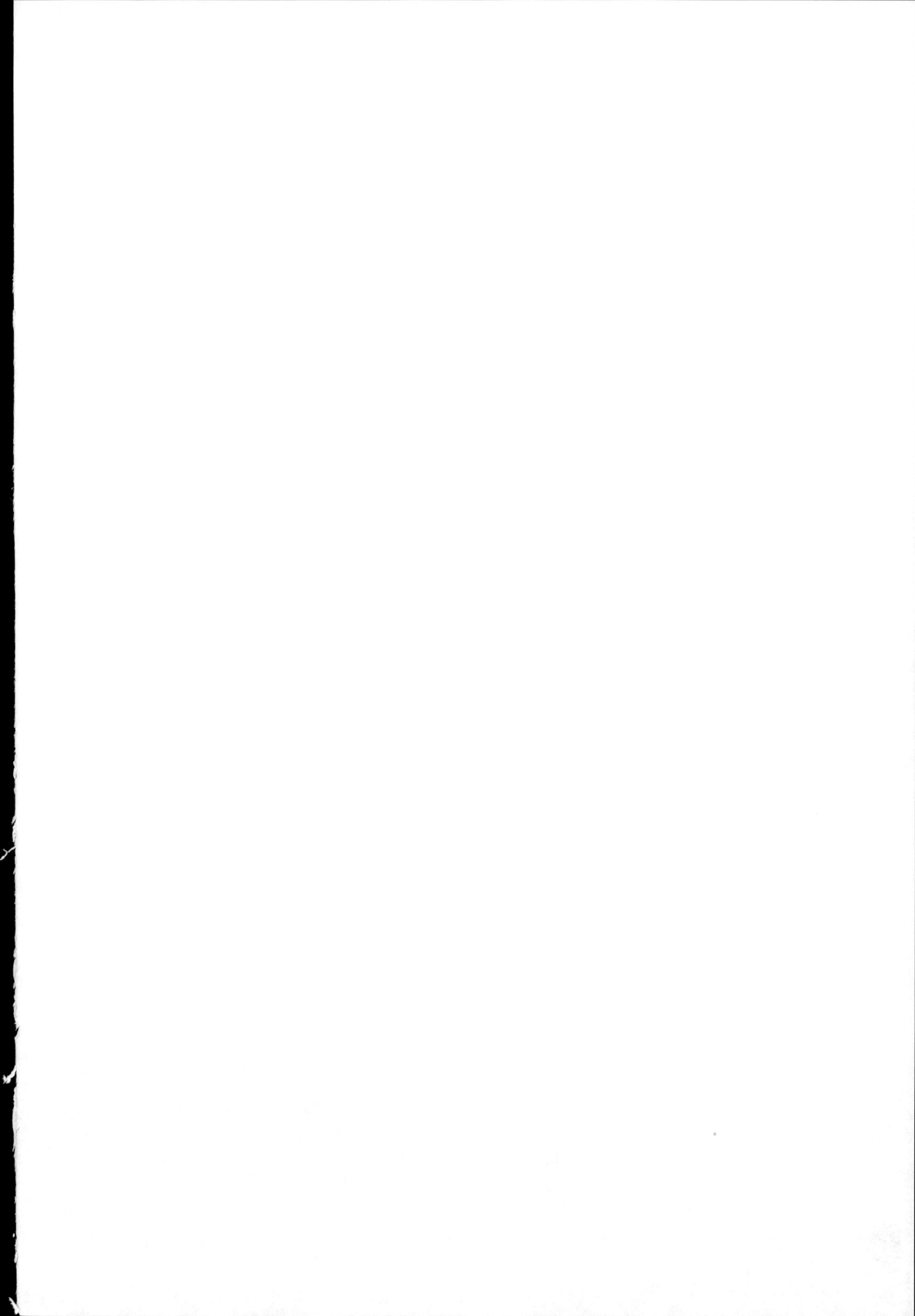

www.ingramcontent.com/pod-product-compliance
Lightning Source LLC
Chambersburg PA
CBHW082027190326

41458CB00010B/3295